Louis Bell, Oscar Terry Crosby

The Electric Railway in Theory and Practice

Second Edition, Revised and Enlarged

Louis Bell, Oscar Terry Crosby

The Electric Railway in Theory and Practice
Second Edition, Revised and Enlarged

ISBN/EAN: 9783744649865

Printed in Europe, USA, Canada, Australia, Japan

Cover: Foto ©berggeist007 / pixelio.de

More available books at **www.hansebooks.com**

THE

ELECTRIC RAILWAY

IN

THEORY AND PRACTICE

BY

OSCAR T. CROSBY, AND LOUIS BELL, PH.D.

SECOND EDITION, REVISED AND ENLARGED.

NEW YORK
THE W. J. JOHNSTON COMPANY, LIMITED
41 PARK ROW (TIMES BUILDING)
1893

PREFACE.

In considering so widely ramifying a subject as electrical traction two quite distinct methods of treatment present themselves—one the discussion of general principles, methods, and results, the other the detailed study of motors and their peculiarities, the specific parts of the general equipment and their use in daily work. These two are mutually exclusive in any volume of finite size. Believing that the latter course would soon lead us into the mere allotment of a graveyard for defunct apparatus, description of which would be less interesting than a collection of epitaphs and would possess but casual value even to the moralist, we have chosen the former line of treatment.

We have endeavored to present both the elementary theory of the subject and the general features of the best practice, describing in detail particular methods and forms of car machinery only in so far as they are of importance in illustrating the broad principles on which they depend. Specific instructions have been introduced, however, when, in the present state of the art, they seemed necessary to a fuller comprehension of the subject and a more thorough grasp of modern methods.

We have not ventured to forecast the future of the transmission of energy to moving motors, for nothing save prophecy after the fact could be equal to the situation; but have contented ourselves with indicating certain paths of improvement that may lead to important results.

In pursuing this policy we hope and believe that our sins will be found to be those of omission rather than commission, and that the reader will deal gently with them, as well-nigh unavoidable in this first general discussion of a new and important branch of applied science.

Our hearty thanks are due to the many friends who have most courteously aided us in the preparation of this volume by freely giving information of every sort, and particularly to Prof. Elihu Thomson, Mr. Carl Hering, Mr. W. E. Baker, Mr. H. I. Bettis, Mr. Thomas Pray, Jr., and Mr. Theo. Stebbins, for most valuable favors of such kind.

OSCAR T. CROSBY.
LOUIS BELL.

TABLE OF CONTENTS.

CHAPTER X.

CHAPTER XI.

CHAPTER XII.

APPENDICES.

APPENDIX A.

APPENDIX B.

APPENDIX C.

APPENDIX D.

APPENDIX E.

APPENDIX F.

APPENDIX G.

THE ELECTRIC RAILWAY

IN

THEORY AND PRACTICE.

CHAPTER I.

GENERAL ELECTRICAL THEORY.

IN common parlance men speak of electricity as "that mysterious force;" and indeed it is mysterious, but not more so than the "force of gravity," "capillary attraction," "chemical affinity," or such like familiar phenomena. "Scratch a fact and you find a mystery" is a homely phrase the truth of which is as a seal of bondage upon the human race.

Two portions of iron attract each other simply by virtue of their mass (itself a word impossible to define absolutely), and the force between the two bodies diminishes as the square of the separating distance increases. This phenomenon we say is *explained* by the law of gravity. But why does the force vary inversely as the square of the distance? Why not as the cube, or the fourth power? Why, indeed, does it vary at all? Why is there attraction or force at any distance? Is not the whole matter, ultimately, as little understood as the fact that under certain conditions the same iron bodies may exhibit the action of another force, called another because its relation to other phenomena—its law—is different from that of gravitation? We call that other force magnetic. Its law seems to be somewhat less simple, but ultimately is not more mysterious than that of the force of gravity.

In furtherance of efforts to simplify our conceptions of—that is, our working hypotheses for—all the operations of nature, it is convenient to consider:

First, that all bodies of appreciable magnitude are composed of minute bodies of inappreciable magnitude.

Second, that these smaller bodies—atoms or molecules—are perpetually in motion.

Third, that the varying states of any body as to hardness, mechanical strain, temperature, light, etc., correspond to varying modes of motion of the constituent molecules. These may change their paths, their velocities, or both, or their groupings.

Fourth, that the resultant force between a given body and the surrounding medium will generally change when any change occurs in the motion of its molecules. The change of state in the body may thus be recorded by corresponding change in a sentient organization, such as the human being.

It seems possible to conceive that all differences of material in the universe may correspond to variations, from point to point in space, of the motion of particles themselves homogeneous.

As the very nature of these supposed ultimate particles cannot be accurately conceived, this mechanical conception of the universe rises up, as do all other fundamental conceptions, from an inexplicable mystery. It does, however, serve a good purpose in correlating phenomena.

Magnetism, or the property of developing magnetic force, has been discovered in but few substances. Iron, nickel and cobalt show this property in specially marked degree. Of these, iron and its compounds—steel and cast iron—are by far the most important. No iron can be said to be wholly non-magnetic, although some of its alloys are nearly so.

To define magnetism completely would be to describe in full its relations to other phenomena. These are not all understood. Expressions have, however, been reached for the chief characteristics of magnetic action.

To-day the most important relation of magnetism is that with the flow of electricity. This latter term, like magnetism, can, of course, be only approximately defined, and a complete definition cannot be made without a lengthy recitation of its characteristics. We will, however, be helped by some such formula as this: There is said to be a flow or current of electricity in a given medium when—due to an initial disturbance which may be a magnetic, chemical, frictional or thermal change—the particles of that medium so move as to (1) produce the sensation called heat, or (2) light, or (3) certain chemical changes, or (4) magnetic change, or (5) other electrical flow, and (6) finally—appealing again to bodily sensation—the peculiar effect experimentally associated with the phenomenon described.

This definition may seem to take much for granted, may seem to be only a circular reasoning, but substantially all definitions found themselves on things taken for granted, otherwise every definition would be a complete explanation of the universe.

Such a mode of molecular motion—a current of electricity—may be produced in any known substance, but it is most readily produced in the metals and certain liquids. When the disturbing force must be very great in order to produce an appreciable electrical effect in a particular medium, that medium is called an insulator or non-conductor.

The earliest method of producing a sustained electrical current was that familiar even to-day in primary batteries. In these it is found that if a metal be brought in contact with another different metal, or with certain non-metallic bodies, and both be then immersed in a fluid producing unequal chemical reactions upon them, a current of electricity will flow across the surface of contact between the two bodies, or along a third body (as a wire) if it be a conductor and joined to the two immersed bodies. Zinc has been found to be generally the best metal for use in batteries, as that which is to be most acted upon by the fluid—usually an acid.

A primary battery is very convenient and useful for many cases in which an electric current is desired. When, however, the supply of power (or current) is required to be considerable the method is expensive to a prohibitory degree, as zinc must constantly be consumed in proportion to the energy supplied by the battery. The maximum possible supply of energy from a pound of zinc is about 1,808,000 foot-pounds. Zinc usually costs about six cents per pound. Hence if the energy be utilized by a process having 95 per cent. efficiency, the cost per foot-pound = 0.0000035 cent.

The maximum possible supply of energy from a pound of good coal is about 11,120,000 foot-pounds. The coal, at $5.00 per ton, costs then 0.25 cent per pound. The efficiency of utilization in a good steam plant is about 13 per cent., giving to the fly-wheel about 1,445,600 foot-pounds per pound of coal. The dynamos of modern make return 90 per cent. in electrical energy of the mechanical energy received from the fly-wheel of the engine. Hence the cost per foot-pound delivered from the dynamos = .0000002 cent, or only one seventeenth part of the cost in the case of the primary battery.

Ignorance of these facts has resulted in the wasteful expenditure of much honest effort and money.

The possibilities in the matter of the direct production of electrical energy from heat seem to be worth considering, and some day may hear the proclamation of a success that will revolutionize present methods in electrical work. For the present, however, the familiar steam engine or water-wheel, connected to drive an electro-magnetic machine, is the only combination fitted to produce electrical energy for general distribution to lamps or motors.

(A) Faraday observed some sixty years ago the fact that if a conductor, say a loop of copper wire, be moved in certain ways in the neighborhood of a magnet, an electrical current is caused to flow in the loop. (B) It was further established that if a current be made to pass around a magnetic body—as when a piece of iron is surrounded by a coil carrying a current—the magnetic action in the body would be affected thereby.

These two facts or laws are involved in the operation of all modern dynamos. The armature of such machines consists of a number of conducting loops, the circuit of which is made complete through the brushes and any wire or other conductor to the ends of which the brushes may be connected.

For simplicity the armature is represented in Fig. 1 as a simple loop. When rotation of this loop begins under the action, say, of a steam engine, we have at once the case referred to in (A). The magnets and their extensions, called the pole pieces, are generally of wrought or cast iron, and these

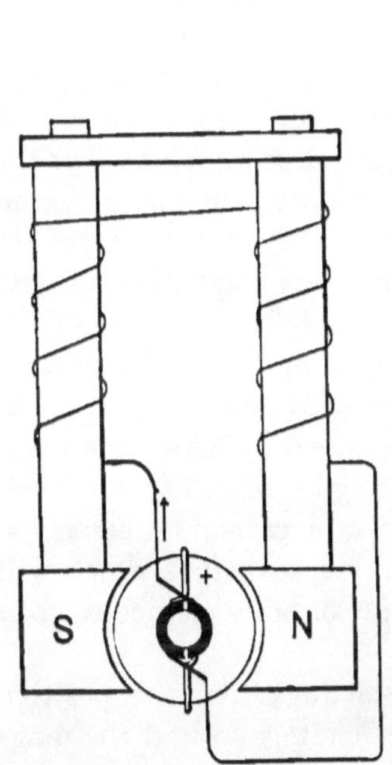

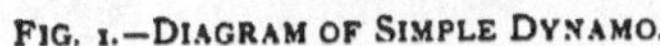
FIG. 1.—DIAGRAM OF SIMPLE DYNAMO.

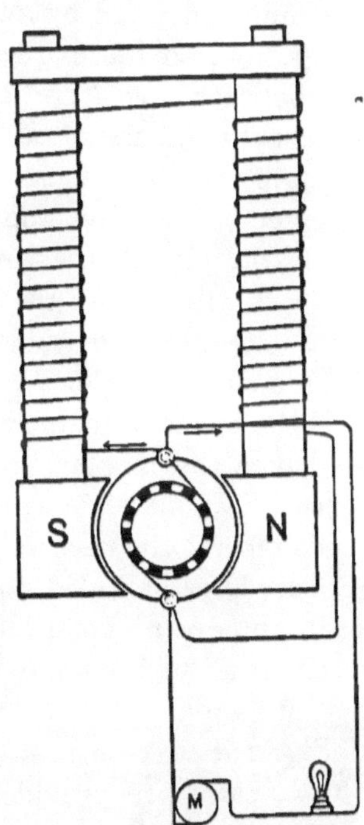

FIG. 2.—SHUNT WOUND DYNAMO

substances are often, loosely, said to be not permanently magnetic, while steel is credited with that property. But in fact these field magnets are always slightly magnetic. They thus create around them a "field of force," that is, a region in which magnetic force may be shown to exist, as by the collection of iron filings around the poles. Assuming some unit of force which need not here be defined, the strength of the field is conveniently expressed by referring to the number of lines (units) of force passing through a unit area, taken at right angles to the direction of the force, as shown by the iron filings. The slightly magnetized pole pieces cause a few lines of force to traverse the space within which the armature revolves, the result being that a very weak current flows to one of the brushes, called the positive, thence through the coils around the magnets and returns to the other, the negative brush By virtue of the flow that has taken place around them, the magnets become (case B) more strongly magnetic, the current generated by and in the armature wires becomes stronger, and so there goes on an interactive, cumulative effect between armature current and magnetic strength, until, usually in about 20 to 30 seconds, the full normal strength of the magnet has been attained. If the current be interrupted, either by stopping the driving engine or by breaking the circuit, the wrought or cast iron magnets and pole pieces will lose all the strength gained from the flow of current, but will retain enough permanent magnetism to start again in a similar cycle.

If all the electrical energy developed by the armature were expended in the magnet coils, no useful outside work could be done. Two methods will be explained by which the magnet coils receive the necessary exciting current, while a much larger supply is provided for heating a carbon to incandescence, or producing rotation in a motor armature which will do mechanical work on any machine suitably connected to it by belt or gears.

In Fig. 2 it will be seen that starting from the positive brush two paths are provided, one leading around the magnets, the other through a lamp and a second armature, representing a motor. This second path is generally called the "working circuit." The first is called the "magnet" or "field" circuit.

In any conductor there is found to be a certain resistance to the flow of an electric current; for any given substance the resistance is found to increase directly as the increase of length, and decrease directly as the increase of cross-section. The relative resistances of circuits fed from the same source, as in Fig. 2, can thus be easily regulated. If the resistance of the magnet coils be ten times that of the working circuit, the latter will receive ten times the quantity of current in the former. Instead of having only a single wire for each of the two circuits, there may be any number, but they may always be imagined as forming a single wire, the resistance of which is just equal to that of the combination of wires; and in such a discussion as this we need consider only this representative wire.

Looking at Fig. 3 it is seen that the current has but one path; leaving the positive brush, it passes around the magnets, thence to the outer working circuit, and returns to the negative brush. If in this case the total output of current be the same as in the previous case, the number of turns in the magnet coil may be very much less than before, the magnetizing effect remaining equal. It has been experimentally proved, and may now be called a law, that the magnetizing effects of different coils carrying varying currents may be compared by comparing the *product* of the *number of coils* by the *strength of the current* flowing through them. Thus ten turns of wire carrying one unit (called an ampère) of current will produce the same effect as one turn carrying ten units—

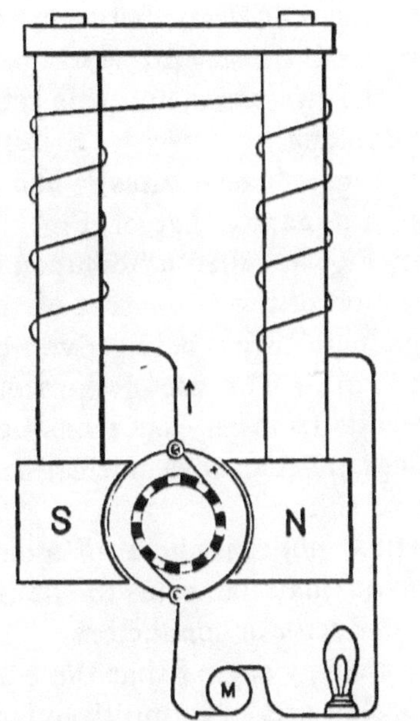

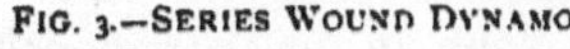
FIG. 3.—SERIES WOUND DYNAMO

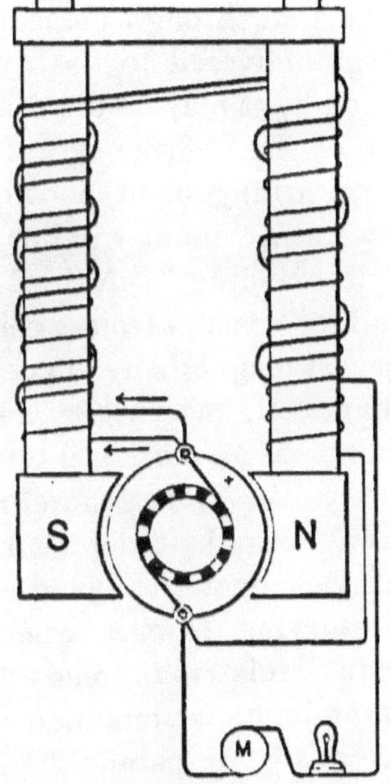

FIG. 4.—COMPOUND WOUND DYNAMO

or ampères. One hundred turns carrying half an ampère ($100 \times 0.5 = 50$) will produce one-half the effect of fifty turns carrying two ampères ($50 \times 2 = 100$) and thus generally, multiply the number of turns in the magnet coils by the number of ampères flowing through those coils, and the products formed show the relative magnetizing effects of the respective coils. This product of the ampères by the turns measures what may be called the magneto-motive force.

Referring to Figs. 2 and 3, if we suppose the total current to be the same, and that in Fig. 2 only one-tenth of the total amount goes through the magnet coils, the number of these coils, for equal magnetization, would have to be ten times as great as in Fig. 3, in which there is but one circuit, containing the magnet coils, as well as the lamps or motors for which the supply is given. In Fig. 2 the work of the external circuit is done by a part of the whole current generated (say 0.9) flowing under the whole of the electrical pressure between brushes. In Fig. 3 the external work is done by the whole of the current flowing, under a pressure less than that existing between brushes by the amount required to force the current through the magnet coils. The ratio between the energy delivered to the dynamo and the useful work performed by it may be the same in the two cases, and this ratio is called the "efficiency" of the dynamo.

The arrangement shown in Fig. 2 constitutes what is known as a "shunt winding" for a dynamo, that of Fig. 3 a "series winding." A third type, Fig. 4, called a "compound winding," results from a combination of the two—part of the required magnetizing effect (or ampère turns) being given by a shunt coil, and part by a series coil. The use of the word "shunt" as an electrical term results from an easy transition from its use (more general in England than in America) as a railway term denoting a branch or siding.

Any conductor carrying a portion, not the whole, of a current flowing from a given point may be said to be in "shunt" relation to some other conductor or conductors. The same relation is sometimes indicated by saying that the conductors are "in parallel," "in parallel arc," "in multiple," or "in multiple arc" with respect to each other. It would be best, perhaps, to use no other expression than "multiple." Figs. 5, 6, 7 show various relations of combined conductors.

In speaking of the quantity of electricity flowing in a conductor we have simply used a popular and convenient expression. No direct dimensional measurement of "quantity" can be had. But by observing that in one case a current separates a certain weight of water into its constituent elements, while in another twice or three times the weight is thus decomposed; that now one ounce of silver is deposited on a metal

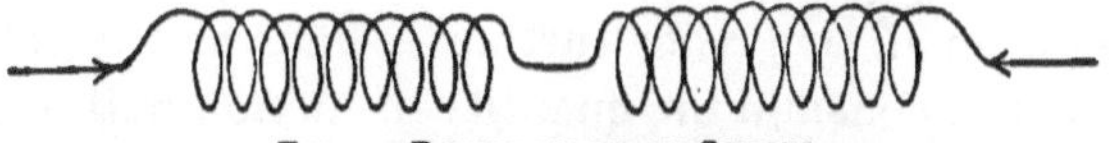

FIG. 5.—RESISTANCES IN SERIES.

surface, and again two or three ounces: that now a lamp carbon is only dull red, while again it is brilliantly white—we are brought face to face with the different quantities of work performed, and are forced to conclude that the molecules themselves of the conductor have different quantities of the

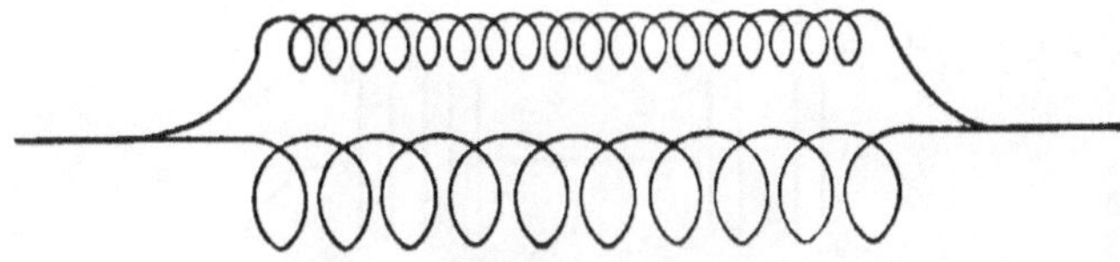

FIG. 6.—RESISTANCES IN MULTIPLE.

energy of motion, due to the original disturbing forces of different magnitudes. The force producing the molecular movement, which in turn presents electrical phenomena, is called an "electromotive" force. If electromotive forces emanating from different sources are applied to a conductor

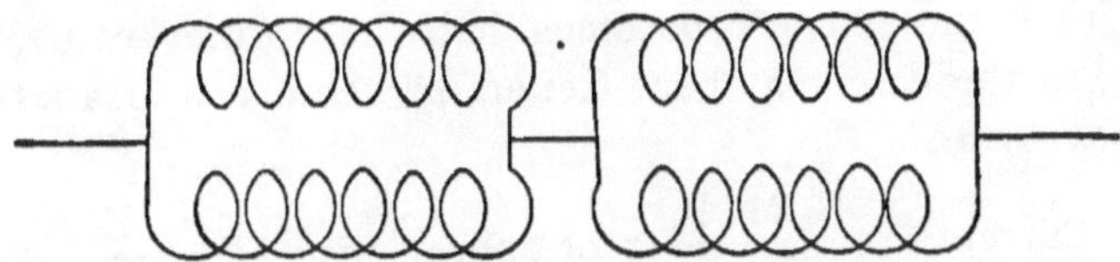

FIG. 7.—RESISTANCES IN SERIES MULTIPLE.

which itself remains unchanged, and it is learned, by measuring their effects, that currents of the same or different strengths flow from these different sources, then the equality or inequality of the electromotive forces is implied. So, if we consider the flow of water in a particular section of pipe, we judge of the relative pressures causing the flow by noting its quantity or measuring the work it can do, the resistance to flow—such as friction on the walls of the pipe—being assumed constant.

Generally it will be readily understood that the quantity of motion produced by a disturbing force acting in any medium will be relatively great or small as the resistance to that form of motion is small or great; further, whatever the resistance may be, if it be constant the quantity of motion will be greater as the force becomes greater. In electrics this relation may be expressed thus: The current flowing in any circuit varies inversely as the resistance, and directly as the electromotive force (usually written E. M. F.). Adopting a system of correlated units, if the E. M. F. be of magnitude expressed by unity and likewise the resistance, then the current flowing will be unit current. If the E. M. F. remain the same and

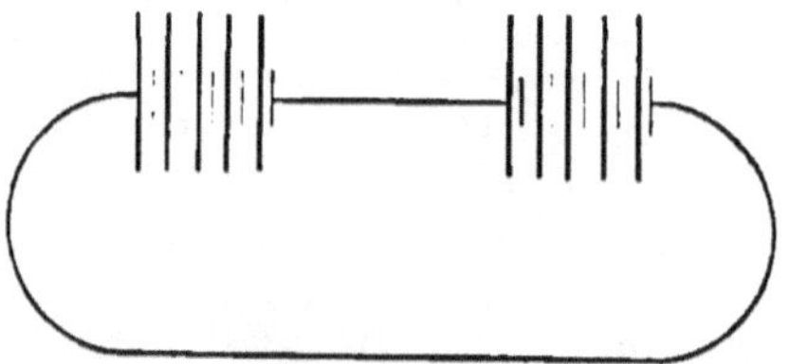

FIG. 8.—ELECTROMOTIVE FORCES IN SERIES.

the resistance become 0.5 or 2, the current would become twice its former value, or half, as the case may be. If the resistance remain the same and the E. M. F. become 0.5 or 2, then the current becomes half its former value, or twice, as the case may be. Generally, then, it appears that we may write:

$$\text{Current} = \frac{\text{E. M. F.}}{\text{Resistance}}; \text{ or abbreviating, } C = \frac{E}{R}.$$

This relation is always known, from the man who demonstrated it, as Ohm's law.

The specific resistances of many substances (that is, the resistances of bars of unit length and cross-section), expressed in units of resistance, have been carefully tabulated.

It would also be fair to assume equality of electromotive forces in different cases if there be the same physical arrangement of parts in the generator. Thus a particular combination of zinc, copper and acid, if applied to any number of circuits, may reasonably be supposed to produce the same electromotive force in all; observed differences in the quantity of current produced would then fairly be attributable to differences of resistance in the various circuits. And, again, if two similar combinations be placed in the same circuit, one following the other, as in Fig. 8, it is fair to assume that we have twice the electromotive force that would be given by one of the combinations acting alone; and that consequently we would have twice the current if the resistance of the circuit were the same in the two cases.

Likewise, if we make two similar dynamos and run them under similar conditions, they will produce equal electromotive forces. As to the force developed in any particular case, experiment shows that it increases with the number of complete changes from zero to maximum in the number of lines of magnetic force passing through the looped conductor of the armature in any fixed interval of time, and with the magnitude of each change. It follows that in an actual dynamo, increase in four separate elements will independently increase the electromotive force produced in its armature, (1) strength of magnetic field, (2) area of revolving loop, (3) number of loops, when arranged each in series with its neighbor, (4) velocity of rotation. If we can determine experimentally the electromotive force produced by a change of one line of force through the loop in unit time, the dynamo may then be proportioned to produce a given E. M. F.

Concerning the strength of field—or number of lines of magnetic force per unit area—it has already been explained that the force called magneto-motive force, producing such effects, increases with the increase of ampère-turns around the body of the magnetic metal. While this is always strictly true, it is also true, as in all other cases of disturbing forces, that the quantity of motion, of any particular kind, depends upon the resistances met, as well as the original forces acting. Indeed, the relation between "lines of force," magneto-motive force, and magnetic resistance may be expressed in terms as simple as those of Ohm's law. Using the term "Induction" in a sense analogous to that in which "Current" is used, we may write

$$\text{Induction} = \frac{\text{M. M. F.}}{\text{Resistance}}, \text{ or } I = \frac{M}{R}.$$

All substances save three (iron, nickel and cobalt) have enormously high magnetic resistance.

Magnetic action takes place along lines connecting one pole

FIG. 9.—BAR MAGNET.

of the magnet with the other, and the resistance to magnetic action along such lines determines the "strength of field" set up by a given magneto-motive force.

Thus (Fig. 9) in case of the straight magnet N S there is a long air space between N and S, and air is susceptible to magnetic action only in a very low degree. We may then increase the strength of field by simply bending the magnet into a U shape, thus shortening the air space between the poles, without change in the dimensions of the magnet or in the number of ampère turns exciting it. An armature revolving between N′ S′ (Fig. 10) would, at the same velocity, generate more than twice the E. M. F. that it would if placed at either A or B (Fig. 9).

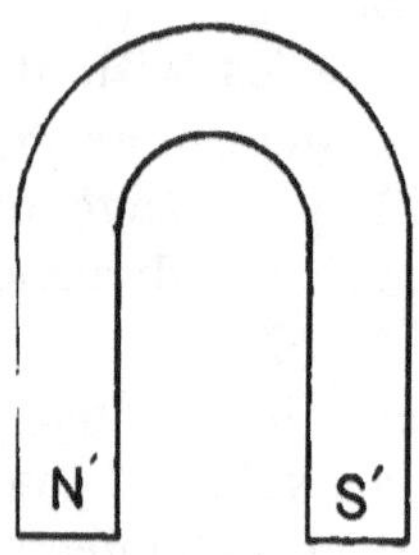

FIG. 10.—HORSESHOE MAGNET.

Further, an armature whose wires are wound on an iron core, occupying as much as possible of the space between N and S, will generate in each coil a greater E. M. F. at same speed than would one on a non-magnetic core, since the total resistance to magnetic action is less in the first case than in the second.

If, starting with the straight magnet, we not only bend it to bring the ends near each other, but also actually shorten the magnet, we again diminish the total resistance in the magnetic circuit, and with the same magneto-motive force obtain more "lines of force" passing through unit area from N to S—or we may again diminish resistance by increasing the cross-section of the magnet, just as we would decrease resistance to an electric flow by using large wires instead of small wires.

It might seem from the foregoing remarks that any desired strength of field could be obtained by proportionate increase of magneto-motive force or decrease of resistance. But unfortunately experiment has shown that after being magnetized to a certain degree, even iron becomes stubborn as air or other non-magnetic bodies to any increase of magnetism, however high we may urge the magneto-motive force. When magnetized to such a degree, iron is said to be "saturated." It has been further found, as might be expected, that in approaching saturation the resistance of iron to further action becomes *gradually* greater until it becomes as great as air.

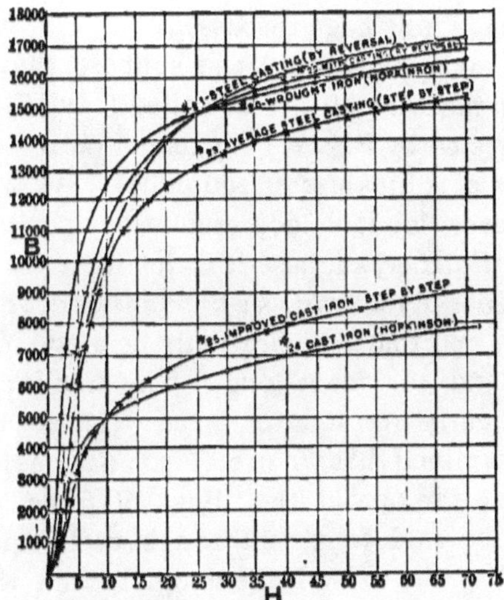

Fig. 11.—Magnetic Properties of Various Irons.

Over a certain range, the increase of magnetic strength in iron seems exactly proportional to increase of the magnetizing ampère turns, but beyond that range this strict proportion disappears. Thus let us say that ten ampère turns (often written A-ts) produce a magnetic strength denoted by unity. Twenty ampère turns may then produce a strength 2, but in going to forty we may find the strength not 4 but 3 only, and in going to 160 we find the strength barely 3.5, and so on, until a point is reached where no perceptible increase of strength follows from any increase of magneto-motive force. This change in the magnetic resistance of different kinds of iron is best shown by the curves (Fig. 11).

On the horizontal axis are shown magneto-motive forces increasing from left to right. On the vertical axis are shown the corresponding strengths of magnetization produced in various materials under like conditions. If the iron did not increase in magnetic resistance the E. M. F. generated in an armature would increase steadily with the M. M. F., as it does, very nearly, until the curves turn sharply to the right—but beyond that point the increase of E. M. F. is less, relatively, than the increase of M. M. F.

The general principles governing the design of dynamos and motors have now been discussed.

To show their application, let us suppose that it is desired to construct a machine of the so-called U type (see shape of letter U suggested by Fig. 10) capable of delivering a current of 80 ampères at a pressure of 200 volts. We will predetermine the speed—let it be 600 revolutions per minute, or 10 per second. Further, suppose it is desirable that a solid core armature—generally known as a Siemens or drum armature—shall be used, and that the core shall be a cylinder 10 inches in diameter and 10 inches long. The area inclosed by one loop would then be 100 square inches. It has been experimentally determined that when soft iron has been magnetized so highly that 120,000 of the "lines of force," above mentioned, may be said to pass through each square inch of metal normal to the axis of the magnet, then the mass has practically reached the point of saturation. To carry the iron to that point requires a very great M. M. F. Let us be content to obtain in the armature core 80,000 lines per square inch. We cannot in practice avail ourselves of the entire cross-section of the armature core as an iron conductor of magnetic lines of force, because this core is not made solid, but generally is built up of a great number of very thin plates of iron separated by paper or other non-magnetic material. This is done to prevent the flow of currents of electricity induced by moving such a conducting mass in a field of force, and which, in a solid iron core, necessarily of very low resistance, would be very great. Their injurious effects would be (1) heating of the armature and (2) consequent loss of energy.

The use of paper reduces the iron to a cross-section about 85 to 90 per cent. that of the area of the loop. Hence if we obtain 80,000 lines per square inch through the armature as a whole, we must obtain about 18 per cent. more through the iron of the armature, or about 95,000 lines per square inch of iron.

Experience has shown that if the total lines of force passing through a section at D T (Fig. 12) be represented by 100,

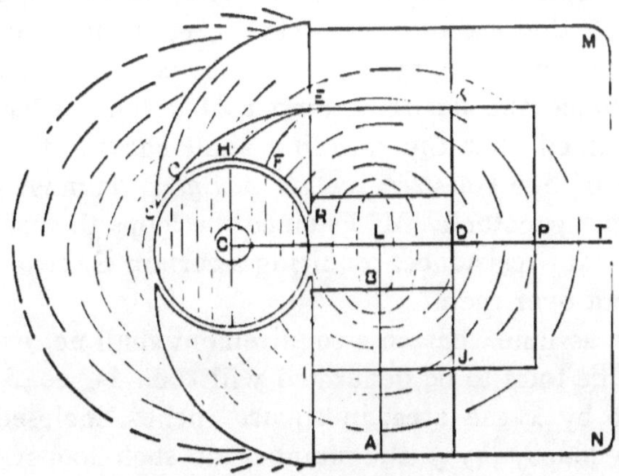

FIG. 12.—MAGNETIC CIRCUIT OF DYNAMO.

then the number passing through the armature will be from 60 to 80, the remaining 40 to 20 passing through the air, as indicated in the figure.

With reasonable attention to shaping our machine we may count upon a loss not greater than 30 per cent. In order, then, to have 80,000 lines per square inch through the armature loops, we must either have the section at D T of greater area than at V W, or must force about 120,000 lines per square inch through the iron of the magnets, and their connecting piece, the "keeper." Supposing the case not to be one which requires the weight of iron to be a minimum, we would prefer to increase the area so that the total lines at D T shall be distributed at the rate of 80,000 per square inch. The section A B of the magnet core may be nearly a mean between the sections V W and D T, the lines per square inch being still 80,000.

To obtain absolute values for these areas we may proceed as follows: It has been experimentally shown that if one loop be rotated once per second in a field such that one unit line of force passes through the loop, then the E. M. F. generated in such a loop will be 0.00000002 volt. As we have chosen a speed of 600 revolutions per minute, or ten per second, we may at once use the larger coefficient 0.0000002 (see page 17). There must be such a number of loops and such an enclosed area through which 80.000 lines per square inch are passing that the number of loops, multiplied by the total lines of force enclosed, multiplied by the coefficient 0.0000002, shall produce the desired number of volts. It has already been assumed that the machine shall show a difference of potential of 200 volts *between the brushes.* It must generate a somewhat greater E. M. F. since the loops themselves will offer a certain resistance, requiring a certain E. M. F. to force the current over them.

Let us assume that this requirement shall not exceed 4.0 volts. The total to be generated will then be 204. Let us represent by x the area, in square inches, inclosed by the armature loops; by y the number of such loops; we may then write this simple equation

$$204 = x \times y \times 00000002 \times 80000;$$

or
$$204 = x \times y \times 0.016;$$

or
$$x = \frac{204}{y \times 0.016} = \frac{12750}{y}.$$

We may readily obtain a working value for y based on these considerations. All the loops represented by y must be connected by the commutator bars, either by making only one loop out of each length of wire, and connecting its ends with adjacent bars; or by making several loops from one length of wire, the number of commutatator bars being lessened to one-half, one-third, one-fourth, etc. Electrical considerations preponderate in favor of equal numbers of loops and bars, and in larger machines, of comparatively low voltage and high speed, it is good practice to produce such equality. The difference of potential between successive bars is thus reduced to a minimum for the given number of loops; this reduces the tendency to spark at the brushes; injurious inductive disturbances in the loops themselves are likewise reduced by decreasing the number of loops per section. Ordinarily, however, it is difficult, with present methods of commutator construction, to produce this equality. Practice has shown that we should limit the difference of potential between successive bars to about ten volts. As we have in the present case 204 volts, we might have about 20 bars in each half of the commutator, or 40 in all. But, as we also know from practice, a considerably larger number of bars may be made into a commutator suitable for a machine of the kind, in view of which we may adopt a larger factor of safety as to sparking. For convenience let us assume 64 bars in the commutator. If we further assume two loops to the bar, the value of y becomes 128, and the value of x practically 100 square inches. If the cylinder of which this figure gives the area of cross-section be taken as ten inches long by ten inches in diameter, we have an armature of convenient and usual size for the conditions to be met. We may now readily determine, from what has before been said, that the section D T should be about 160 square inches; the section A B about 130 square inches. If these areas be circular, the diameters are about 13 inches.

We have now determined the cross-section of the iron in the magnetic circuit, except that of the pole pieces. While passing through these the magnetic lines curve in toward the armature, and those on the upper side of the shaft have a path somewhat shorter than those below. To produce a symmetrical distribution through the armature core, it is desirable to make the lower portion of the pole pieces quite full, *i.e.*, of as large area as practicable. No exact rule need here be given for this shaping of the pole pieces, but, as just indicated, the magnetic resistance along E F should be as nearly as possible equal to that along E G. One other consideration affects the shape of the pole piece and its area normal to the lines of force, that is, the angle between the lips, S C R. Without entering into the rather lengthy discussions which have been heard as to the best angle, we may assume for an ordinary case, without serious error, that each pole piece covers 145° in arc of the armature. The length of this arc on the edge of the pole pieces will then be very nearly 13 inches.

The length of the line E H, which may be taken as the length of the magnetic circuit through the pole piece, depends on the distance between the axes C D and E K. This in turn depends on the diameter A B (13 inches) and the breadth B L of the space left for the magnet coils and the free space required for hand room and ventilation. From general experience we may safely make B L = 4 inches, thus making K D = 10.5 inches, the resulting length of keeper M N thus being about 34 inches and E H about 7 inches.

If we make proper allowance for the air space from surface of pole to surface of armature core (this will be done later) we may, as to length of iron circuit, consider E H and H C as continuous, making a total of 12 inches as to length, and without great error we may consider the area for this length as 100 square inches (it is greater in pole, less in armature). In the keeper we have already a defined length of 34 inches, or for one-half the magnetic path in this part about 17 inches, with cross-section of 150 square inches. If now the length E K of the magnet and the length of the air gap were known, the total resistance of the magnetic circuit would become known. As E K is dependent upon the M. M. F. required (and hence the space needed for winding the magnet coils), and as the M. M. F. is largely dependent on the resistance of the air gap, let us first consider that. Its length from iron to iron is determined by the space needed for laying the loops of armature conductors (of copper), insulation for same, and a reasonable clearance between exterior of armature and pole piece. This clearance we may fix at 0.1 of an inch. We may also allow 0.1 of an inch for laying the armature conductors; or, since copper is practically of equal magnetic resistance with air, we may say that the air space is 0.2 inch in length. Its area is about 130 inches, namely R S (= 13 inches) × length of pole piece (= 10 inches = length of armature).*

The total "induction" or flow of magnetic lines of force, through the armature is 80,000 × 100 = 8,000,000. This number must be forced across the air gap, being equivalent to about 62,000 per square inch. The magnetic resistance of this air gap will increase with its length and decrease with its cross-section: we may therefore, simply for convenience, calculate the M. M. F. needed for forcing 62,000 lines across one square inch; that force will be the same required for total flow across the total section. From experience it is known that to force one unit line through an air path one inch long and one square inch in cross-section we must apply, as M. M. F., 0.25 ampère turn. In our particular case the length of path is only 0.2 inch, while the number of unit lines is 62,000—hence for the total M. M. F. we have 3,100 ampère turns (0.25 × 62,000 × 0.2 inch = 3,100).

Going to the pole piece and armature, as has been said, the magnetic resistance of iron varies with the "induction" maintained. In this case we require 80,000 lines per square inch, and at that density the resistance of air is about 1,200 times greater than that of soft iron, or for one unit line through a path one inch long and one square inch in cross-section 0.000208 ampère turn is required. Hence for total length (12 inches) and total induction the ampère turns = 0.000208 × 80,000 × 12 = 200.

We may now go to the keeper.

The machine being symmetrical, we may take half its length separately, since we are now determining the M. M. F. needed to overcome resistances in one-half the machine, and this force is to be supplied by the current encircling one of the magnets. The induction being 80,000 lines per square inch, we have 0.000208 × 80,000 × M T = about 300.

Let us now assume an approximate M. M. F. for the magnet core. We may use 900 ampère turns as an outside figure

* If the armature conductors be wound in slots, cut in the periphery of the armature core, instead of being laid on the surface of this core, the length of the air gap proper may be considerably diminished. It is then to be remembered that since the iron of the core has, to the depth of the slot, been largely cut away, and the spaces filled with non-magnetic substances, the magnetic resistance of this outer ring of the core is greater than before the slotting.

allowing a considerable margin for joints in the magnetic circuit. Add this to the other figures, we have 3,100 + 200 + 300 + 900 = 4,500. Now this number of ampère turns, as has been explained, may be made up of any two selected factors, representing current and turns respectively.

If we wish to design a series motor, the current is at once determined by the original assumption to be 80 ampères.

In this case the resistance and number of turns would be small. If a shunt motor is under consideration, we should make the current quite small, that the constant loss of power in the field may be small, the resistance and number of turns being relatively large. Assume 1.5 ampère for the field current in this case.

Then the number of turns = 4,500 ÷ 1.5 = 3,000. The resistance of these turns follows from the fact that this current is to be produced by an electromotive force of 100 volts, being one-half the total of the normal E. M. F. for which the machine is calculated. From the rule previously given, namely,

$$C = \frac{E}{R}, \text{ we have } 1.5 = \frac{100}{R}. \quad \therefore R = 66 \text{ ohms.}$$

This then must be the resistance of the 3,000 turns of wire. The average length of each turn, assuming the winding to be two inches deep, will be equal to the circumference of a circle about 13.5 inches diameter—the magnet core being supposed to be of circular cross-section. This length equals 42 inches, or for the 3,000 turns 10,500 feet. Its resistance per foot must be 66 ÷ 10,500 = 0.00629 ohm. From tables prepared for the purpose we find that a wire having a diameter of 0.052 inch will serve our purpose. When covered with the usual cotton insulation such a wire would have an outside diameter of about .075 inch.

Approximately 300 such wires could be laid along one inch of the length of the magnet, the depth being two inches. The length of magnet core would then be 10 inches, in order to place the 3,000 turns.

Making now calculations for the M. M. F. as determined by this length of magnet core we have: ampère turns required = .000208 × 80,000 × 10 = 166.

Our assumed value, namely 900, is thus found to be ample, providing a large allowance for joints in the magnetic circuit, the resistance of a good joint having been estimated to be equal to that of 8 inches of iron having a cross-section equal to the area of the joint.

It remains now to determine the size and number of wires to be wound on the armature, and also the number of separate sections into which the total number shall be divided.

As will be seen by reference to Fig. 13, the current in flow-

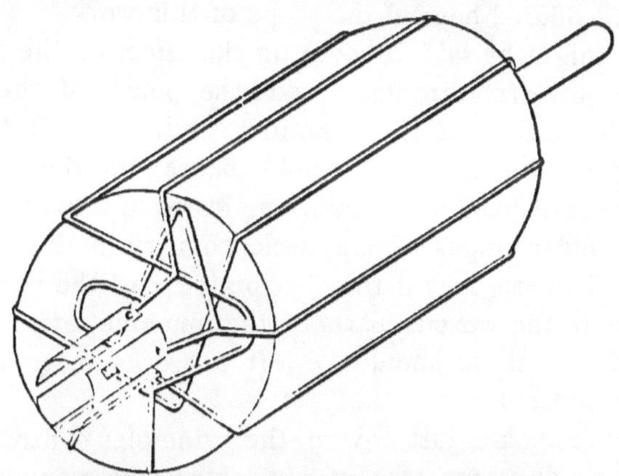

FIG. 13.—DIAGRAM OF ARMATURE CONNECTIONS.

ing over the armature as a whole is divided into two equal parts. In this case we will then have 40 ampères to be forced over a circuit made up of one-half the whole length of wire on the armature, and we have already designated 4.0 volts as the E. M. F. for this purpose. The resistance of this half of the armature wire must follow, $40 = \frac{4}{R}$. $\therefore R = 0.1$.

The total length of a turn around the armature (see dimensions) will be 40 inches, or with a reasonable allowance for the length of the end going to commutator, and for overwrapping at the armature heads, we say 44 inches. For one-half the total length we have, in feet, $44 \times 64 \div 12 = 235$.

The resistance per foot must be $0.1 + 235 = 0.00043$, given by a wire having a diameter of .144 inch (No. 7 B. & S.).

We may use this, or more conveniently perhaps two smaller wires in multiple, and on investigation we find that two wires of No. 10 B. & S. gauge will give equal resistance, while permitting more space, measured along a radius of the armature, for insulation and clearance.

All the principal features in the design have now been determined. The values assumed for speed, E. M. F., number of commutator bars, etc., entering as "known quantities" in the problem, have of course been made to "fit" more accurately than is often the case on first trial, in the practice of design. Certain refinements have not been mentioned, being considered beyond the scope of this work.

Much might be said concerning the effect of the magnetization of the armature itself upon the poles; of the mutual and self-induction of the armature conductors; of the exact shaping of the lips of the pole pieces; of the important phenomena of hysteresis, involving losses of imparted energy due to rapid changes of magnetic polarity in the armature core (or field magnets if they be rotated); of the rise of temperature in the various parts of the machine, etc.; but it is believed that these should be left to works more technical than the present.

We have spoken just now of the principles controlling the design of a dynamo—that is, a machine whose function is to generate electric current which may be used for lighting, heating, or for producing motion in other electric machines, similar in construction to the dynamo, but whose function it is to transform the electric energy received from the dynamo into mechanical energy. Such a secondary machine is called generally a motor. The principles of design are identical, and indeed a machine made for use as a dynamo may instead serve as a motor and vice versa.

In the dynamo the action may be briefly described thus: Given lines of magnetic force and looped conductors located in the resulting field of force, a rotation of such conductors produces electric currents and requires continuously supplied mechanical energy to continue such rotation against the attraction set up between the lines of force and the electric current in the conductors. In the motor a complementary action takes place and may be thus described: Given lines of force and looped conductors located in the resulting "field," a flow of current through such conductors will produce an effort at rotation due to attraction set up between the electric current and the lines of force, and requires continuously electric energy to produce rotation against the mechanical resistance presented by the friction of parts and the useful work or "load," and to overcome electrical resistances in the armature and magnet circuits.

The energy for magnetizing the field is a very definite quantity and may be calculated as explained above. Considering the armature of a motor, it is apparent that when rotating under the influence of a current from an external source it likewise fulfills in all respects the conditions required for action as a dynamo; *i.e.*, for the generation of an electromotive force within its own conductors. Such an electromotive force is indeed set up, as shown by the fact that, save when a motor armature is held mechanically against all possibility of rotation, the current flowing through is not that which would be produced by the given external E. M. F. acting over the given resistance of the armature, but is a less current, such as would flow over such a resistance, under the influence of an E. M. F. equal to the *difference* between the external E. M. F. and the E. M. F. that would be produced by such a machine if driven at the given speed as a dynamo. This is actually the E. M. F. generated by the motor, and is generally called counter E. M. F.

Now the total energy absorbed by the motor is, of course, that measured by the product of the external E. M. F. (called usually the "impressed" or "applied" E. M. F.) and the current that may actually flow; or, to use expressions more familiar in hydraulics and pneumatics, these two elements of pressure and of quantity when multiplied together measure the power or "work."

The mechanical energy developed by the motor armature is measured by the product of the *current* in the armature and the *counter* E. M. F. The ratio between the electrical energy absorbed and the mechanical energy given out by the motor is the measure of its efficiency. We may express absorbed energy by $C \times E$, and developed energy by $C \times E'$ (E and E' being impressed and counter E. M. F. respectively). The efficiency then becomes $\frac{C \times E'}{C \times E} = \frac{E'}{E}$. This efficiency is found in many motors to be as high as 93 to 95 per cent., leaving little to be desired. The same machine may be caused, by overload, to work at lower efficiencies, since the great currents required for the work require greater proportions of the absorbed energy simply to force these currents over the internal resistance of the motor, leaving less for external useful work. This point will be more fully developed later.

CHAPTER II.

CONCERNING PRIME MOVERS.

By prime movers we mean such mechanisms as are fitted to utilize somewhat directly natural sources of energy for the production of mechanical power. So far as electrical purposes are concerned, the number of available prime movers is very limited, and when we take into consideration the particular demands to be met in the supply of electrical power for traction purposes, we find the only generally available prime movers to be steam engines and water-wheels. One or two very small electric roads in England are operated by gas engines, but in this respect they are quite alone, and the practice, for various reasons, is seldom to be advised. It does not befit our purpose here to enter into any exhaustive discussion of the theory of the steam engine, or to describe in detail the very numerous modifications of the few types that are met in modern engineering practice. We shall merely attempt to give in a very brief form, without having recourse to mathematics, the general principles involved in the construction and operation of modern engines, and to give some account of the means by which their action and particular properties may be most conveniently investigated. Having done this, it will only remain to describe in a manner necessarily condensed the general principles of the utilization of water power.

The steam engine is to-day our most universal means of obtaining mechanical power, and although the theory of its action is somewhat complicated the main facts are easily understood. The fundamental principle of heat engines, of which steam engines are the most familiar variety, is the production of pressure in a gas by the transference of heat energy to it, and the utilization of this pressure to recover, in the form of mechanical motion, the energy thus given. The primary source of energy is that which exists as potential chemical energy in fuel of various sorts. The gas almost universally employed is the vapor of water; in other words, steam. It is a singular fact in physics that all gaseous bodies are remarkably alike in their dynamical properties. They expand and contract, if free to do so, in almost exactly the same amount when the temperature varies; or, if the volume be kept constant, the pressure varies with the temperature very uniformly for all gases and all temperatures.

If, then, we are to utilize steam pressure practically in an engine cylinder, the higher the temperature of the steam the greater will be its pressure. The amount of work that can be gotten out of a given quantity of steam at a certain pressure obviously depends on the extent to which we are able to utilize that pressure. If we could employ it all in pushing on a piston we should be able to get very efficient engines indeed; but since we are living in an atmosphere that has a pressure, and in a world that has sometimes a rather high temperature, we are in practice totally unable to avail ourselves of all the pressure produced. In other words, if we give a certain amount of heat to a volume of steam, creating thereby a high pressure and temperature, we shall be unable to get all the heat energy back in the form of mechanical motion of the engine simply because of the necessary limitations imposed by our conditions with respect to pressure and temperature. Aside from these the energy lost is small in amount.

It is found by experiment that if we were to have a unit volume of some gas, like air, at the temperature of melting ice, one degree Fahrenheit rise of temperature would increase its pressure by 1-493d part. Consequently, there might be some temperature so low that if it were possible to reach it the gas would have no pressure; in other words, it would have lost its elasticity and given up all its energy. This point is 461 degrees below zero Fahrenheit, as appears from the rate of increase just given. This particular temperature is known as the absolute zero. If we were able to start at this point, and by applying heat to raise the temperature of any gas up to our present working temperature, we should give to that gas a certain amount of heat energy, all of which could be recovered if we were able to let it expand and cool itself off by expending its heat as mechanical work until all its pressure was exhausted, when its temperature would be back to the starting point of minus 461 degrees. Unfortunately, we are unable either to start with a gas or vapor at such a low temperature, or to reduce its pressure after heating anywhere nearly to zero.

Obviously we cannot use all its pressure unless we get down to the absolute zero, and in practice we are compelled to cease utilizing its expansive power when the temperature goes down to the every-day temperature at which we live; so only part of the whole amount of heat can be recovered as mechanical work, the exact proportion depending on the extent to which we are able to cool by expansion the working gas, allowing it to do work and lose pressure. If, for example, we work with gas at 400 degrees Fahrenheit, and let it expand until it cools down to zero Fahrenheit, we shall utilize that portion of the pressure of the gas that corresponds to a change in temperature of 400°. All the pressure which corresponds to the temperature from zero down to the point where pressure ceases is unavailable.*

The efficiency of such a transformation of energy is, then, somewhat less than 50 per cent. To be exact, the efficiency of the heat engine is equal to the fraction $\frac{T_1 - T_2}{T_1}$, where T_1 is the highest temperature to which the gas or vapor has been heated, and T_2 the temperature at which we are compelled to stop utilizing its pressure, T being always reckoned from the absolute zero. In practical engineering, T_2 is the temperature of exhaust or condensation. The general law of the efficiency of the utilization of heat by en-

* In ordinary engines a nearly saturated vapor, such as steam, is used, and in producing it much heat is taken to drive the substance from the liquid to the gaseous state, so that, although we begin to give heat to the gas at a comparatively high temperature, the total heat required is much as if we had started at or near the absolute zero. As various liquids require different amounts of heat to vaporize them, one might think that there could be a great gain by properly choosing the liquid In fact, however, the liquids that vaporize with little heat per unit weight give a considerably smaller total volume of vapor, so that, so far as working in engines is concerned, the particular liquid used makes little or no difference.

gines is that the heat transformed into mechanical energy is to the whole heat received by the working fluid as the range of temperature is to the absolute temperature reached by the fluid. The perfect efficiency theoretically possible can never be reached because of our inability to command sufficient range of temperature, but the best modern engines come quite near to the maximum efficiency *possible between the temperature limits attainable in practice.* Taking nature as we find it, and even condensing our working gas at ordinary temperatures, we can really recover as mechanical work a very respectable proportion of heat. Sufficient has been said, however, to show that the higher the initial pressure given to the steam and the lower the pressure at which it is exhausted or condensed, the more efficient will be the engine in converting the heat energy of coal into mechanical power.

Now in every-day work the pressure of steam is utilized in only one way, that is, by employing it to drive a piston along a steam cylinder, thereby working a driving wheel of some kind by means of the piston rod. The rudimentary steam engine, then, consists of a closed cylinder containing a movable piston, with a piston rod reaching into the outside air. The pressure of the steam, admitted to the cylinder by valves, drives the piston to and fro. In some of the earliest engines the steam was admitted to the cylinder through a valve operated by hand, and when the end of the stroke was reached the steam, instead of being allowed to escape, was condensed by a spray of water thrown inside the cylinder; afterward the steam was allowed to escape at the end of the stroke through an exhaust valve, or else to flow into a separate chamber, where it was condensed. Finally, about one hundred years ago, the engine was provided with valves to admit automatically the steam to the cylinder, and to let it escape at the end of the stroke. With various modifications, that is the form of machine used to-day.

Fig. 14 shows a diagram of the working parts of a modern engine very extensively used for electric light and power purposes. Its cylinder is provided with a steam port A, at each end, communicating with the steam chest S S. This latter is placed immediately next the cylinder, and receives the steam at full boiler pressure through the supply pipe. The entrance of the steam into the cylinder is controlled by the valve V, which is a hollow cylinder closed at the ends and contracted somewhat in the middle. This piston valve has a slight to-and-fro motion communicated to it by an eccentric on the engine shaft, and slides backward and forward in a cylindrical cavity in the steam chest; at each end of this is an exhaust pipe communicating with the outer air, or with the condenser.

Observe that in this arrangement the valve is surrounded on all sides by live steam, and consequently is not pressed against any of the wearing surfaces: in other words, it is what is called a balanced valve. In the position shown the valve is just admitting steam to the piston end of the cylinder; a very even flow is secured, not only around the shoulder of the valve as its enlarged portion passes beyond the

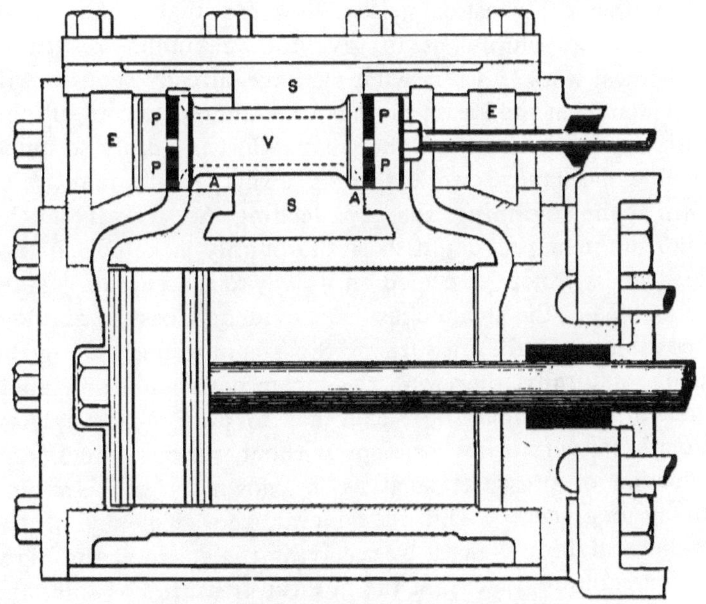

FIG. 14.—WORKING PARTS OF SIMPLE ENGINE.

steam port, but also through the ports P P at the opposite end of the valve and through the inner space. At the same time it will be seen that at the other end of the cylinder the valve has moved in its bearing completely past the steam port, allowing the steam to flow out freely into the exhaust; when the piston has moved a little way on its return, the port A, through which the steam has been flowing, will be closed as the solid part of the valve slides over it, and open for the exhaust at the other end of the stroke when the valve slides clear beyond. It will thus be seen that the same mechanism admits the steam to the cylinder or allows it to flow into the exhaust, according to the relative positions of the valve. This rudimentary description of the steam engine will be unnecessary to most of our readers, but will prove useful in connection with points to be presently mentioned with reference to the economical use of steam.

Suppose, now, that we have communicated heat to a steam boiler to produce a quantity of steam at an ordinary working pressure—such, for example, as 100 pounds to the square inch in excess of the ordinary pressure of the atmosphere—a very practical question arises: how can that steam be so utilized in an engine as to give the maximum return in mechanical work? From what we have already seen, it will be evident that the steam must not be allowed to cool off and lose pressure, because all the heat so lost is simply so much possible mechanical work thrown away. Consequently, to begin at the beginning, the pipe leading the steam from the boiler to the engine ought to be thoroughly jacketed, so that the steam will not get cooled on its way to the engine. Having arrived at the steam chest, it is evidently best to employ, if possible, the full pressure of the steam to push along the piston; naturally, therefore, the steam ports and other apertures through which the steam has to pass must be large enough to permit its passage without serious check and reduction of pressure, such as is known as wire-drawing among engineers. And in general, the efficiency of the engine will be increased by receiving the steam at the highest temperature, that is, at the highest pressure possible, and discharging it at the lowest temperature possible; in other words, after having obtained all the heat available from it in the form of mechanical work.

The fundamental principle of the efficient working of steam depends on following out this very obvious suggestion, that has been admirably stated by Rankine as follows:

"Between given limits of temperature the efficiency of a thermo-dynamic engine is the greatest possible when the whole reception of heat takes place at the higher, and the whole rejection of heat at the lower limit."

It is desirable, therefore, in admitting steam to the cylinder to let it in, so far as it is admitted at all, at full boiler pressure, and to exhaust it as completely as may be at the lowest available pressure. A vast amount of ingenuity has been spent in arranging the steam engine so as to produce this result; in other words, to move the valves in such a way that the steam admitted may come in at a uniform high pressure and be exhausted at a uniform low pressure. The difference between a good and a bad engine lies largely in the success with which this has been accomplished. In the early days of the steam engine it was usual to leave the admission valve open, allowing the steam to rush in direct from the boiler, until the piston rod reached the end of its stroke. Such a plan permitted a given engine to do a great deal of work, because the piston was all the time subjected to the full pressure of the steam, but at the same time it is obviously very uneconomical, since the steam when we are through using it will have nearly the same temperature and pressure as at first, the deficiency, which has gone into work in pushing the piston along, being kept up by fresh supplies from the boiler during the stroke. An engine of this character requires, evidently, enormous amounts of fuel, so one of the early improvements in engineering practice was to use the steam expansively; in other words, to admit to the cylinder a comparatively small amount of steam, and let it finish up the stroke of the piston by its own expansion.

Under these circumstances, steam enters at full boiler pressure only, perhaps, for a quarter of a stroke; the supply valve being then closed, the steam expands and uses up its heat quite completely in doing work on the piston, so that when the stroke is completed the steam exhausted will be low in temperature and pressure—or, in other words, working steam expansively enables us to increase the range of temperature through which our heat engine works, and hence increases its efficiency. In so using steam, however, the pressure produced on the piston is not uniform, and should we desire to compute the rate at which the engine is doing work, it is necessary to know the average pressure that the expanding steam has exerted. An instrument for finding this average will be described later.

If an engine is allowed to condense, that is, to exhaust its steam into a vessel that is kept cool, it is possible still further to increase the working range of pressure by very nearly the pressure of the atmosphere, and hence the efficiency. Wherever plenty of water for keeping the condenser cool is available condensing engines are very frequently used, and are especially useful when the engine is but lightly loaded. By so getting rid of the steam, there is a very perceptible saving. From what has been said we are prepared to pass at once to the practical discussion of every-day engines.

Following Rankine's general principle, we may begin at once by saying that the ideal valve gear for an engine is that which can open instantaneously when steam is wanted, admits it to the cylinder at full boiler pressure, and closes instantly, to avoid letting any steam in at a lower pressure when it is desired to shut off the steam supply. When the expansion of the steam is completed the exhaust valve should open instantly and let the pressure down to that of the outside air, or the condenser, at once. From the diagram of the engine just given you will see that these conditions are but partially fulfilled; for so long as the same ports are used for admission and exhaust, and are controlled by a single sliding valve, the engine is not always capable of proper adjustment to the conditions under which it is working, because the four functions performed by the single valve—the opening and closing of the supply ports and the opening and closing of the exhaust ports—are not independent of each other, and, consequently, have to be so related as to produce the best average effect, not always the most desirable one in a particular case. It was for the purpose of avoiding the difficulties met in these single-valve engines that more complicated valve gears have been introduced. The best type of these is the Corliss, which has been used for over forty years.

In this form of engine, shown in Figs. 15 and 16, four separate valves are employed, two to admit the steam to the

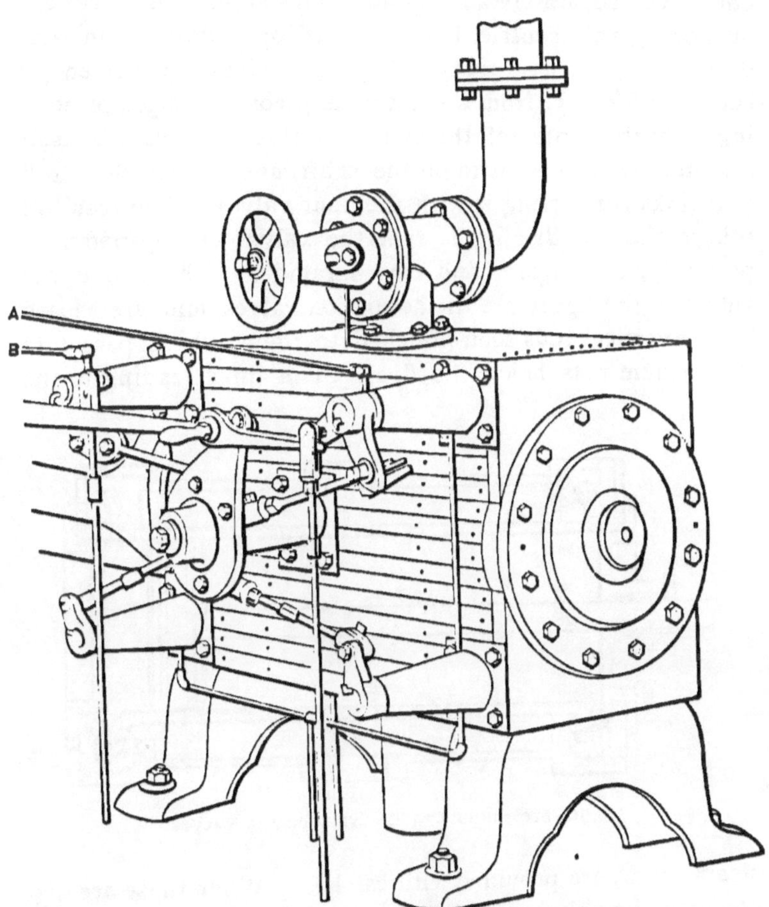

FIG. 15.—CORLISS VALVE MOTION.

cylinder and two others to let it out when it has been used: they operate by a slight turning motion about their axes, and are usually in the form of slightly tapering cones, having the steam ways cut as long longitudinal slots. It is apparent that valves of this sort can be thrown wide open very quickly and closed as speedily, and that they are capable of any sort of independent adjustment; the result is apparent in the superior economy that is obtained by such mechanisms.

The Corliss valve gear has various modifications, but all retain the distinguishing characteristic of four independent valves separately adjustable. The admission valves are opened by an eccentric, but are closed by gravity or, in later designs, by vacuum pots. All four valves are driven (as shown in Fig. 15) from a wrist plate pivoted on a pin projecting from the center of the cylinder—this wrist plate is itself actuated by the eccentric on the shaft, and has attached to it four links connecting it with the four valves. The result of this method of driving is that the valves are opened very promptly, held open, and closed rapidly. The two upper valves in the figure are the admission valves, and are closed by the vertical rods seen attached to them, which pass into the vacuum pots below the floor. The links leading to the

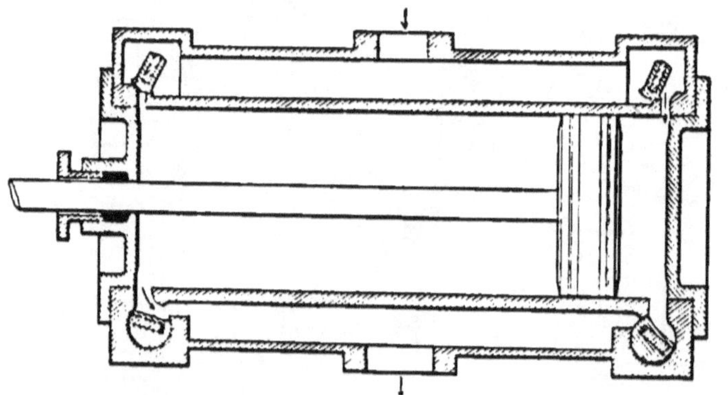

FIG. 16.—SECTION OF CORLISS CYLINDER AND VALVES.

steam valves are provided with catches. When these are disengaged the valves are free to close, and each catch takes hold on its backward stroke ready to open its valve. The rods A and B control the position of these catches, so that the valve may be released and closed at any point in the motion of the piston up to nearly half the full stroke. These rods are controlled by an ordinary centrifugal governor, usually belted to run much faster than the engine in order to make it sensitive. This Corliss valve gear is used only on comparatively low-speed engines, 140 revolutions per minute or less, because the vacuum pots or weights that close the valves do not work rapidly enough to operate at a much higher speed.

For most engines running two or three hundred revolutions per minute a simpler valve gear, much like that shown in Fig. 14, is almost universally employed. This may take many forms, most of them, however, retaining the same general characteristic of a balanced valve simply and strongly made, and controlling a pair of ports that serve both for admission and exhaust. The particular diagram shown refers to the Armington & Sims engine, which is used to a very large extent in electric light and power work. But something more than proper valve motion is necessary to secure the best results from an engine. One of the most serious losses met, particularly in engines driving electric railway plants, is cylinder condensation, or loss of heat due to the steam being cooled by the metallic portions of the cylinder; the result is that a certain amount of heat which ought to be utilized on the piston is wasted.

Considerable loss of this kind is almost certain to take place where an engine is underloaded. The steam is admitted at a high pressure and expanded altogether too much, during which process a portion of it gets condensed. There is for every steam pressure a particular amount of expansion that gives the best results; if more than this be employed, the engine generally suffers from condensation and excessive cooling of the steam by radiation of heat through the sides of the cylinder; while if less expansion be used, the steam is rejected while it still has a considerable amount of valuable energy which is thus lost.

When high steam pressures are to be used, as is most desirable on the score of economy, it is best that the expansion should not be carried out in a single cylinder, but consecutively in two or more. Engines so arranged are called compound. The advantage of this arrangement lies largely in avoiding the losses incidental to changes of temperature that take place during expansion; the more nearly uniform the temperature of the cylinder can be maintained, the less loss from radiation and condensation. Where the temperature range of steam in the cylinder is great, as in the case where high initial pressures and high expansions are used, it is a difficult matter to keep the cylinder at a steady enough temperature to allow economical expansion of the steam. If, however, the expansion is distributed over two or more cylinders, only one of them will communicate with the condenser, in itself a ready source of cooling, and each cylinder can be kept at a temperature much more favorable to escaping loss than if the entire temperature range of expansion were in a single cylinder. Engines in which this useful plan is carried out are coming into very extensive use. The compound engine, either non-condensing or condensing, has found its way into a large number of electric central stations, and the triple-expansion engine, in which the work of the steam is still further subdivided, is gradually displacing the compound engine where large powers are required. The compound construction is generally more advantageous in condensing than in non-condensing engines, and under favorable circumstances will save something like 40 per cent. of the fuel that would otherwise be required.

A very important part of every engine is the governor. In its original form it consisted simply of a pair of balls supported by arms hinged to a vertical shaft put in rotation from the main shaft of the engine; links from each ball led to a collar capable of vertical play up and down the shaft. This collar gave motion to a lever controlling the main steam valve, and when, owing to the high speed of the engine, the balls were thrown outward by centrifugal force, the collar was raised and the steam partially cut off. The objections to this form of governor are numerous; two of them being lack of sensitiveness, owing to the considerable amount of work that has to be done by the governor in operating a steam valve, and the fact that engines equipped with this throttling governor work at a fixed cut-off, that is, the steam is always admitted to the cylinder for a definite part of the stroke, but at a pressure depending on the amount the steam is wire-drawn in passing the half-closed valve. By thus lowering the initial pressure the efficiency of the engine is decidedly diminished. One of the great improvements made in the introduction of the Corliss valve gear was a successful way out of both of these difficulties. In the first place, the work done by the governor is very little, only sufficient to release the catch on the link that works the admission valve and permit the latter to close. This gives such sensitiveness that a very slight change of speed will suffice to move the governing levers. The second advantage is that instead of admitting steam to the cylinder for a fixed portion of the stroke, but at varying pressure, the governing is accomplished by changing the cut-off. In other words, when the load grows heavy and the engine begins to slow down, the catch that holds the admission valve is shifted so as to keep the valve open during a greater portion of the stroke, thus furnishing a higher average pressure during the stroke. The steam is thus used at nearly the full boiler pressure and temperature, but in amounts depending on the work to be done.

Although the Corliss form of governor is capable of very excellent work, and is a vast improvement over any of the throttling governors, it still leaves, in its usual form, much to be desired as regards keeping the engine absolutely constant in speed.

We may as well state at once that so far as most electric railway service is considered an engine cannot be kept at really constant speed by any device yet invented. There are good and bad governors, but there is no perfect governor. One of the reasons for this is the enormous rapidity of the variations in load encountered in driving railway generators for a small road. Quadrupling the load will slow down any engine with any kind of governor. The decrease in speed is, of course, only temporary, and in a few seconds the engine recovers itself. But even with the most reliable and sensitive forms of governor now constructed the momentary slowing down is quite noticeable, although counted minute by minute the speed may be almost uniform. In the Corliss arrangement for governing the faults are two. First, the speed of the engine in revolutions per minute is so slow that opportunities to govern are too infrequent, for the drop cut-off used in Corliss engines is seldom worked above 100 revolutions per minute, and if forced beyond this point it is apt to fail in promptitude. Hence a very perceptible fraction of a second elapses between the action of the governor and the next ensuing opportunity to shift the valve. Besides this, the inertia of the governor itself causes some delay. The second fault is one that is common to all fly-ball governors of the kind described; they act too uniformly. In other words, the force tending to pull the balls downward is uniform, and the position taken by the balls for any given speed of the engine is always the same, so that there is continually a fixed relation between the action of the governing balls and the position of the valves.

Now, if instead of pulling the balls down by gravity a force can be substituted for it that will vary with the change in the position of the balls, so that the shifting of the valves may go on rapidly until the engine comes to the required speed, we shall have as the result a governor much more sensitive than the ordinary form, especially as regards changes in steam pressure or in load. On all modern high-speed engines such a governor is in use. Most of the smaller engines nowadays are high-speed machines, running from two to four hundred revolutions per minute according to size. The governor is, in such machines, usually placed where it will act directly on the eccentric to change the throw of the valves. Fig. 17 shows a governor of that class. Its operation will be seen at a glance. Two weighted arms pivoted near the periphery of the fly-wheel are thrown outward by centrifugal force, and their motion is resisted by a pair of powerful spiral springs, acting in this case in tension; the motion of these arms is transmitted through a pair of levers, clearly shown in the cut, to lugs on the eccentric, and tend to shift the latter's position on the shaft, and thus vary the valve motion. The engine can be conveniently adjusted for various speeds by varying the tension of the springs and the position of the weights on the arms, or both. These governors are much wider in range of action and far prompter in motion than the fly-ball governors previously described.

Some of them permit the admission of steam for over four-fifths of the entire stroke, thus taking care of an overload very effectively. We thus see that there are two quite distinct types of engine in use to-day for power purposes. In the first place, engines with Corliss or equivalent valve gear, having separate admission and exhaust valves and running at a relatively low speed, seldom over 100 revolutions per minute; and, second, the great group of single-valve engines in which the admission and exhaust are not independent of each other, running at speeds usually over 200 revolutions per minute, and furnished with fly-wheel governors of a

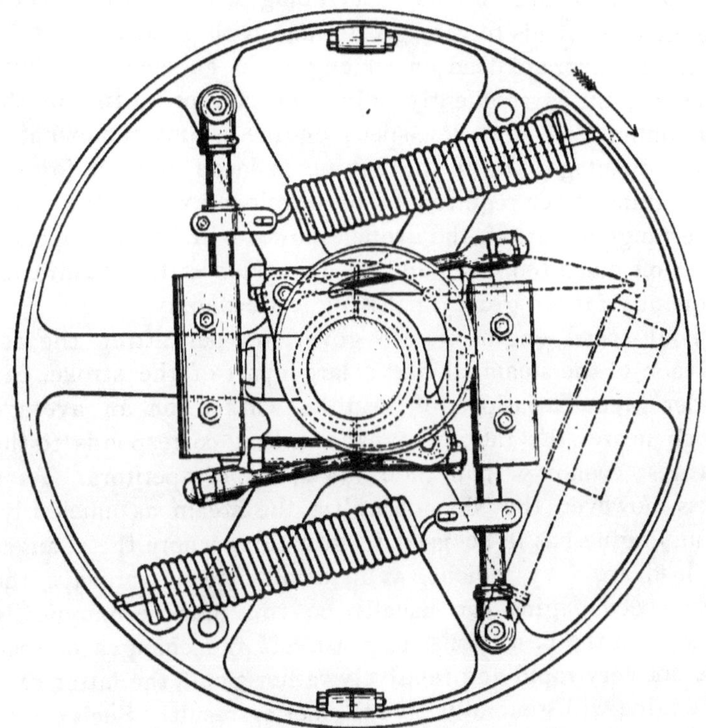

FIG. 17.—GOVERNOR OF HIGH-SPEED ENGINE.

very effective character. The choice between these two kinds of engine is somewhat difficult. There are a few excellent transition forms between the types, but nearly every engine with which the reader is likely to have to do belongs quite distinctly either to one or the other. As will be shown later, the slow-speed engines with the Corliss valve gear, or its equivalent, utilize the steam rather more efficiently than any single-valve engine has yet been able to do. On the other hand, they suffer more from condensation than high-speed engines of the same capacity for doing work, on account of the greater size of cylinder necessary to compensate for the decreased speed. They do not govern quite so well, and, as a class, do not permit of admitting steam throughout a portion of the stroke anywhere nearly as great as that which can be reached with the other mechanisms. For electric-railway service these are serious disadvantages, because the load is apt in small roads to vary greatly and with great rapidity.

As the average load on an engine in an electric-railway power plant is frequently only a quarter or a third of the maximum load, the slow-speed engine is put somewhat at a disadvantage, for if the engine is large enough for the maximum power required to be developed within the available range of cut-off, the average power will be developed at a cut-off much too short and an expansion of the steam consequently far too great for the highest economy.

High-speed engines, with governors permitting the admission of the steam during a large part of the stroke, can under great variations of load be worked on an average much nearer the ratio of expansion that corresponds to the greatest economy than their low-speed competitors. As a class, however, they do not utilize the steam as efficiently. If an engine has to be used in situations where the changes of load are very violent, as in small electric railways, the high-speed engine can usually be run more economically than even the best Corliss engines. If the changes of load are not very rapid and relatively rather small, the latter type of engine will generally give the better result. Such conditions are found now and then in large electric roads where the number of cars is so great that the starts and stops of individual cars are not important factors in change of load. This subject will, however, be more thoroughly discussed in the chapter on station design.

It is important that every man who is to use a steam engine should understand thoroughly how the power it produces can be measured. Ordinarily an engine is designated by its available horse-power, but this practice really gives very little useful information about the machine, and at present makers either do not specify the horse-power, but give the dimensions and speed, or in mentioning the horse-power mention also the corresponding point of cut-off of the steam. By one horse-power, of course, the reader will understand a rate of doing work equivalent to 33,000 foot-pounds per minute. This power is derived from the pressure on the piston of the engine, and the following almost self-evident rule gives the method of finding the horse-power of any given engine.

Take the area of the piston in square inches, multiply it by the number of feet that the piston travels each minute, multiply this product by the mean pressure on the piston during its stroke, divide the entire product by 33,000, and the result will be the horse-power of the engine in question. The area of the piston in square inches can be immediately obtained from the bore of the cylinder as given by the maker, and the number of feet of piston travel per minute from the stroke and the number of revolutions. The only uncertain and troublesome factor, then, is the mean pressure on the piston.

If the steam entered the cylinder throughout the entire stroke, the mean pressure against the piston would be very nearly the pressure of the steam in the boiler, and the problem would reduce itself to a push equal to the number of square inches in the piston multiplied by the pressure of the steam per square inch, and working as many feet per minute as the piston travels. But inasmuch as steam is always used expansively, entering the cylinder at nearly boiler pressure and leaving it at only a very slight pressure, the mean push of steam against the piston is not so easy to find. It is evident, however, that if we could measure the pressure of the steam at each point of the piston's travel, we could get the mean pressure at once, for we should merely take the pressures at a sufficient number of points and average them. We have fortunately a method of doing this and ascertaining at the same time a vast variety of other valuable facts concerning the performance of the steam engine. This is found in the indicator, which is really nothing more nor less than an instrument for registering the steam pressure against the piston at each point of its path. So important an instrument is the indicator that no engineer is fit to care for a modern engine who does not understand its principle and use, and it is worth while devoting some little space to the consideration of the information it can give us.

But first let us follow roughly the actual cycle of operations that goes on in the cylinder of a steam engine, beginning with the moment when the valve that admits steam to the cylinder opens. Let us suppose that we are possessed of an eye quick enough to follow the fluctuations of a steam gauge attached to one end of the cylinder. The first effect, of course, is that steam at practically full boiler pressure rushes in, and we should find the steam gauge darting up to that pressure; then, as the piston started and the steam from the boiler continued entering, the pressure would remain almost exactly constant up to the point of the stroke where the admission valve began to close. The pressure would then fall, as the steam in expanding and pushing on the piston loses its temperature and pressure together. We should thus find the steam gauge sinking gradually down until very near the end of the stroke. Then as the piston neared the extreme limit of its motion the exhaust valve would open and the pressure would fall immediately very nearly down to that of the outside air. Meanwhile the piston would be on its backward stroke, driven by the steam at the other end of the cylinder, and the pressure at the end we are investigating would remain, so long as the exhaust port remained open, very small in amount and almost uniform. Finally, as the piston almost reached the point where it started the exhaust port would close, and the small amount of steam left in the cylinder would be slightly compressed, until at the end of the stroke the admission valve would open again, and the pressure would rise as at first.

This is the very cycle of operations that goes on in either end of the cylinder of a steam engine, and it is the duty of the indicator to register it; the instrument, in fact, is nothing more nor less than a registering steam gauge arranged in such form as to be convenient for the purpose. Fig. 18 shows an excellent modern indicator, and Fig. 19 a sectional view showing its exact construction. It consists, primarily, of a cylinder, seen at the left of Fig. 19, containing a snugly fitting piston, the upward motion of which is resisted by a spiral spring; this forms a true steam gauge, as the compression of the spring is exactly proportional to the

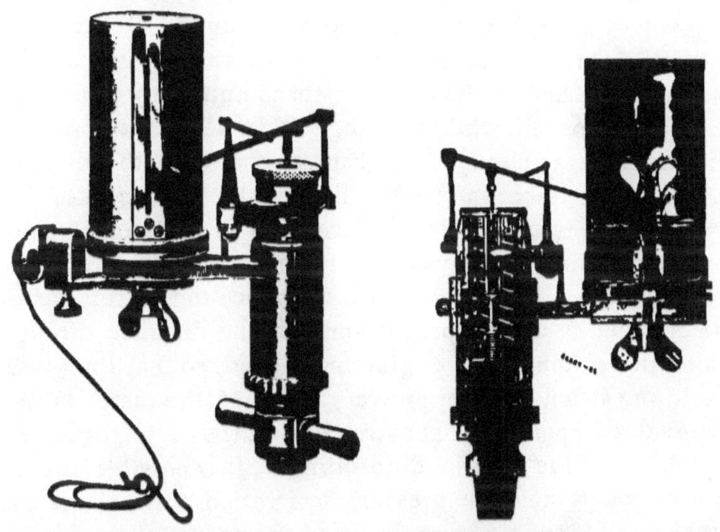

FIG. 18.—THOMPSON INDICATOR. FIG. 19.—SECTION OF THOMPSON INDICATOR.

pressure applied to it. If we screw this indicator cylinder vertically into the end of a steam-engine cylinder, or to a pipe communicating with it, and then allow the engine to run, the position of the piston in the indicator cylinder will depend on the pressure of the steam applied to the engine piston, and at each moment of the stroke the compression of the spring will be exactly proportional to the pressure of the steam in the cylinder at that same instant. A little piston rod passing through the top of this indicator cylinder moves a lever carrying at its extremity a pencil point. The lever is not a simple one, but a combination of levers so arranged that the pencil shall travel vertically up and down instead of in an arc. The point of this pencil rests, as seen from Fig. 18, on the surface of another cylinder, over which, in practice, a piece of paper is tightly stretched and held in place by the pair of slender springs shown in the cut.

If, now, the steam be admitted to the engine, the pencil point will fly up as far as the tension of the spring will allow the indicator piston to push it, and as the steam expands and the pressure falls the position of the pencil in its downward path will be at every point exactly proportional to the steam pressure in the cylinder. Now the cylinder is made to revolve through nearly an entire revolution; by the cord shown it is attached to the piston rod of the engine, so that as the piston moves the cylinder turns uniformly under the point of the pencil, while a spring within keeps the cord taut and takes up the slack on the back stroke, so as to bring the cylinder exactly back to its starting-point. The effect of this arrangement is as follows: The position of the pencil point in a vertical line is always proportional to the pressure in the steam cylinder; its position with reference to the rotation of the cylinder is always proportional to the distance through which the piston of the engine has moved, so that the pressure in the steam cylinder at every point of the piston stroke is registered upon the paper stretched on the revolving drum. On taking off the paper and unrolling it, it is possible at once to see exactly how the pressure has varied throughout the stroke; and if the pressure required to produce a given compression of the spring in the indicator be known, we can also tell by the heights at various points of the indicator diagram just the steam pressure in pounds per square inch that was exerted on the piston.

With every indicator are furnished two or three springs of strengths accurately known, and graduated so that each inch of motion of the pencil point shall correspond to an exact number of pounds pressure on the spring. A valve is inserted between the indicator and the steam cylinder, so that the piston of the instrument will not be in motion except at the moment when it is desired to trace an indicator diagram. As an added precaution against injury to the indicator a little handle, shown at the top of the indicator cylinder, enables the upper portion of it to be turned so as to draw the pencil away from the paper except when it is wanted. As the stroke of most engines is greater than the periphery of the indicator drum, the latter is usually attached to the piston rod, or its cross-head, through the medium of a reducing motion of some sort, frequently a pendulum arrangement, of which the lower extremity is operated by the cross-head of the engine, while the cord to the indicator is attached to an arc further up toward the center of motion. It should always be remembered, however, in arranging anything of this kind that it must be so put together that the motion imparted to the indicator drum shall be exactly proportional to that of the cross-head. For details of the practical adjustment of the indicator it is best to refer to one of the excellent treatises upon it, or to the plain working directions that are furnished by the makers of indicators, as an elaboration of this sort of thing is hardly within the scope of the present chapter.

On admitting the steam to our indicator and pressing the pencil down lightly against the drum during a single complete stroke by means of the little handle before alluded to, a tracing of the indicator diagram will be obtained, and its general shape will be that shown in Fig. 20. In this diagram the line F A represents the rise in pressure as the admission valve is opened, and the line A B shows the

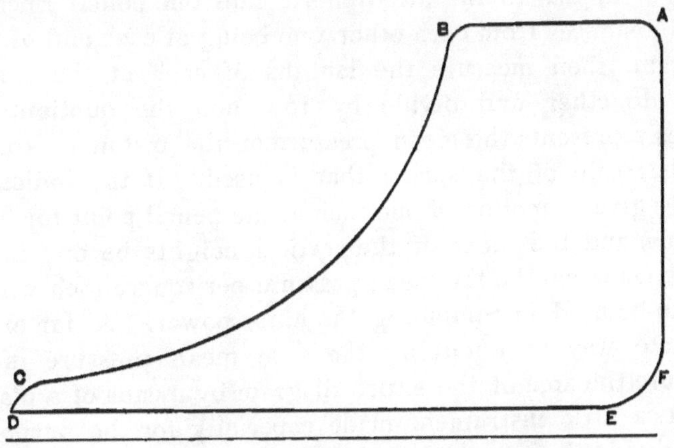

FIG. 20.—SPECIMEN INDICATOR CARD.

period during which steam is freely admitted to the cylinder and the pressure kept constant; the smooth curve B C shows the gradually falling pressure as the steam expanded up to the point where, at C, the exhaust port begins to open; from C to D the pressure falls to a minimum, and continues constant along the backward stroke up to E, where the exhaust port closes, and the small amount of steam that remains is slightly compressed until the opening of the admission valve. The object of this compression is to cushion the motion of the piston slightly, so as to prevent too violent strains upon the engine. In taking an indicator diagram it is customary to turn one of the valves of the instrument so as to admit the outside air to the indicator cylinder before steam is turned upon it; then, the drum being in motion, the pencil point is swung against it, and a single line showing the position of the spring for atmospheric pressure drawn near the bottom of the paper. This furnishes a point of reference both for showing the exact length of the indicator diagram and the pressure of the steam above that of the atmosphere.

Now, having the indicator card taken, the first use to which we can put it is the determination of the average pressure of steam on the piston for the purpose of finding the horse-power by the process indicated previously.

The simplest way of obtaining this mean pressure is to erect at right angles to the atmospheric line ten pencil lines at equal distances from each other, one being at each end of the diagram, then measure the lengths of each of these, add them together and divide by 10; then the quotient in inches represents the mean pressure on the piston reckoned on the scale of the spring that is used. If the indicator spring gives a motion of one inch at the pencil point for fifty pounds, and the mean of the vertical heights be one inch, fifty pounds will be the mean pressure per square inch which has to be used in computing the horse-power. A far more accurate way of obtaining the true mean pressure is to measure the area of the entire diagram by means of a planimeter, a little instrument made especially for the purpose; then divide the area in inches by the length in inches, and the mean pressure is at once found. The planimeter is not always at hand, however, and the other method gives a tolerable approximation.

In the next place, suppose the piston of the engine does not fit well and the steam leaks past it: the result will be, evidently, that the pressure behind the piston will lessen very rapidly, and consequently on the indicator diagram the line B C will fall toward the atmospheric line much more rapidly than in the figure given. If, on the contrary, the piston is tight but the steam admission valve leaks, the pressure will not fall by any means rapidly enough. If the steam is only admitted for a very small portion of the stroke, and there is considerable condensation in the cylinder, the pressure will again fall too rapidly.

If the valves do not open and close at the proper time the result will be at once seen in the indicator card. For example: Fig. 21 shows a card taken from a well-built modern

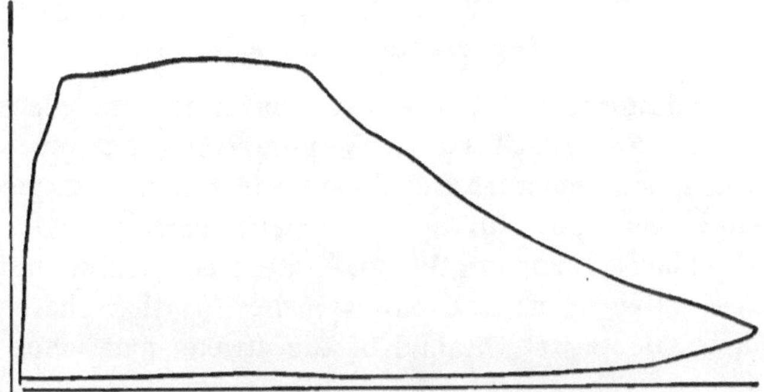

FIG. 21.—INDICATOR CARD—VALVES WRONGLY SET.

engine in which the valves were not set correctly; the vertical line at the left shows on the scale used for the diagram the full boiler pressure. The admission valve, as will be seen from the sluggish rise of pressure, only opens after the piston has started on its trip; the steam pressure never catches up effectively with the piston, and steam does not enter the cylinder as freely as it should, although after the valve is wide open the pressure has risen a little. The line showing the expansion of the steam does not look promising, and the pointed toe of the diagram shows that the exhaust port was very late in opening, for the pressure did not fall anywhere near the atmospheric pressure until the piston was well started on its backward stroke. Besides opening too late, the exhaust port closed too late, so that there was not the slightest cushioning at the point where the admission valve began its rather slow opening motion. Fig. 22 shows the same diagram after the valves had been reset by reference

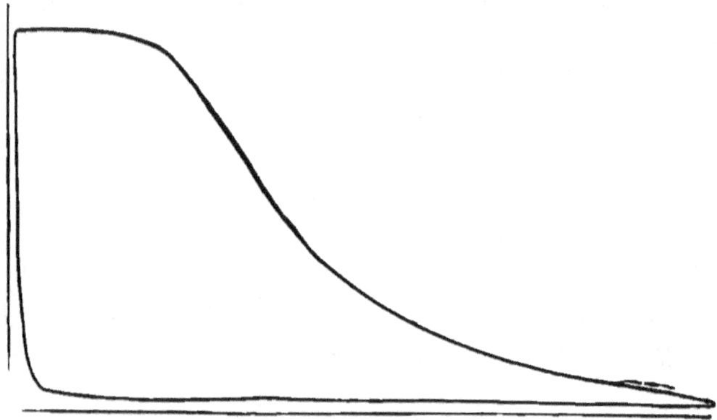

FIG. 22.—INDICATOR CARD—VALVES READJUSTED.

to the indicator card. You will see that in the first place the boiler pressure is utilized much more fully, the expansion gives a much smoother curve, and the steam is expanded far more nearly down to the atmospheric line.

The exhaust is apparently not quite free enough, but the exhaust valve opens and closes earlier, so that there is a slight cushioning at the end of the stroke, and when the

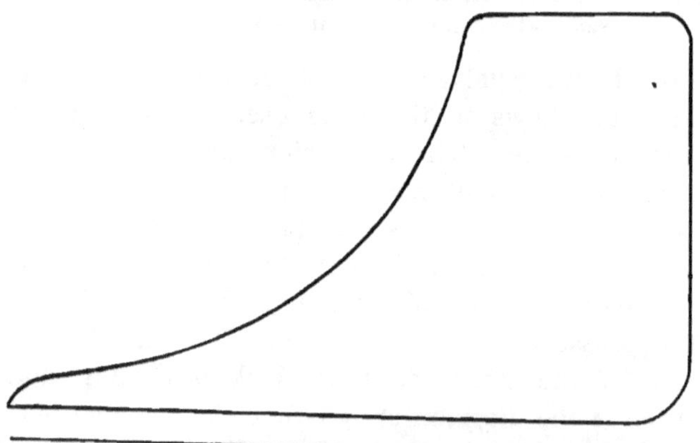

FIG. 23.—CARD FROM COMPOUND CONDENSING ENGINE, HIGH-PRESSURE CYLINDER.

admission valve opens it moves rapidly and opens fully, so that the steam pressure rises sharply and is fairly well maintained. Figs. 23 and 24 show two very perfect indicator diagrams taken from a compound Corliss pumping engine. Fig. 23 is the high-pressure diagram and Fig. 24 the low-pressure. In the latter, of course, the atmospheric line cuts through the diagram as the engine was condensing, and the pressure consequently fell below the atmospheric pressure. In the former of these diagrams the proper working of the valves is very elegantly shown. The steam pressure rises almost up to full boiler pressure instantly, and is maintained almost absolutely constant until the admission valve closes. The expansion line of the steam is almost perfect, and when, just at the end of the stroke, the exhaust valve opens, it opens fully and drops the pressure to a nearly constant

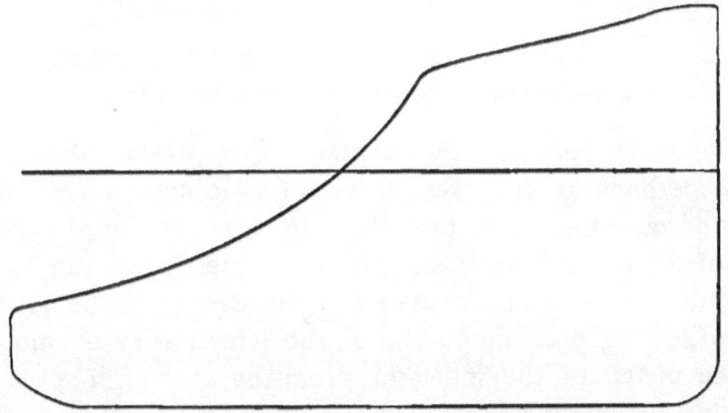

FIG. 24.—CARD FROM COMPOUND CONDENSING ENGINE, LOW-PRESSURE CYLINDER.

amount, where it remains until the exhaust port closes, and produces a slight cushioning at the end of the stroke.

The diagram from the low-pressure shows almost the same beautiful characteristics. The admission line is vertical and turns sharply at the top into the steam line. The steam admitted, however, falls off somewhat in pressure, inasmuch as it comes not from the boiler, but from the high-pressure cylinder. The expansion line is as perfect as in the previous diagram, and the exhaust works with the same beautiful promptness. Altogether, these are model diagrams. The engine from which they were taken has a high-pressure cylinder 15 inches in diameter and a low-pressure 30 inches in diameter, both of 30-inch stroke. The high-pressure cylinder takes steam at 125 pounds. The engine makes 46 revolutions per minute, and the consumption of coal is but 1.8 pounds per horse-power per hour, a result very seldom approached in stationary engines of this size. Figs. 25 and 26 show a pair of excellent cards taken from an Armington & Sims engine with a single piston valve, such as was shown

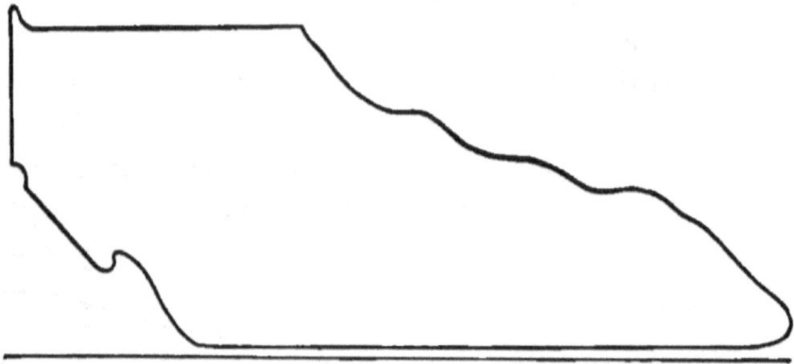

FIG. 25.—CARD FROM SINGLE-VALVE ENGINE. HEAVY LOAD.

in the early part of this chapter. The present diagrams show perhaps as good results as can be obtained from this class of valves in actual practice. In this case the admission line of the steam is vertical, and runs almost up to full boiler pressure; the expansion curve will be seen to be somewhat waved, owing possibly to the slight irregularity of motion in the pencil of the indicator, from the very high speed at which the apparatus is worked.

The cylinder was 9 1-2 by 12 inches, the speed 200 revolu-

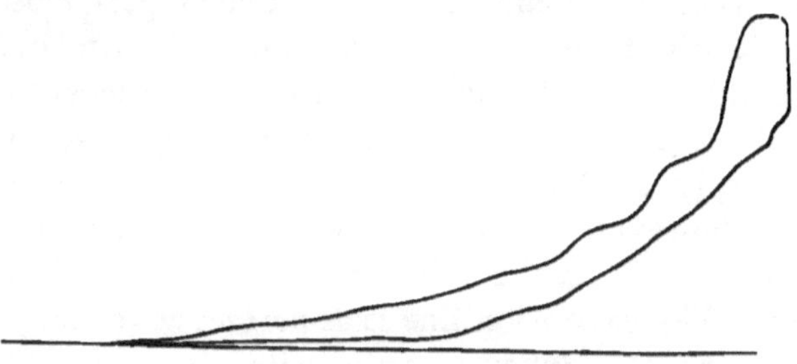

FIG. 26.—CARD FROM SINGLE-VALVE ENGINE. LIGHT LOAD.

tions per minute, with steam at 85 pounds. The noticeable feature of the cards from this and nearly all other high-speed engines is the relatively large amount of compression, the exhaust port closing rather early in the stroke;

especially is this true when, as in Fig. 26, the load on the engine is exceedingly light, hardly more than the friction of the moving parts. In this case the compression begins almost at half stroke, the cut-off of the steam admitted being very early in the stroke. These cards are most excellent for a high-speed engine, but do not show the economy of steam of the pair of cards just given. The expansion of the steam and the operation of the exhaust valve fail to give the good result that is obtained by the use of separate admission and exhaust valves.

Fig. 27 shows an indicator card of a somewhat less desirable kind taken from an engine actually engaged in driving an electric-railway plant. In this machine the valves were not adjusted anywhere nearly as well as in the example just given. The compression begins very early in the stroke and runs insensibly into the admission line; the steam line falls

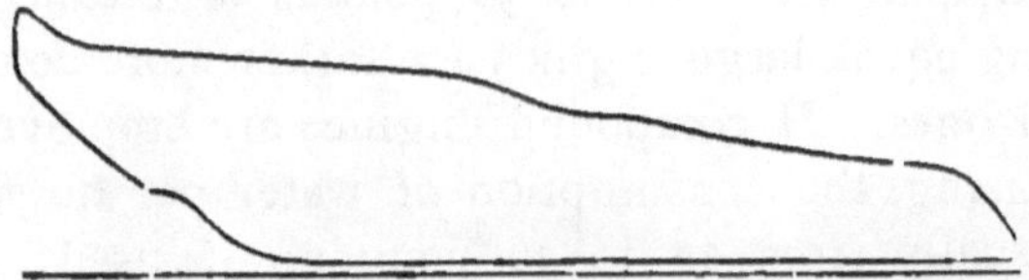

FIG. 27.—CARD FROM SINGLE-VALVE ENGINE IN RAILWAY POWER STATION.

off in a curve, leaving the point where the admission valve closed rather indefinite. The expansion curve is bad and the opening of the exhaust valve rather gradual. All this means lack of economy, for neither is steam admitted at as high an average pressure as is desirable, nor is it expanded properly before the exhaust begins to open; besides this, the compression is much larger than is ordinarily desirable.

High-speed engines seldom give indicator diagrams showing anything like the efficiency found in using some other forms of valve gear; nevertheless, the practice in this respect is improving, and some of the modern engines give very good results. At all events, the difference between the two classes is not nearly so formidable as it once was. The weak point of the low-speed engine has already been mentioned—the relatively large cylinder and consequent internal waste, lack of range of cut-off, and sluggish governing. The weak point of the high-speed engine with its single valve is that the steam is not used as effectively because of the interdependence of all the movements of the valve. If the cut-off is changed the action of the admission and exhaust is changed throughout, with the result that in most cases the steam is not admitted freely up to full boiler pressure, the exhaust is apt to take place at the wrong time, and there is often an objectionable amount of compression. All these mean that the steam is not used in the way to secure the maximum efficiency, and the difference, rather evident from an inspection of the indicator cards, is often seen in the respective coal bills met in using these two types of machine.

When economically run, the modern high-speed engine will produce one horse-power per hour with a trifle over 30 pounds of steam, seldom or never below that figure. An engine fitted with Corliss or equivalent valve gear, and of similar size, will, under favorable conditions, do the same work with the expenditure of 27 to 30 pounds of steam. Other things being equal, large engines are rather more economical than small ones. If compound engines are employed, used non-condensing, the consumption of water per horse-power hour is usually from 22 to 26 pounds; if used condensing, from 16 to 20. Condensation in either simple or compound engines is likely to save something like 15 per cent. of the fuel. Very large triple-expansion condensing engines may produce the horse-power hour on as low as 12 or 13 pounds of steam, but such figures are rather unusual. These data can, of course, be regarded only as approximations, as the results obtained are influenced by the care exercised in setting the valves of the engine to do the most economical work. For example, in Figs. 21 and 22 the changes made in the valve motion reduced the amount of fuel per day from 16 tons to 12 tons for the same work, which of itself is a sufficient lesson in the importance of using the indicator and working the engine in the very best manner of which it is capable.

In choosing an engine for operating an electric-railway power station very many things have to be taken into consideration; the general principle to follow, however, is to employ an engine of such size and character that it can be worked with its average load at nearly its maximum point of economy. With large roads, where the load does not vary very much, these conditions can be fulfilled admirably by large Corliss or similar engines, either simple or compound. If, however, the load is likely to vary excessively, and the maximum load will probably be three or four times the average load, such engines if used will be so seriously underloaded as to be really less economical than the high-speed engines, which, although intrinsically less effective in their use of steam, more than make up for this loss in their greater range of power and consequent ability to handle heavy loads on occasion, while on the average working near their point of maximum efficiency. This principle will be elaborated further in the chapter on station design.

A very large portion of this chapter has been devoted to the steam engine because it is at present, and is likely to be for some time to come, the principal prime mover for electrical power stations, especially electric railways. Gas engines, which have been before alluded to, and water-wheels, of which we will now speak, are occasionally used for such purposes. The difficulty of governing under a very variable load is so great that the advantage of cheap power rendered available by these means is materially lessened. In speaking of water-wheels, text-books and popular treatises generally begin by a formal mention of overshot and breast wheels, both vertical wheels on which the water acts partly by weight and partly by impulse, and finally give some brief mention of turbine water-wheels. At the present time the latter constitute the only class worth the slightest discussion for our particular purposes, with the exception of a peculiar form of impulse wheel frequently used for enormously high heads of water.

The overshot, breast and undershot wheels are almost as obsolete to-day as single-acting engines and steam pressures of 15 pounds per square inch. The turbine is a water-wheel generally, but not always, arranged on a vertical axis, and receiving and discharging water in various directions around its circumference. It differs from all other water-wheels in that all the vanes or paddles provided are acted on simultaneously instead of portions of the periphery being used in succession. The wheel consists in general of a drum or annular passage containing a set of curved vanes shaped in such a manner that the water, after glancing from them, is left behind with a minimum amount of energy. They have the great advantage of being of very small bulk for their power and being equally efficient for quite high and low heads of water. On such wheels the water acts by virtue of its weight and velocity, and the supply is received either directly from a reservoir, in which case the wheel is placed close to an opening in its bottom, or through a supply pipe and wheel case. The former method, involving, as it does, no loss from friction in pipes, is preferable for low falls; the latter for considerable heads. In most turbines the opening through which the water rushes upon the wheel is furnished with guide blades to give the stream the most effective direction. Turbine water-wheels are generally divided into three classes, according to the manner in which the water acts upon them. First, the parallel-flow turbines, in which the water is supplied and discharged in a current parallel to the axis of the wheel; second, outward-flow turbines, in which the water is supplied and discharged in currents radiating from the axis; and, third, inward-flow turbines, in which the water is supplied at the periphery and discharged through the center. To these may be added the modern class of horizontal turbines, in which the turbine principle is used in wheels upon a horizontal shaft, thus avoiding the necessity of gearing. These may belong to any one of the three classes above mentioned. Turbines frequently work "drowned,"

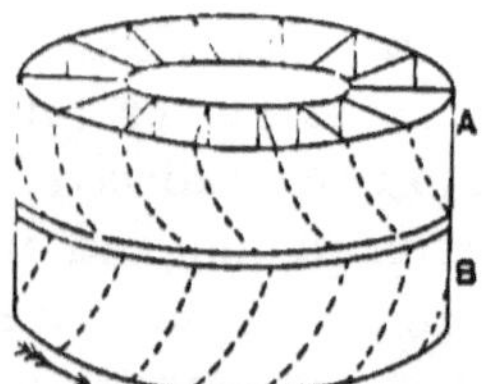

FIG. 28.—PARALLEL-FLOW TURBINE.

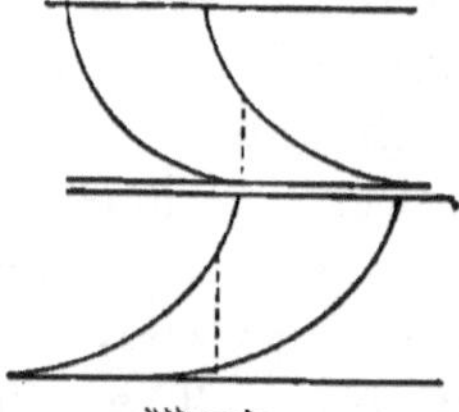

FIG. 29.—ARRANGEMENT OF GUIDE VANES. PARALLEL-FLOW TURBINE.

that is, entirely submerged; but inasmuch as some efficiency is lost in this way, the practice is not to be generally recommended.

Fig. 28 gives a diagrammatic view of a parallel-flow turbine, and the accompanying Fig. 29 shows the way in which the guide blades and vanes are arranged to produce the proper reaction. The water enters the guide blades around the periphery of the casing, and from these passes to the wheel itself, putting it in powerful rotation. The drum A is a supply chamber, containing the guide blades; B is the wheel proper. Fig. 30 shows a horizontal portion of an outward-flow turbine wheel; A is a supply chamber in the interior, being in the form of a vertical cylinder, with a ring of openings around its lower end; C shows the guide blades corresponding to those in the figures previously shown, and D the

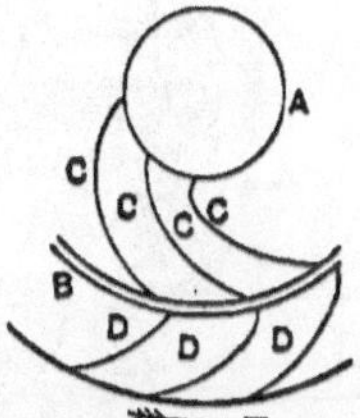

FIG. 30.—DIAGRAM OF OUTWARD-FLOW TURBINE.

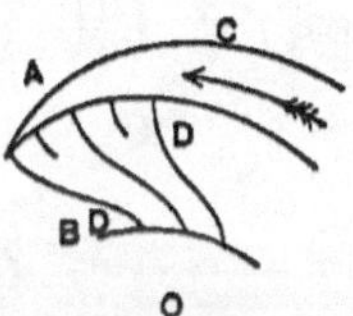

FIG. 31.—DIAGRAM OF INWARD-FLOW TURBINE.

buckets of the wheel. Passing to the inward-flow turbine, Fig. 31 gives a similar diagram for it. The water is led, as in the outward-flow turbine, by a vertical chamber; B is the wheel, provided with vanes D D and free to discharge its water through the centre; C is one of the guide blades to direct the water against the vanes. Fig. 32 gives a view of a horizontal inward-flow turbine, showing how these general principles are applied in a modern wheel often used for electrical purposes. The efficiency of all these wheels is about the same, and equals say 80 per cent. of the theoretical energy of the water as it strikes the wheel. No simple rules can be given for computing the water-power of a given fall, as the data regarding the velocity and amount of water cannot be readily ascertained without a careful survey, which should be made by a qualified engineer. The general principle of supply, however, is to utilize as high a fall as practicable, with as little intervening pipe as possible. The horizontal turbine is rather a favorite on account of the convenient position of the main shaft. Turbine water-wheels not only are enormously powerful in proportion to their bulk, but have a very high velocity of rotation; so great, in fact, that in some large electric stations the dynamos are belted direct to the water-wheels, but furnished with larger pulleys, the speed of the wheels being actually greater than that desired.

Where water-power can be cheaply had it is generally

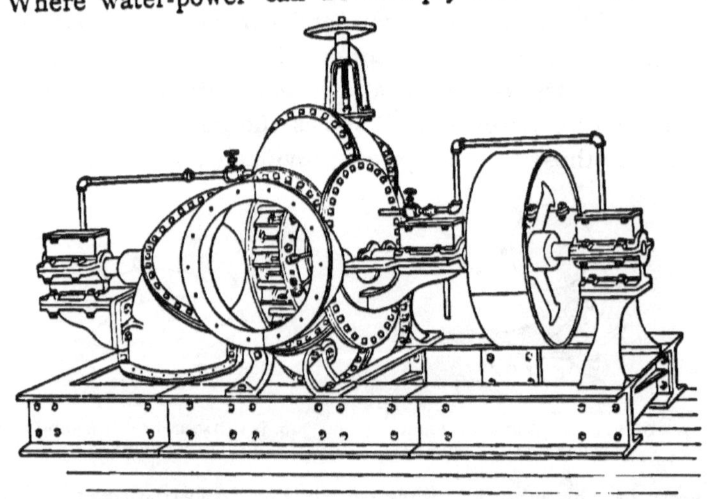

FIG. 32.—A TYPICAL HORIZONTAL TURBINE (LEFFEL FORM).

advantageous to use it, and in certain localities, where fuel is very dear, the use of water is almost imperative; but it must not be forgotten that there are circumstances under which no advantage can be gained over steam. It is probable that the water-wheel will be used more, rather than less, for electrical power purposes, as there are many parts of the country where fuel is particularly expensive, and by the transmission of power for a few miles, water-falls can be utilized with the greatest advantage. With the electro-motive force at present used for railway work, the distance of practical transmission is somewhat limited; but there are cases in which it might even be worth while to supply power from a distant dynamo at a very high potential for driving a railway power station at the usual electro-motive force. The principal disadvantage of the water-wheel is the difficulty of governing it. This difficulty is real and great, and arises from the large mass of falling water that has to be controlled, and the consequent mass and inertia of the gates. The result of this is that a water-power governor lacks sensitiveness, and it is only with great difficulty that the wheel can be held at constant speed. This is of course a serious objection for electric uses, and more especially so where the load, as in electric railways, may vary with great rapidity. All the disadvantages of bad governing in engines are exaggerated when these are replaced by water-wheels, because the governor does not act so quickly nor is it so effective a regulator

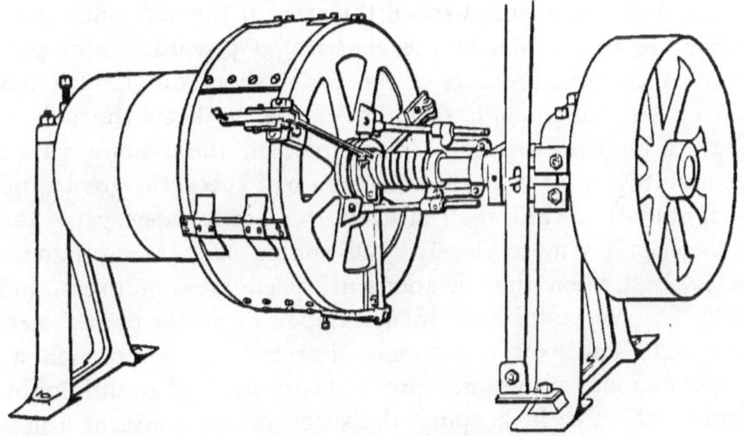

FIG. 33.—FOOTE SPEED REGULATOR.

as the isochronous governor of a steam engine. Nevertheless, on occasion, good work can be done with water-wheels.

There are ingenious electrical devices for controlling the flow of water at a distant point by a governor at the water-wheel; and there is at least one speed regulator that obviates the necessity for a sensitive governor at the wheel by regulating the speed of the driven dynamo, irrespective, within certain limits, of the velocity of rotation of the driving pulley.

The Foote speed regulator, shown in Fig. 33 is an apparatus for accomplishing this end that has met with considerable favor in electric light and power stations. The prime mover is belted to a pulley, shown on the left-hand end of the regulator, while the dynamo is belted to the larger pulley at the other end. The condition for operation is that the driving shaft shall have a speed greater at all times than that of the driven shaft. The large central drum has around its periphery friction brakes adjusted by the centrifugal governor. The brake shoes are of sheet steel, with a layer of paper pads that fit closely on the surface of the friction pulley; they are pressed down upon it by means of the levers shown, and actuated by the coiled spring on the shaft just at the right of the drum. The tension can be changed by means of a screw collar, and the controlling device is the centrifugal governor, the pressure of which counteracts that of the spring when the machine is in motion. When the driving pulley is running faster than the required speed the arms of the automatic governor are thrown out by the centrifugal governor, and partially release the pressure of the brake shoes on the friction wheel, so that enough slipping ensues to keep the driven shaft at its proper speed. If the load on the dynamo varies enough to cause the slightest decrease of speed the governing weights fall toward the shaft and the brake shoes gripe the friction pulley more closely. Of course such a regulator is at its best when the variations of speed are not great, and it is in successful use in a number of water-power electric light and power stations. For railway work such an apparatus may sometimes prove desirable, and would doubtless be effective in keeping the speed nearly constant unless the changes of load were exceptionally violent.

Before closing this brief mention of water-wheels, one should be mentioned that has made a remarkable name for itself in service under certain peculiar conditions. This is the Pelton water-wheel shown in Fig. 34. It is purely a jet wheel, acting by the tremendous impulse of a water jet under a powerful head. The buckets, as will be seen at a glance, are numerous, small, and doubly concave, sweeping gradually away from a ridge in the center; this is for the purpose of allowing the water jet to divide and move smoothly off to the sides instead of spattering, as would be the case if it impinged on a flat or smoothly concave vane. These wheels are made in a single massive casting, with buckets bolted firmly to the periphery, and have proved very successful where an immense head of water is to be utilized. They are not employed for replacing turbines under ordinary conditions, but where a small quantity of water under a very large head is available the Pelton wheel is almost the only one that has in practice given a thoroughly good account of itself. It is in use already in a number of power-transmission plants in and about the Rocky Mountains and on the Pacific coast, and replaces the turbine with admirable effect when the head of water amounts

FIG. 34.—PELTON WATER-WHEEL.

to a hundred feet or more. Its efficiency is quite equal to that of the best overshot or turbine wheels.

Wherever good water-power is available it is certainly well worth investigation as a source of power for electrical enterprises in the vicinity. To decide on its ability to compete successfully with steam may not always be easy, but it is worth while to make careful estimates whenever there is a reasonable chance of good results from the use of water-power. Under favorable circumstances its advantage of great cheapness is obvious; its principal disadvantage is the difficulty of proper governing, although this can be somewhat reduced by proper design of the station and by governing appliances such as those mentioned.

CHAPTER III.

MOTORS AND CAR EQUIPMENT.

To drive a shaft when speed and load are desired to vary both widely and independently is the work of the electric-railway motor.

As to variation of speed, it has already been explained that in the design of a dynamo electric machine, or in the comparison of one motor with another, variation in the number of armature loops must be considered as affecting the speed, other things being equal. But in studying the action of any particular motor, when finally in operation, this cause of speed variation may be neglected, since the armature, once wound, carries in the magnetic field at all times the same number of conductors.*

The other quantities, changes in which are connected with changes in speed, are (1) strength of field, (2) the rate of work, *i.e.*, the quantity of work done in a given time. As has been already shown, the rate of work is measured by the product of (3) the current, and (4) the counter electromotive force. The current, however, is readily expressed in terms of the counter E. M. F., the resistance of the machine, and the applied electromotive force, since it is always such a current as will flow over the given resistance under a pressure equal to the difference between the two opposing pressures, —that applied or impressed by the dynamo through the line, and that generated by the motor armature itself. (5) As seen from the above, change in the applied E. M. F. is also connected with change in the speed. The quantities (1), (2), (3), (4) and (5) are interdependent, but are separately men-

* It is of course possible to devise mechanism such that, in connection with the commutator, it would afford a means of changing at will the number of active loops, but, thus far, practice has not justified such a complication. We may, therefore, in examining the action of a railway motor, take the number of armature wires moving in the field as constant.

tioned, since convenience requires reference first to one, then to another.

Of these five quantities, perhaps that which in practice is most constant is strength of field. As has been shown (Chapter I.), the efficiency of a motor depends upon the relation of the counter E. M. F. to the applied E. M. F. High efficiency, or relatively high counter E. M. F., is desirable at any and all speeds. But high counter E. M. F. goes with a large product of the three factors, (*a*) number of armature loops, (*b*) speed of rotation, and (*c*) strength of field. As noted, (*a*) the number of loops is fixed; (*b*) the speed is limited by practical requirements, hence (*c*) great strength of field is constantly desirable. It is therefore good practice so to wind a street-railway motor, by putting a relatively large number of turns around its magnets, that a maximum strength of field is attained, even when the current flowing is small as compared with the maximum current.

This is equivalent to saying that the magnetization given by a relatively small current is yet sufficient to saturate or nearly saturate the iron of the magnetic circuit. If, however, the field be kept below saturation and be varied in strength, this variation may be used to accomplish a certain degree of speed regulation. Let us suppose a car moving on a level (or on a uniform grade), and at such a speed as to produce a counter E. M. F. of 400, the applied E. M. F. being 500. For convenience assume the internal resistance of the armature circuit to be 10 ohms. Then the current flowing will be $\frac{500-400}{10}=\frac{E-E'}{R}=\frac{100}{10}=10$ ampères. The mechanical work done would be $400\times 10=E\times C=4{,}000$ watts.

Now suppose it is desired to run more slowly. Increase the field strength by 10 per cent. The counter E. M. F., at the same speed, would be 440 volts, the current $\frac{500-440}{10}=$ 6 ampères, and the work $440\times 6=2{,}640$ watts. But since the car requires 4,000 watts to maintain the previous speed, it is evident that it must now decrease its speed until there shall be an equality between the work required to maintain the lower speed and the work done by the motor at the lower speed due to the greater field strength. Let us learn what this speed is.

It was seen above that work at the rate of 4,000 watts = $\frac{4,000}{746}$ = 5.36 horse-power = 176,880 foot-pounds per minute, must be performed in order to maintain the speed existing before the change of field strength. Suppose the car to weigh eight tons, and suppose that on the particular track in question a horizontal effort of 25 pounds per ton is required to overcome all resistances, including those of gears or other mechanism between the armature and the axles; then, a total horizontal effort of 200 pounds must have been exerted. This quantity, multiplied by the number of feet traveled per minute, must be equal to the number 176,880, representing the total foot-pounds of energy utilized per minute. Hence, the travel per minute $= \frac{176,880}{200} = 884.4$ feet = 10.05 miles per hour. At the new speed we must have a similar relation—

$$\frac{\text{Work done in foot-pounds}}{200} = \text{feet traveled per minute; or,}$$

since 1 watt-minute = 44.24 foot-pounds—

$$\frac{\text{Work in watts}}{200 \div 44.24} = \text{feet traveled per minute; whence, the}$$

work in watts = 4.52 × feet traveled per minute.

The number of watts may always be expressed as the product of the counter E. M. F. by the current, or thus, watts = $E'\ C'$.

Let Z' = strength of field before change. Then,
1.1 Z' = strength of field after change.
Let S' = speed of armature before change.
S'' = speed of armature after change.
C' = current in armature before change.
C'' = current in armature after change.
E' = counter E. M. F. before change.
E'' = counter E. M. F. after change.

Now, $E' = Z' \times S'$. (We may omit the number of armature turns.) The speed of armature has a fixed ratio to speed of car in all practice of to-day. Hence, we may now assume some convenient ratio, such that, for example, the speed of car above calculated, *i.e.*, 884.4 feet per minute, shall correspond to 1,768.8 revolutions per minute of the armature (1 revolution = 0.5 of a foot travel). Then we may write $E' = Z' \times 1,768.8$. Also the work in watts = $4.52 \times 0.5 \times$ revolutions per minute = 2.26 S′.

$$C' = 10 = \frac{500-(1,768.8\ Z')}{10}, \quad C'' = \frac{500-1.1\ Z'\ S''}{10}$$

$$Z' \times 1,768.8 = 400. \quad \therefore Z' = 0.226.$$

This is a purely relative expression for the strength of field, resulting from the assumed values and ratios as given above. Then $1.1\ Z' = Z'' = 0.2486$.

$$(1)\ E'' = 0.248 \times S'',$$

$$(2)\ C'' = \frac{500-(.248 \times S'')}{10},$$

$$(3)\ E'' \times C'' = 2.26 \times S''.$$

Placing in (3) the values taken from (1) and (2), we have

$$(4)\ (0.248 \times S'') \times \frac{(500-0.248 \times S'')}{10} = 2.26 \times S'';$$

$$\frac{500-(0.248 \times S'')}{10} = 9.11.$$

$$\therefore\ (5)\ 408.9 = 0.248 \times S''. \quad \therefore S'' = 1,649.$$

Feet per minute = $\frac{1,649}{2} = 824.5 = 9.4$ miles per hour approximately. The counter E. M. F. will then be found thus: $E'' = 0.248 \times 1,649 = 408.9$. Also, $C'' = \frac{500-408.9}{10} = 9.11$, and the work in watts = $408.9 \times 9.11 = 3,725$.

In like manner, if the field be weakened 10 per cent., we may find: $(0.203 \times S''') \times \frac{(500-0.203 \times S''')}{10} = 2.26 \times S'''$; $500 - .203 \times S'' = 111$. $\therefore S''' = 389 \div .203 = 1,916$. Feet per minute = 958 = 10.9 miles per hour. Counter E. M. F. = 389, current = 11.1 ampères, and the work in watts = $389 \times 11.1 = 4,317$.

It should be noted that these results, increase of speed with decrease of magnetic strength, and *vice versa*, will be reversed if the changes be made from a state in which the counter E. M. F. is less than one-half the initial E. M. F. When the counter E. M. F. of any motor is just one-half the initial E. M. F., the rate of work, that is, the output per unit of time, is greater than at any other efficiency. Since this output is proportional to the product of counter E. M. F. by the current, and the current is proportional to the difference between the fixed E. M. F. of the line and the counter E. M. F., we have the output varying as the general expression $E' \times (E - E')$, the resistance being supposed to be constant. Now, the product of a number as expressed by E′, and of another number which is equal to the difference between the first number and a second fixed number, will always be a maximum when the first number is just half the second. Suppose that $E = 500$, for all values of E′.

Take $E' = 249$, then the product $= 249 \times 251 = 62,499$;
Take $E' = 250$, then the product $= 250 \times 250 = 62,500$;
Take $E' = 251$, then the product $= 251 \times 249 = 62,499$;

and any greater departure in either direction from 250, as the value for E′, will be followed by a further departure from the maximum value for the product. While E′ is greater than the difference between E and E′ (that is, when it is greater than one-half E), any change of absolute value made in E′ is less, proportionally, than the corresponding change in the smaller quantity, $E - E'$. Thus when E′, as above, was taken at 400, and the magnetic strength was increased, E′ also increased to 409, a change of 2 per cent., tending to increase the product E′ C′, measuring the output. But this change of 9 volts was necessarily followed by an equal change in the difference $E - E'$, reducing it from 100 to 91, a reduction of 9 per cent., and as the current flowing is directly determined by the quantity $E - E'$, its value must also change by 9 per cent. Since, therefore, we have increased one factor, E′, by 2 per cent., and decreased the other, C′, by 9 per cent., the product representing the new output must be less than the former. But if the original value of E′ were 200, instead of 400, then the difference $E - E'$ must be 300, and any addition to E′ resulting from an increase of field strength would increase this factor, E′, in greater proportion than it could decrease the other factor, which depends on the larger number, 300. The net result would therefore be an increase in the product E″ C″, and the car would consequently run faster.

The fact that the rate of work of a motor is a maximum when its efficiency is 50 per cent. is an important one. A motor which has been properly rated as a 15 horse-power machine should do that work at, say, 90 per cent. efficiency, since that is an easily attainable figure. It then receives about 17 horse-power of electric energy to perform 15 horse-power of mechanical work. It is possible to obtain a rate of 20 horse-power from such a machine, but only by lowering its efficiency, say to 60 per cent., in which case about 33 horse-power of electric energy must be supplied for the 20 horse-power of work.

Having discussed the relation between changes of field strength and of speed, we now pass to changes of load and field strength. It is evident that if in the case above assumed the weight of the car or condition of track or steepness of grade be changed, then the horizontal effort, 200 pounds, required to maintain uniform motion, will also change. Let us suppose it to be increased. Then if nothing be done to change the magnetic strength of field, or the impressed E. M. F., there must be, in a series-wound motor, a *decrease* of speed, and that whether the original efficiency be above or below 50 per cent. In the former case, increase of output (due to the lowering of the speed, and hence of the counter E. M. F.) will meet the larger demand, which is itself less than if the speed were maintained at the previous rate, and an equilibrium will again be established, when E″ C″ = K × revolutions per minute.

In this, K is constant, corresponding to 2.26 in the previous calculations, and depending on the particular value of the horizontal effort required. If, before change of load, the motor be working at less than 50 per cent. efficiency, the decrease of speed must be greater, since the equality between supply of and demand for power can be re-established only because the one decreases less rapidly than the other. Thus, the supply, varying with the product, E′ C′, will tend to decrease with decrease of E′, which, in a constant field, will vary directly as the speed; but decrease of E′ means increase of C′, though this increase will be less proportionally than the decrease of E′, as was explained above when considering field strength. The product of these two factors will then evidently diminish less rapidly than in the direct ratio to the speed. The demand for power will, on the other hand, decrease in exact ratio to the speed, since we have supposed the horizontal effort, after the first change considered, to remain constant. Another change would produce another lowering of speed, another adjustment between supply and demand; and this process may continue until the horizontal effort becomes greater than the rolling friction between wheel and rail, in which case the wheels will "skid;" or greater than the motor with maximum current in the armature and fields is capable of overcoming, in which case no motion whatever will be produced.

If we suppose these changes of load to take place when the field is not saturated, we may, of course, use the changes of field to counteract, if so desired, this speed reduction which would otherwise follow. Diminution of field strength would tend to increase or decrease the speed, as the efficiency is greater or less than 50 per cent.

The second important method of regulating speed under varying loads is found in the variation of the E. M. F. applied to the armature. From what has been said concerning the measure of work done, it will be readily seen that, other things being equal, decrease of applied E. M. F. must be followed by decrease of output, and increase of E. M. F. by increase of output. Let us again consider the case of a motor developing 400 volts counter E. M. F., the applied E. M. F. being 500 volts, the current being 10 ampères, and the resistance to motion being uniform. If now we reduce the applied E. M. F. (E) to 450 volts, and if we conceive that for a moment the counter E. M. F. (E′) remains at 400, the current would be reduced to 5 ampères thus: $\frac{450-400}{10}=5$. Such a change would reduce the output to one-half its former value. The speed would then necessarily drop to one-half, unless this decrease of output be checked in some way. This check would come through the fact that a decrease in speed would be followed by a corresponding decrease of E′, which would at once increase C′. To learn what the change in speed would actually be, let us proceed as before.

Using the new value for E, and assuming other conditions as before, we may write

$$E' \times C' = E' \times \frac{(450 - E')}{10} = 2.26 \times S.$$

or $\quad E' \times (450 - E') = 22.6 \times S.$

We suppose now that the field will remain constant. There will then be a very simple relation between E′ and S, that is, they will vary directly each with the other. It was proved above, under the conditions assumed, that

$S = 4.42\ E'$. Substituting above, we have

$E'(450 - E') = 22.6 \times 4.42 E'$,

or $\quad 450 - E' = 22.6 \times 4.42. \quad \therefore E' = 350.11.$

Calling this for convenience E′=350, we have,

$$C' = \frac{450 - 350}{10} = 10 \text{ ampères, as before.}$$

Output = 350×10 = 3,500 watts. Speed of car, in feet per minute $= \frac{3,500}{4.52} = 774 = 8.8$ miles per hour.

The original speed, it will be remembered, was 10.05 miles per hour. Now, let us suppose a much greater change in applied E. M. F., say to 250 volts. Then

$250 - E' = 22.6 \times 4.42 = 99.9. \quad \therefore E' = 150.$

$C' = \frac{250 - 150}{10} = 10$ ampères, as before.

Output = 150×10 = 1,500 watts.

Speed of car $= \frac{1,500}{4.52} = 332$ feet per minute = 3.8 miles per hour.

Again, suppose E = 200 volts, then

$200 - E' = 100$ (taken for 99.9). $\therefore E' = 100$ volts.

$C' = \frac{200 - 100}{.10} = 10$ ampères, as before.

Output = 100×10 = 1,000 watts.

Speed $= \frac{1,000}{4.52} = 221$ feet per minute = 2.5 miles per hour.

Generally, we see that $E' = E - 100 = E - C'R$. R is a constant by the construction of the machine (usually), and C′, under a constant load, also remains constant, while E varies, as seen above. The output = E′ C′ = (E — C′ R) C′ will then vary in proportion to the difference (E — C′ R): and this will vary the more rapidly as E approaches the constant value C′ R. Thus, in dropping from 500 to 450, the change of 50 volts is applied to an original value of (E — C′ R) = 450—that is, a change of one-eighth. But in dropping from 250 to 200, the difference of 50 volts is applied to an original value of (E — C′ R) = (250 — 100) = 150—that is, a change of one-third. Looking again at the expression E — C′ R = E — 100, we see that if E drops to 100 volts, the output will be zero—that is, the car will not move, unless the resistance to motion be in some way diminished.

We may say then, generally, that in any motor, the strength of field being fixed, a certain current, C, is required to produce a certain torsional effect in the armature (or a certain corresponding horizontal effort through the car axle). To produce this current over the internal resistance of the motor, R, a certain E. M. F. measured by C R is required. If the impressed E. M. F. be *just equal* to C R, then the rotating force in the armature and the resistance to rotation will just counterbalance each other, and there will be no motion. Excess of pressure over that measured by C R will produce motion, and in the ratio of this excess. The internal resistance will cover that of the armature and field windings, in case of a series motor, but only that of the armature in the case of a shunt motor. During changes of impressed E. M. F. the constancy of field strength supposed above will follow in the case of a series motor, if the number of turns and resistance of the field windings remain constant (since we have seen that C will then be constant).

Having seen the effects produced by changes in field strength and in impressed E. M. F., we should now consider the convenient and customary methods of producing those changes.

Suppose three coils of wire to be wound around the magnet cores of a motor, as shown in Fig. 35. Let the resistance of each be one ohm. Connect D to B and E to C. Suppose a difference of potential of 100 volts be maintained between the terminals A and F, then the total resistance of this field circuit would be three ohms, the three coils being arranged in series. The current flowing would be $\frac{100}{3} = 33.3$ ampères. The number of turns would be two for each coil (one above, one below) or six altogether. The magnetizing effect would be measured by $C \times T = 33.3 \times 6 = 199.8$.

Now suppose the connections be changed by disconnecting D and B, E and C, and connecting A, B, and C together, and D, E, and F together. The resistance of the three coils, now in multiple, will be only one-third of an ohm, or one-ninth

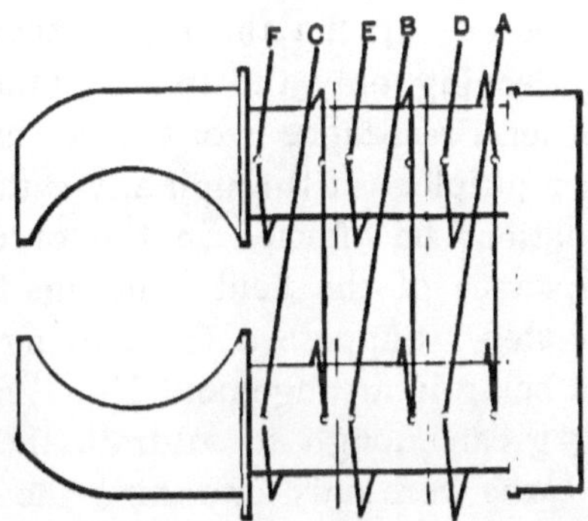

FIG. 35.—CONNECTIONS FOR COMMUTATING FIELD COILS.

the former value. The current will be nine times its value, or 299.7. The number of turns which this whole current makes around the magnet is only two, one above and one below, or one-third the former number. The magnetizing effect will be measured by $299.7 \times 2 = 599.4$, or three times the former value. If the magneto-motive force, 199.8, were itself sufficient to saturate the magnets, then the increase to 599.4 would be useless, so far as the strength of field is concerned. But if a force of say 2,000 ampère turns be required for saturation, then increase from 199.8 to 599.4 will be followed by a proportional increase in the field strength. In either case, the drop of potential required to force a given current over the field windings will be much less, with the second arrangement of coils, in which the resistance is relatively low, than with the first arrangement, in which it is relatively high. If, therefore, these coils be placed in series with the armature, we have a ready means of varying the E. M. F. applied to the armature itself, and also of varying the magneto-motive force applied to the field. The method of regulation thus afforded is usually known as that by "commutation of field circuits."

In its practical application, provision has been made for other arrangements, intermediate in effect between the two above described. Thus, (1) starting with the three coils in series, (2) one may be cut out or short-circuited, the other two remaining in series; (3) two may be placed in multiple with respect to each other, the third being in series with this multiple combination; (4) the third may then be cut out or short-circuited, leaving only the two in multiple; (5) the arrangement of least resistance is of course that in which all the coils are in multiple. If through any suitable switch the various combinations be effected in the order above mentioned, the resistance of the field windings becomes less at each successive step. Suppose a car moving at a constant speed, the coils being in arrangement No. 1, and the current required being great enough to saturate the fields, even if the number of turns were only one-third the number given by this arrangement. If we now pass through No. 2, No. 3, etc., successively, and there be no change of load, then the E. M. F. applied to the armature will increase, the speed will correspondingly increase (as explained above); the current will remain constant, for, by supposition, the field is continuously saturated. The changes, then, do not affect it, and affect the armature simply by varying the applied E. M. F., as would be the case if the field circuits were independent, as in a shunt motor, and some resistance external to the motor were used to vary the drop of potential on the lines connecting the dynamo and motor. The increase of output without increase of current is due to a progression in motor efficiency. In the first arrangement the resistance of the field circuit (hence the waste of energy therein) is greater than would be permissible in good design, were it not plainly set forth as a means of regulation. Only in the last arrangement does the resistance become as low as good design would give were efficiency alone in question.

If the current flowing before the changes be less than that required for saturation with any number of turns less than that given by the first arrangement, then the increase of speed in motors in practice will be more rapid than in the case just considered, and the current will not remain constant, but will increase. For there are now two causes tending to increase the speed: first, the increase of applied E. M. F., as before; second, there is decrease of the field strength, due to the decrease of turns, which tends to increase the speed (provided the efficiency of operation be greater than 50 per cent.).

As explained, the increase of speed due to this cause requires an increase of current. These changes tend, indeed, to check each other. Thus, increase of current tends to diminish the E. M. F. applied to the armature, just increased by lowering resistance, and to strengthen the field, just weakened by decrease of turns. To determine the constant speed and current which would be established with arrangement No. 2 or No. 3, etc., it would be necessary to know the relation between the number of turns and the resistance of the three coils. Assuming them to be equal in all respects, the changes of turns and the resistances may be seen from the following table:

ARRANGEMENT.	TURNS.	RESISTANCE.
1	6	3
2	4	2
3	4	1.5
4	2	0.5
5	2	0.3

In moving from step to step, the rate of change is evidently not uniform, but is continuous in the direction of higher speed. By making the coils unequal, it is possible more equally to divide the total change effected from first to fifth, but in the very limited space usually available for windings, such inequality may require dangerously small cross-section for the maximum currents required.

In the early motors of the Sprague Electric Railway and Motor Company, rated at 7.5 horse-power, at ordinary street-car speeds, say ten miles per hour, and for 500-volt circuits, the turns and resistances were as follows:

Switch No...	1	2 and 3	4	5 and 6	7	Positions 2 and 3, and 5 and 6, are effectively the same.
Turns........	8	5	5	3	3	Proportional numbers, not absolute values. Armature R = 1.4, constant.
Res...............	20.0	12	9.5	5.0	3.2	Actual R in ohms, including armature.

In a later machine, rated at 15 horse-power, under the same conditions the field resistance of each motor was 8.42 ohms, total, made up of three coils of 2.16, 2.65, and 3.61 ohms, respectively, the number of turns being as 11.5, 11, and 15, respectively.

In using this method, the principal difficulty has been met in disposing of the excessive heat necessarily generated in the compact mass of field windings. Many ordinary forms of insulation have been found to deteriorate under the influence of the high temperatures maintained. When, however, the incidental difficulties have been met by use of proper material, the method is a convenient one for securing a considerable range of regulation. The mechanism required, external to the motor, consists only of such a switch as will in a simple manner produce the changes of connection, such as described above. In the practice of the Edison Electric Company (successors in this business to the Sprague Electric Railway and Motor Company, who most widely introduced the method considered), these changes of connection have been produced by taking all the coil *terminals* to a series of spring contacts, pressing against metal surfaces arranged on a cylinder. These surfaces are so shaped that rotation of the cylinder produces the desired combinations. It is of course possible to use a smaller or larger number of independent coils than three. In the former case, the number of combinations giving different effects is not sufficient for a smooth control. In the latter, the complication of switches becomes excessive.

The control by external resistance, thus lowering or raising the E. M. F. applied to the motor, has been widely used and scarcely needs extended description. The practical problem has been to secure a convenient rheostat, made of such material as will give the high electric resistance required, and at the same time withstand the heat produced. Open coils of iron or German-silver wire seem to offer the most ready means of attaining the end in view, but the space demanded is found excessive as compared with the meagre dimensions required by the condition that the whole mechanism shall go under a car floor. Plates of thin iron, bent into a "U" shape, insulated from each other by mica, and arranged in a semi-circular frame, have composed the rheostats used by the Thomson-Houston Electric Company, which has widely used this method of control. An arm moving around the center of the semi-circular frame carries contact pieces over the convex surface of the resistance plates, leaving all, a part, or none of them in circuit, according as a low, or higher, or maximum E. M. F. is desired to be applied to the motor. When this method is used, the field winding should be permanently of low resistance, but sufficient to nearly saturate the cores at all loads.

One of the two methods of control thus described, or a combination of them, has been used in almost every effort at electric-railway work. The chief exception is found in storage-battery cars, in which the batteries themselves may be thrown into various combinations, thus changing the E. M. F. applied to the motors. There are few examples, if any, of the exclusive use of either method for motors now being manufactured. Thus, the Edison Company uses an external resistance to prevent too great a rush of current into the motors when starting; their proper resistance, even with all the coils in series, being so low when the two motors are connected in multiple on a 500-volt circuit as to cause a current of about 100 ampères to flow. This is usually much in excess of that actually required for a comfortable start of the car, and is of course highly objectionable in respect both to its effect on the motor and its demand upon the station. This "starting coil," as it has been called, is cut out of circuit as soon as the start has been made. The switch then produces the first combination of field coils, as above described, and as it rotates further, produces the others successively—this commutation of the windings being the chief means of control. On the other hand, the Thomson-Houston Electric Company, which has used the external rheostat as the chief means of control, in the last position of the switch, when the full line potential has been applied to the motors, short-circuits or cuts out one of the two parts of the field winding. Up to this last point these two have been connected in series, acting, indeed, as one coil. Now, in order to increase speed, a large part of the turns are made ineffective, hence the field is weakened and the increase of current and increase of speed take place, as already explained. The total resistance of the rheostat is fixed at such a value as will allow to flow the maximum current required for starting from rest, the line potential being taken, as in practice, at 500 volts. This total resistance may then be so subdivided as to produce any desired rate of change.

In case two or more motors be used on one car or train, there remains a third method of producing a considerable range of control; namely, by changing the machines from the multiple to the series arrangement, with respect to each other, and *vice versa*. The applied E. M. F. is just one-half in the second case what it is in the first, if there be two motors, one-third if there be three, and so on. A variation of this method might be had by changing the field windings, leaving the armatures permanently in series or permanently in multiple. The series arrangement of the two machines is of special value in starting a car and in maintaining low speeds. In both cases it may become the equivalent in effect, and the superior, in economy, of either of the above-described methods of control. Through this method it is indeed possible to make the same motors fulfill widely different service conditions. Thus, suppose cars are to begin their trips in the most populous portions of a city, where many stops and relatively slow running are unavoidable, while, on reaching the suburbs, long runs at high speeds are desired. We may then design motors which, if placed in multiple and permitted to run at say 25 miles per hour, will develop each 25 horse-power. If placed in series, they would each develop 12.5 horse-power, at 12.5 miles per hour, and at both speeds their efficiency may be high. To meet speeds lower than 12.5

miles per hour, the load being such as would permit only that speed if the impressed E. M. F. be 250 volts (half of 500 for each), we may resort to either of the two methods of general control, the motors as a whole being kept in series. So, for speeds between 12.5 and 25 miles per hour, the torsional effect required being still the same, we may in like manner control the two motors placed in multiple. In case of any heavy and long grade, this latter arrangement would be used.

Early in 1888 multiple-series switches were placed by the Sprague Electric Railway and Motor Company on many of the cars installed by it. They were subsequently abandoned, not because of failure to produce the anticipated results, but because at that date, when many other things were giving much trouble, it seemed wise to sacrifice variety of effect to simplicity of mechanism.

Others, when studying the problem, have desired to make the change from series to multiple connection a step in the progression constantly taking place after and during every stop of the car, and hence, to be produced when the car is in motion and current flowing. This may involve a quick jump of the car, in one case, or a corresponding drag in the other, both occurrences being objectionable. The greatest value of the method seems to be as pointed out above, in which cases the breaking of the circuit and operation of a switch, separate from that used for general control, would be permissible though shown by recent experience to be needless.

Now that many previously troublesome features of railway practice have been made easy, while at the same time there is constant increase in the variety of service called for, attention is actively devoted to this method. It appears in several important installations recently made, and its use will doubtless become more general. Its advantages are evident. See Appendix F.

More than nine-tenths of all the installations thus far made have shown two motors on a car, and these two have almost invariably been placed in multiple. Each is therefore wound so that it may produce the maximum speed required, having between its terminals the full line potential. If they are placed in series, there is, of course, no harm done, the effect being only to produce a lower speed in this way rather than by high-resistance field coils, or an external rheostat. When the motors are placed in multiple, each is nearly independent of the other. One may have its circuit broken while the other is propelling the car.*

If, on the other hand, the motors are so wound as to give maximum required speed when placed in series across the line, and this be considered the normal relation, it then becomes practically impossible to place them in multiple. If one machine must, by reason of injury, be cut out, then the other must be protected by some external resistance, or left, at best, to run with a high-resistance field. The advantage to be gained lies in this, that for a given quality of workmanship there will be fewer failures of insulation, since the pressure on each motor is only half that due to the line potential. Further, this prevents effectively, save in a special case, that widely unequal division of labor which has often been observed between two motors on the same car. This is generally due to a difference of resistance in the two magnetic circuits.

Its effect can be best understood by considering the fact that the current flowing over an armature is due to the difference between impressed E. M. F. and counter or self-induced E. M. F. The speed of the two armatures must, in car service, be practically equal. The number of turns, save by gross carelessness in manufacture, will also be equal. Now suppose that by reason of either or both of the causes mentioned, the two armatures are moving in fields of force differing in strength by 5 per cent. of the stronger, the number of ampères being the same. Consider armature (A) as generating a counter E. M. F. 490. Then armature (B) would generate $490-(490 \times 0.05)=465.5$. The effective E. M. F. would then be 10 volts for (A) and 34.5 for (B)—the line E. M. F. being as usual taken at 500 volts. The current would

* A trouble that has sometimes occurred with alarming frequency, however, consists in this: that the commutator of one motor begins to "throw solder"—*i.e.*, melts the solder at the connections with armature wires. In such case, look for a loose connection in the *other motor*. Such a loose connection, being of high resistance, may have thrown an undue load on the motor which has shown the trouble at commutator, this load causing an excess of current resulting in the melting of the solder.

be $\frac{10}{R}$ for A and $\frac{34.5}{R}$ for B, or, since the resistances of the armatures are practically equal, we have for a 5 per cent. difference in the field strength a difference in the current of 345 per cent. This supposes the currents in the fields of the two machines to be identical, and, indeed, such has generally been the case, the connections of the two machines, however controlled, having been usually as shown by Fig. 36, the effect mentioned not having been fully appreciated until lately.

The field windings being simply so many turns of wire having equal resistance in the two motors, the division of

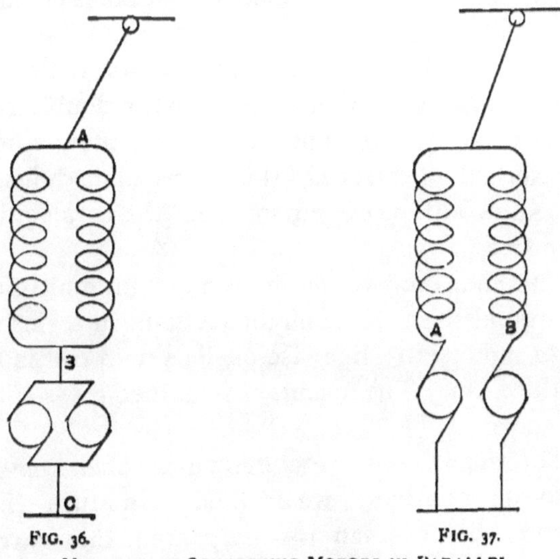

FIG. 36. FIG. 37.
METHODS OF CONNECTING MOTORS IN PARALLEL.

current from A to B will be equal, while from B to C the division will be determined by strength of field, as just explained.

If the connections are as shown in Fig. 37 the possible differences are very largely reduced.

In this case, each field winding with its armature constitutes a separate circuit, and the excessive current which would tend to flow in B, under the condition of unequal reluctance of field, will be checked by its own effect in increasing the ampère turns in circuit B over that in circuit A, thus increasing the counter E. M. F. The current in B will,

6

unless saturation for both machines has been reached, be greater than that in A, since by supposition the field strength is less for a given current in B than in A, and the counter E. M. F. must then be less. But the difference will be only proportional to the difference of permeability below saturation. The connections as shown in Fig. 36, in which the machines are cross-connected, have the advantage of offering greater facility for reversing the current. Thus, one switch, reversing the current over both fields or both armatures, through the terminals A and B, or B and C, controls both motors. If the machines be not cross-connected, then a separate switch must be used for each machine. This, however, is a matter of small consideration compared to the importance of the result secured by it. The connection of the two motors in series of course renders unnecessary the double reversing switch, and makes the current in the two motors identical. Another method of meeting this particular difficulty was used, indeed, is still used, on some of the cars equipped by the Edison Company.

In this method, the two motors remain in multiple: a coil is placed around the field of motor (A) and in series with the armature of motor (B): likewise on B, a field coil is in series with armature (A). The machines remained cross-connected as shown above.

These additional coils were known as "balancing coils." If, due to its relatively weak field, armature (B) tends to take more current than its companion, this current traversing the coils around the magnets of A, and in a direction opposite to that of the principal windings of A, diminishes the field strength of that machine. Likewise, the current in armature A circulates around B and diminishes its field strength, but by a less quantity than that subtracted from A. The tendency, then, is plainly toward equality. The method has been applied with considerable but varying success. Thus, suppose one machine, A, to be over-saturated; the other, B, to be below saturation. Its armature current, though greater than that of A, may have a less effect in demagnetizing A than will the current of A have in demagnetizing B, and the inequality would thus be increased.

This may be taken to be an exaggerated case, but yet it is possible; and since it shows an actual reversal of the intended effect, it is plain that in cases of smaller and more probable differences, the device may at least vary widely in its effect without actual reversal.

If, further, each pair of machines before being mounted were carefully studied, the adjustment might certainly be made very accurate. The practical necessities of manufacture and installation prevent this. The chief objection to this method may be stated in this: that while it introduces a complication of wiring equal to or greater than that demanded by the double reversing switch, it does not cure the evil so efficaciously or so simply as by independence of circuit and double reversing switch, or by placing the motors in series.

This unequal division of current becomes an evil only when occurring under heavy loads, such that one machine may be required to do much more than its normal work, in which case the efficiency of its performance may be considerably lowered; or the over-load may go so far as to injure the insulation by excessive heat.

Before leaving the subject of speed control, it should be stated that numerous designs have been made (and a few executed) looking to continuous and approximately constant speed of armature, mechanical devices being used to effect a varying speed reduction between the armature and the axle. In some cases, usually by the use of friction plates or hydraulic gearing, the change of ratio between the armature and the axle speed is effected by some continuous movement, carrying the ratio by changes of insensible value from unity to infinity or the reverse.

In other cases, usually by the use of some form of the so-called sun-and-planet or external and internal gearing, the changes are abrupt and limited in number. This variation of leverage, however obtained, is certainly in itself an incontestable advantage. By means of it an armature of minimum weight and working at a maximum efficiency may be made to serve for a given range of work.

If used in connection with a shunt field, the armature speed might be as constant as in the case of stationary motors doing variable work. Change of work requirement would be followed by change of current consumption, which would correspond to slight changes of field strength, the motor efficiency and the speed of armature remaining nearly constant. The proper determination of relations in such case would be as follows: First consider the maximum allowable speed of armature, as determined by electrical and mechanical effects in the armature, and its directly attached devices. Calculate the maximum output that might be required for such a length of time as would cause the efficiency of the motor during that time to be of importance. Assume that this output should be at an efficiency of, say, 90 per cent. The current required then becomes known, likewise the resistance of the armature conductors. Their cross-section results from making proper allowance for the heating effect of such a current over such a resistance. Their length is determined by the relation between the cross-section and the resistance. The size of the loop, or of the body of the armature, is calculated from the conditions that the given length of conductor rotating in a field of assumed strength, say one-half that of saturated iron, shall generate a counter E. M. F. of the required magnitude. The other dimensions of the motor follow from ordinary calculations.

The average output, and those below the average, are all obtained at high efficiency, the field strength being increased, the counter E. M. F. being slightly increased, and the current largely decreased thereby.

In case series motors are used with this method, it would seem best so to wind the field magnets as to produce saturation with a relatively small service current. The armature would then for the most part rotate in a constant field; slight changes of the speed of the armature would be followed by considerable changes of current, corresponding to changes in the required output, practically the whole speed regulation being effected through the particular mechanical device used. Should the field strength become constant only at relatively large currents, there would be a considerable variation of the armature speed, which might or might not be in the same direction as that desired to be effected.

Summarizing concerning this general method, it may be said that if variable gearing could be had, in which the complications would be insignificant, while its reliability and efficiency of transmission would be great, then its use would become general. Active minds are at work on the problem, and it may soon be solved. Meanwhile, the controlling demand for simplicity excludes the devices thus far presented.

The action of shunt motors has just been compared to that of series motors in the particular case of mechanical regulation of speed. It remains to inquire why they have not been used in the already extensive application of electric motors to traction work. At the outset it may be stated that the particular quality named by Silvanus P. Thompson as advantageous to the series motor, in comparison with the shunt machine, is in fact of no practical value. On p. 539 of his great work, "Dynamo Electric Machinery," Thompson says:

"The fact that the torque of a series motor depends only on the current, is of advantage in the application of motors to propulsion of vehicles (such as tram-cars) which at starting require for a few seconds a power greatly in excess of that needed when running. To start, a large current must be turned on. One convenient way of arranging this is to use two motors, coupled habitually in series. When starting, they are, by moving a commutator, coupled in parallel. This doubles the electromotive force for each, and at the same time halves the resistance. For a few seconds a very strong current flows—much stronger than that which the motors would stand for any prolonged work—and so provides the needful additional torque."

As has been pointed out above, the practical difficulty met with in electric-railway design is to get sufficiently high, not sufficiently low, initial resistance. The normal resistance, even of two series motors, placed in series with each other, is not high enough to prevent needlessly large starting currents from flowing. The absence of opportunity for any considerable experience in this direction offers ready explanation for so slight an error in what is, perhaps, the most satisfactory all-round practical electrical treatise in the English language.

A disadvantage which was early discovered in the use of shunt motors is stated thus by Kapp ("Electric Transmission of Energy," pp. 316, 317), speaking of the little Blackpool electric tramway: "It has been explained that the speed of a shunt motor, when running light, can never exceed a certain limit; whereas a series motor may, under the same conditions, assume a dangerously high speed. On purely theoretical grounds shunt motors are, therefore, more suitable for tramway work. But a serious practical difficulty was soon encountered. It arose from the uncertainty of electrical contact between the wheels and the rails. When a current of electricity has to pass through two pieces of metal in contact, the first condition is that the surfaces should be clean, and that is precisely the condition which cannot always be fulfilled in a tramway exposed to the weather and overrun by other traffic. It would thus occasionally happen that the current was interrupted for a very short time, perhaps only a fraction of a second, but the interval was sufficient to cause the field of the motor to lose its magnetism. The consequence of this was, that when contact was restored and the current began again to flow, the armature was not able to offer any counter electromotive force, and an abnormal rush of current took place before the field magnets had had time to again become excited. It will be noticed that the injurious effect here described will be the greater, the lower the resistance of the armature; that is to say, the more efficient the motor, the more will it suffer from an occasional interruption of current. Since it was impossible to absolutely avoid these interruptions, the use of shunt motors was discontinued, and series motors were substituted. In a series motor the intensity of the field and, therefore, the counter electromotive force of the armature are at once restored when the current begins to flow, and no abnormal rush of current can take place. To prevent racing when lightly loaded, variable resistances, placed below the platform at either end of the car, have to be used. These resistances are also employed for regulating the speed when the motor is doing a fair amount of work."

It seems that this difficulty could readily be overcome by

a simple automatic device, which would open the armature circuit unless the field circuit is closed. Such a device would doubtless have been used, were there any marked advantages in other respects of the shunt over the series winding. The reasons for the use of the latter are: 1. The early prejudice founded upon the facts related by Kapp. 2. That in a series motor the regulation of field strength and applied E. M. F. may be effected through one mechanism, the current in the armature and the field coils being the same. 3. The facility of using the field windings as a rheostat, as explained above. 4. The fact that the insulation of the field coils is somewhat easier in the series machine, the difference of potential between the terminals being much less than in the shunt motor. 5. The fact that up to saturation there is an automatic field regulation.

These reasons are good and sufficient to determine present practice, though it is certainly true that good service would be had from a shunt-wound motor, speed control being effected mechanically or through rheostats placed in the armature and the field circuits, to which may or may not be added commutation of field windings.

In considering the motor, in place, in relation to the other necessary parts of the car, we are led to discuss:

(A) The manner of connecting the armature shaft with the axle.

(B) Convenience of position with respect to inspection, removal and repairs.

The means used for transmitting power from the armature shaft to the axle have been:

1. Spur gears. 2. Sprocket chains. 3. Beveled gears. 4. Connecting rods, as from one locomotive driver to the other. 5. Worm gears. 6. Ordinary belting. 7. Ropes and pulleys. 8. Friction plates, etc., as explained, for variable speed reduction. 9. Centering of armature directly on the axle.

1. *Spur Gears.*—Perhaps 99 per cent. of all the railway motors now running are connected by spur gearing. From this general use, it follows that the burden of proof lies upon any other method before it may be considered as being practically in the field. The efficiency of this method

may be roughly stated as 90 per cent. for each couple of intermeshing gears, if exposed to grit, and provided the load be reasonably large. When the load is very small, the fixed loss in friction causes the efficiency to be low.

If the gears be inclosed in oil-tight cases, as is now largely practiced, the couple efficiency is from 90 to 95 per cent. at medium or heavy loads. In mechanical simplicity and reliability, under reasonably good design, there is little to be desired. The serious problems that have been met are:

(*a*) Choice of material, regard being had to strength, freedom from noise, and cost. Cast iron for the heaviest gear (that on the axle), with machine-cut teeth, has been almost exclusively used. For the smaller gears, running at higher speeds, cast iron, steel, brass, bronze, wood (for the teeth), rawhide—all have been tried, with a gradual settling down to steel or gun-metal. The evolution from the train of four gears, with intermediate shaft, to the train of two gears (armature pinion and axle gear) has removed almost entirely the principal cause for endeavoring to depart from the metals for gear material. That cause has been, still is, on earlier apparatus, the excessive noise of the high-speed, much-worn teeth. It seems probable that on the single-reduction gear motors, now so widely introduced, the best practice will use steel wheels with machine-cut teeth both on axle and armature shaft.

(*b*) Diminution of noise. This is connected with the matter of (1) speed, (2) material, (3) relative exposure, (4) suspension method for the whole machine, (5) tooth-design, (6) alignment and (7) load transmitted.

(1) In designing for a speed of about 500 armature revolutions per minute, geared to produce a car velocity of about ten miles per hour, it is probable that the manufacturers have gone as far as is wise, so long as gears are used at all. The noise at this speed may be inconsiderable. The next step after passing the single-reduction motor, geared as just suggested, seems to be the gearless or axle motor. Of its desirability, notice will be taken later.

(2) As to material, we have only to add to what has been said above, that the quietest substitute for metal thus far

found is wood. This material, however, is not strong enough to be used for the armature pinion, the diameter of which scarcely exceeds five inches. Rawhide has been the most successful rival of metal for this hard service. In this, as in other substitutes, it has been difficult to obtain uniform quality in the finished product. Consequently, widely different opinions have been held in regard to its value. The contest, we think, will cease with the passing away of the older high-speed machines, leaving metal undisturbed in the field.

(3.) It is easily practicable tightly to cover the gears of a single-reduction motor. Nor would the problem have been difficult for the four-gear machines, had it been held in view during the original design. That, however, was not the case in respect to the motors most largely used up to date, those of the Thomson-Houston and of the Sprague or Edison companies. Efforts have been made to add suitable covers, but every mechanic knows the difficulty of a "patch job."

Unless the cover be very firmly attached, and of material not particularly resonant, its use may be of no benefit in the matter of diminishing noise, but its great usefulness in excluding dust, pebbles, etc., will of course remain.

(4) That form of motor suspension which places the whole weight of the machine on springs will readily be understood to be that which best conduces to soft-running gear wheels. Until very recently, it has been an almost invariable rule to place one-half of the motor weight dead on the axle, as appears in several of the figures illustrating familiar railway motors. The other half rests on springs which permit a rotation of the motor around the axle. This method has doubtless been of more value in saving the teeth from breakage by sudden strains, than in diminishing noise.

(5) As to tooth-design, there has been little variety, an approved epicycloidal tooth having been most largely used. By using two sets of such teeth on the same gear, "staggered" with respect to each other, the Edison Company has been able largely to reduce the rattle often noticed in well-worn or badly-adjusted gears of the double-reduction type. The design is shown in Fig. 38.

(6) Good alignment is of value in this as in all similar cases.

(7) In the matter of load transmitted, it is observed that a car having been put in rapid motion by the work of the motor, the gears, grinding noisily while the work is being done, become almost noiseless if the current is cut off, though the car may continue its rate of motion unchanged, as on a down grade.

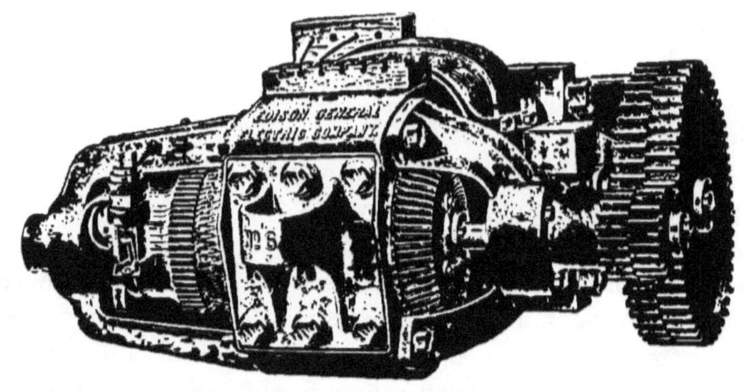

FIG. 38.—SPRAGUE-EDISON MOTOR.

Lubrication of Gear Journals.—For convenience, we may refer here to armature shaft journals also, and state that the best results have been had from the use of grease rather than fluid oil. The self-oiling bearings, such as are so widely used on dynamos, are not so well adapted to service on a much-jolted car motor. Oil is lost rapidly from the basin, and it has seemed, further, that the feeding rings or chains, supposed to revolve eccentrically around the shaft while dipping in the oil, are prevented from traveling by the constant shocks, which tend to keep them playing vertically rather than to rotate them. The grease, placed in a very simple box or cup, should be followed by spring pressure applied to the cover or to some disk inside the box. A pin resting on the shaft (but pressed lightly, so that it will not cut) and passing up through the body of the grease, causes the melted particles to follow easily down its sides.

For the journals of the armature shaft, globe or conical bearings, as shown in Fig. 39, have been much used, while for intermediate and axle shafts straight bearings have been preferred.

There are, however, good authorities in favor of straight, babbitted bearings throughout. If babbitt or other soft metal be used, the *greatest care* must be exercised as to its quality.

2. *Sprocket Chains.*—Four or five years ago, and in the practice of Mr. Charles J. Van Depoele, one of the earliest workers in the field, sprocket chains were used on perhaps seventy-five cars. Examples of quite tolerable service from these early machines may still be found. The sprocket chain was perhaps the best medium of transmission, as long as the motor remained above the car floor, on the platform or inside the car. Nothing better was available for covering the considerable distance between armature shaft and axle. But the speed was high, and the art of making sprocket chains for such trying service was not far advanced. Hence this device disappeared when the motors came to be placed under the car floor (as was done by the Sprague Electric Railway and Motor Company in its earliest work), and the spur gear came to the front. There seems good reason to suppose that the chain may still have its sphere of usefulness, as in transmitting power from the directly-driven axle to a second one, the speed between axles being less than elsewhere in the train. Useful and apparently successful applications of

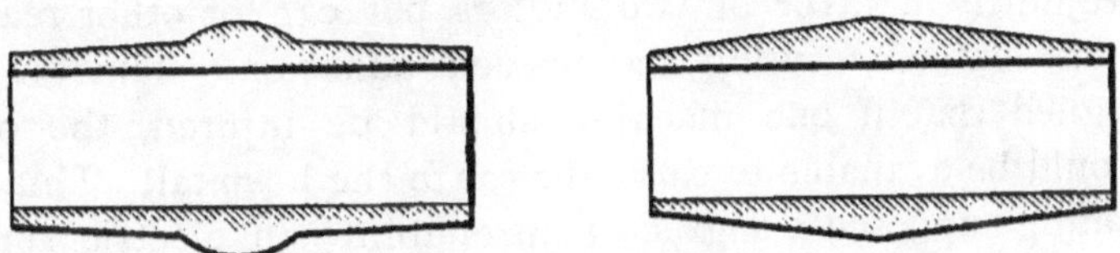

FIG. 39.—FORMS OF BEARINGS.

sprocket-chains have been found in the construction of electric mining machinery, improvements having been made in the manufacture of the links of such chains.

3. *Bevel Gears.*—In using these, the armature is placed parallel to the axis of the car. The only reason for endeavoring to do this is that one motor may thus be connected to two axles of a truck, thus utilizing the whole weight on each

truck for adhesion, or for "traction," as it is frequently but improperly expressed. On grades lower than 3 per cent. this double connection is of relatively small importance, since on such grades one-half the weight of the car, if resting on the driven wheels, is found sufficient to prevent skidding, on average tracks, even when wet with snow—a reasonable supply of sand being available. For an ordinary 16-foot car, operating on such grades, a single motor of such capacity as are those generally rated at 15 horse-power (as by the Thomson-Houston, Edison, Short, and Westinghouse companies in the United States), and geared to one axle of an ordinary four-wheel truck, is sufficient, both as to adhesion and traction. Under favorable track conditions, and unless unusual speeds be required, the same motor will handle, without slipping and without overheating, one trail car of about the size of the motor car. Where greater effort is required, or under exceptionally bad track conditions, it is best to apply the driving power in some way to both axles of a four-wheel truck; if a long car be in question, requiring a six-wheel truck, or two four-wheel trucks, then the power should be applied to those axles, which carry from 70 to 90 per cent. of the whole weight. As is familiarly known, this requirement has been met in the designs most widely used by using a separate motor for each driven axle.

There was, in the early days, four or five years ago, some prejudice in favor of two motors per car for other reasons than that of securing the needed adhesion. Thus it was argued that if one machine should be injured, the other would be available to carry the car to the hospital. This reasoning supposed a degree of unreliability in electric service which would, if permanently, inherently true, rule it out of good practice. The proper reliance in case of accident is upon the next car following, and accidents should be, indeed are, rare enough to make this occasional double duty unobjectionable.

In face of other manifest advantages of a single motor, this prejudice would soon have disappeared, and with it the double motor equipment for ordinary cars, except for this difficult problem of sufficient adhesion on heavy grades.

The device of beveled gears was early considered and tried, especially by Bentley & Knight, who thus equipped a car in 1886. But the mechanical difficulties are great. The truck framing must be rigid to secure alignment of gears at the opposite ends of the armature shaft; the weight of the motor must be practically uncushioned; the efficiency of beveled gears is low; it is difficult to get sufficiently strong teeth in a beveled pinion, of such diameter and pitch as required for use in connection with a 30-inch car wheel, which limits the

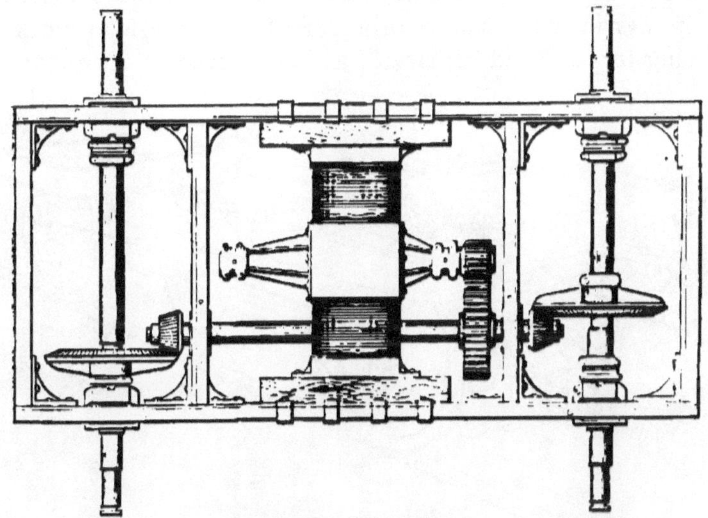

FIG. 40.—RAE MOTOR TRUCK.

diameter of the axle gear to about 24 inches, thus leaving very little clearance.

Many students of the problem thought these difficulties such as to warrant, without trial of the beveled gears, the adoption of other methods. Recently, however, the method has been given some prominence by the efforts of Mr. Rae, electrician of the Detroit Electrical Works, Detroit, Michigan. There are now running perhaps two hundred cars equipped by that company. Curiously enough, they have been most largely used on comparatively flat roads, where this peculiar merit is of least value, since on such roads a single motor could be geared in the ordinary way. The weakness of gear teeth has given some trouble in cases of the heavier work undertaken. It remains to be demonstrated whether satisfactory work will be had from similar installations on very heavy grades. The progress of these efforts will certainly be a matter of great interest. (See Appendix F.)

4. The use of connecting rods instead of gears was tried some years ago by Bentley & Knight, later by Mr. Leo Daft. But it has remained for Mr. Rudolph Eickemeyer again to call attention to this method, by bringing out at the same time a motor of such slow speed that there is no reduction between armature and axle. The connecting rod transmits the power to the two axles of a truck from a single armature, driving forward and backward, as in the case of three driving axles of a locomotive coupled together. (See Fig. 41 for further information.)

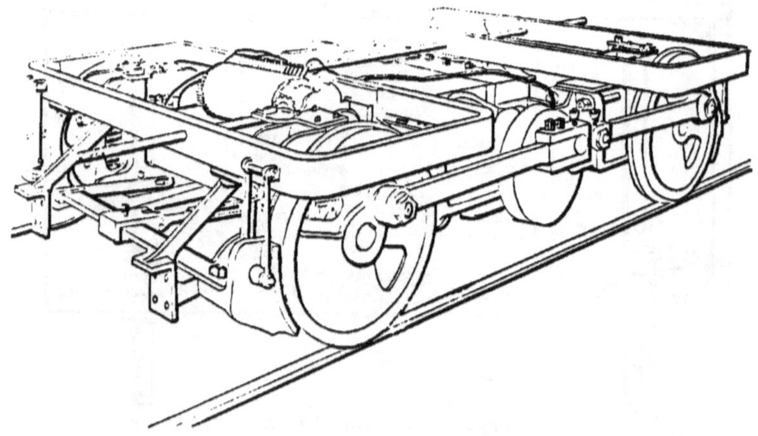

FIG. 41.—EICKEMEYER GEARLESS MOTOR TRUCK.

It will at once be inquired why, having reduced the armature speed to that of the axle, Mr. Eickemeyer does not then center the armature on the axle and avoid any intermediate mechanism. His reasoning is thus: That it is generally better to use one motor rather than two; that two motors would be needed for grade work if the armature were concentric with the axle; that when the motor is placed midway between the two axles its whole weight may be very conveniently placed on springs, a matter of grave importance, by reason of the cost of the track repair due to the pounding of so much dead-weight; and that such position for the motor offers greater facility in the matter of removals than if on the axle.

The few equipments thus far made by Mr. Eickemeyer have been in satisfactory operation for some months. It seems quite clear that the weight of these machines is small, capacity and speed being duly considered. But in comparing this or any other motor of only axle speed with motors requiring reduction gearing, it must be remembered that the same excellence of design that may be applied to the production of a machine of say 20 horse-power at 100 revolutions, and weighing 5,000 pounds, will produce a machine of 20 horse-power at 500 revolutions, weighing about 2,000 pounds, or two motors of 10 horse-power, weighing about 1,200 pounds each. Now, suppose these two motors to be entirely spring-supported, and geared, one to each axle, through spur gearing, in dust-proof, oil-tight cases. We then have, in the one case, one large motor (more difficult to handle), 5,000 pounds weight, spring-supported; four connecting rods, both axles driven. In the other, two small motors, 2,500 pounds weight, spring-supported; four spur gears; both axles driven. The future can best determine which of these designs will be generally preferred.

In comparing the Eickemeyer equipment with other motors of axle speed, we are brought to 9—the direct centering of the armature on the axle. The most notable example of this method now in operation is the City and South London Electric Railway. While, indeed, the particular feature of centering the armature on the axle is here illustrated, yet in many ways it is not a case properly comparable with the efforts now being made in the United States. He who has felt the larger freedom which comes in designing a separate locomotive, instead of a machine placed out of sight under a car floor, will appreciate the fact that gearless motors for ordinary street-railway service cannot be copied from the English example. We shall, then, for the present, return to those designs which have just been built in this country.

Two companies, the Short Electric Company, of Cleveland, Ohio, and the Westinghouse Electric and Manufacturing Company, of Pittsburg, Pa., have already manufactured such gearless motors.

The Short motor has a "disk armature" with pole pieces on the sides, familiar in Brush arc-light dynamos. The armature is attached to a hollow sleeve, concentric with the axle, but supported, through a framework, on springs, and of large enough inside diameter to permit vertical play of about an inch. The field magnets and pole pieces are supported in like manner and move with the armature. The sleeve transmits its power to the axle through six cushions or springs connecting two pairs of disks, one keyed fast to the axle, the other fast to the sleeve (Fig. 42).

The design of the Westinghouse Company shows a Siemens or drum armature keyed directly to the axle. The field is made by four pole pieces connected through an external cylinder. The whole weight of the motor is directly upon the axle (Fig. 43).

It will be noted that the Short design covers the important point of spring suspension, which will be better appreciated as time relentlessly forces attention of electric-railway owners to track repair. It is intended to use one motor for light grades, and two for heavy ones, as above explained. The relative merit of this and the Eickemeyer design will then be determined, in time, by the peculiarities in service—of connecting rods on the one hand and spring connection between pairs of disks on the other.

Ideally simple, the Westinghouse design will be called upon to demonstrate whether or not some departure from such simplicity is warranted by consideration for the track.

We do not here enter into discussion of the relative merits of the three motors viewed simply as electric machines of greater or less efficiency and weight per unit of output. Little is known concerning these points. That it is difficult to obtain high efficiency at such low speeds and in such limited space goes without saying. The necessary limitation of weight also raises a serious question as to whether gearless motors will not eventually be restricted to comparatively light service, unless the speed be considerably higher than in ordinary surface tramway service. If the connecting

FIGS. 42-44.—TYPICAL MODERN MOTORS.
FIG. 42.—SHORT GEARLESS MOTOR. FIG. 43.—WESTINGHOUSE GEARLESS MOTOR.
FIG. 44.—EDISON SINGLE-REDUCTION GEAR MOTOR.

7

rods of the Eickemeyer design, or the sleeve, springs, and disks of the Short design, are indeed necessary to save the track, we have at once a complication fairly comparable with the simple spur gears now used, and a total weight, for a given output, considerably greater.

In regard to (5) worm gears, (6) ordinary belting, (7) ropes and pulleys, (8) friction plates, etc., little need here be said. They have been used sporadically and with no marked success. The worm gear is apparently the best of these four methods. Its low efficiency has prevented many from seriously studying it. Mr. A. Reckenzaun has constructed a motor operating a worm gear, and has run the car by storage batteries. Details of his results have been published, claiming remarkably high efficiency as compared with that usually given by mechanical authorities. It is perhaps unfortunate that an untried gearing should have been experimentally associated with storage batteries. It is hardly fair toward the gearing.

(B) Convenience of position with respect to inspection, repair and removal. In the earlier electric-railway practice numerous examples are found in which the consideration of accessibility seems to have been treated as important. Motors were placed on the platform of the car, in the body of the car, or in a separate cab. The desire to use spur gears, and to leave untouched the space available for passengers, caused the Sprague Company to place motors under the car floor; and there they remain in the general practice of to-day. The position is one of the greatest exposure to dirt and to foreign objects that may cause immediate destruction; it also makes repairing a difficult matter. Nevertheless, it is the right position, pointed out alike by mechanical and commercial reasons. Recent improvements in the reliability of motor windings, in the mechanical construction of motor frames, in the simplicity and protection given the gearing — all combine to diminish the troubles resulting from this difficult position. If separate locomotives are used, the armature must still be concentric with the axle, or removed from it only by one train of gears. The general body of the motor may, however, be extended vertically,

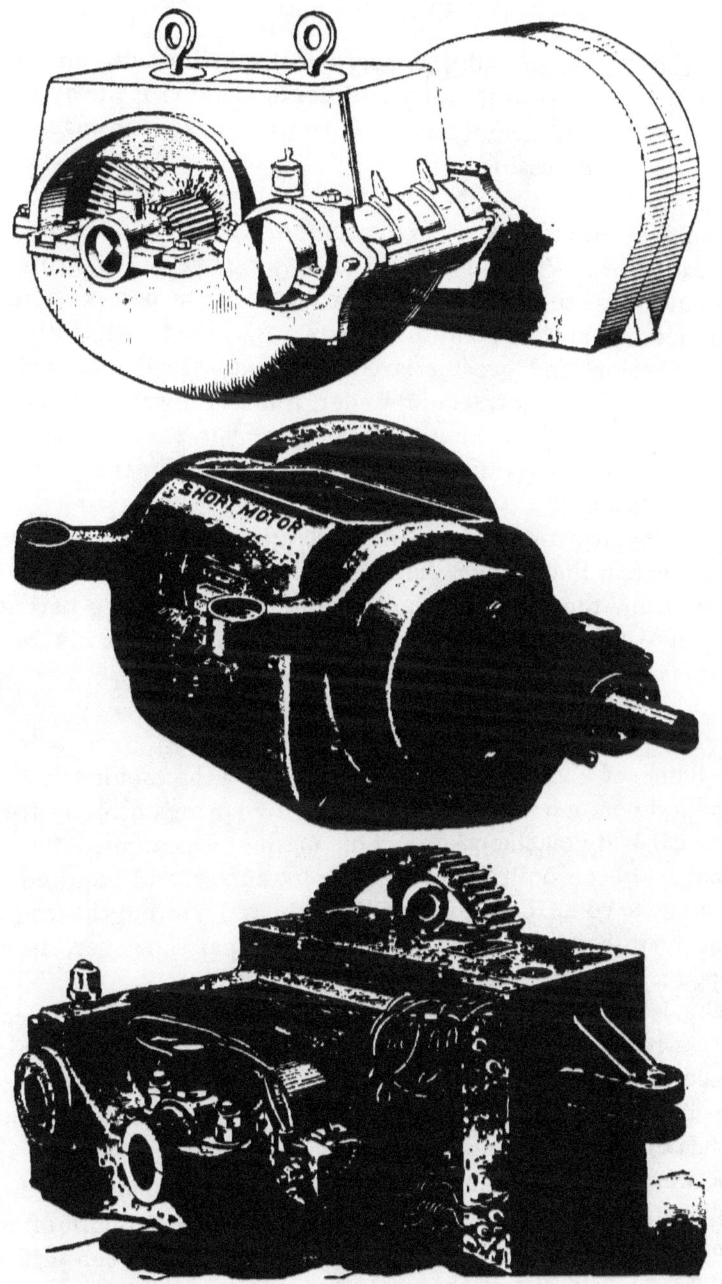

FIGS. 45-47.—TYPICAL MODERN MOTORS.

FIG. 45.—THOMSON-HOUSTON "WATER-PROOF" MOTOR. FIG. 46.—SHORT SINGLE-REDUCTION GEAR MOTOR. FIG. 47.—THOMSON-HOUSTON SINGLE-REDUCTION GEAR MOTOR.

instead of horizontally, as now; or it may lie between these two positions, as in the City and South London Railway locomotives. This arrangement, of course, makes the machines much more accessible.

It has been thought by some that separate locomotives would eventually be introduced into the street service of great cities. This does not seem probable. To bring it about means to introduce greater dead-weight per passenger carried, greater expenditure of power, greater difficulty as to adhesion, and greater street obstruction. For suburban service, or for elevated or underground railways, in which several cars are to be run in a train over easy grades, the problem is widely different, and seems to call for the separate construction.

It may also appear in a sort of compromise case as follows: a suburban line feeds a main cable line; cars are brought to and from the latter by an electric motor car having perhaps a few seats for smokers. This method has already been proposed for one of our large cities.

CAR TRUCKS.

Some of the earlier installations show the motor hung on the axle on one side, and suspended by springs directly from the car body on the other. This method was applied to cars that had been built for propulsion by horses, and required no change save an enlarging of the axle and a strengthening of the floor beams. Subsequent practice has, however, developed a different method, now almost universally used. The wheels and axles are mounted in a frame which carries a cross-bar for the support of the motor, and springs which receive directly the weight of the car body and cushion its oscillations. This separate truck offers facilities for mounting and dismounting motors, for exchange of car bodies (as between open and closed), and in a convenient manner supplies the strength of frame needed for the support of the motor. A great variety of designs may now be seen, differing rather in detail than in any important principle. It is gratifying to note in the later designs a simplicity of construction which was not observable in the earlier forms.

FIG. 48.

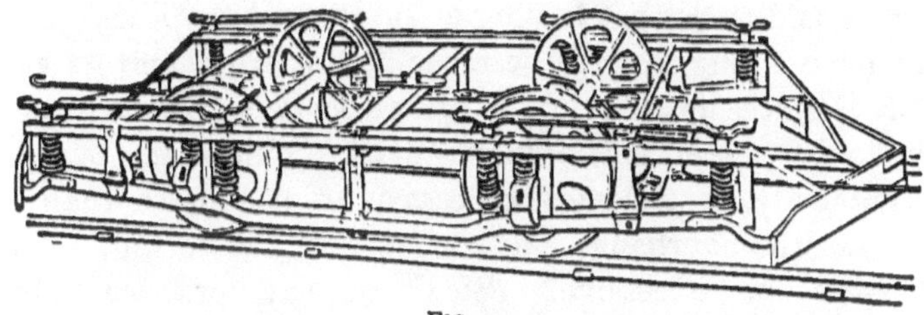

FIG. 49.

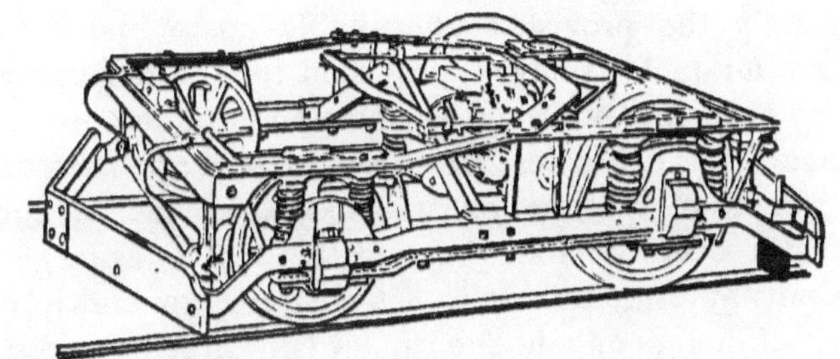

FIG. 50.

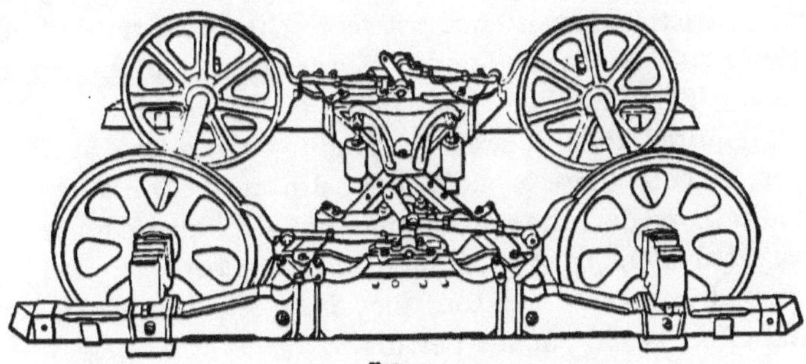

FIG. 51.

TYPICAL ELECTRIC STREET-CAR TRUCKS.

FIG. 48.—BRILL TRUCK FIG 49.—MANIER TRUCK FIG. 50.—BRILL MAXIMUM TRACTION BOGIE TRUCK. FIG 51.—STEPHENSON BOGIE TRUCK.

Nuts, bolts, screws, even rivets, should all be considered as evils—sometimes necessary, but to be kept down to a minimum in numbers. There is room for endless discussion concerning the details of truck design, but the scope of this book permits that only some of the more familiar types shall be shown, as in Figs. 48 to 56.

During the last year, there has been a very rapid increase in the use of larger cars than the standard horse-car, which may be said to be that having a 16-foot body. These larger cars have brought with them the double truck and its variations. Since the maximum allowable wheel base may be said to be 7 feet, it is evident that in going much beyond a 16-foot body, it becomes necessary to use more than two axles, as the overhang, and consequent "teetering," would otherwise be excessive. Steam-railway practice affords abundant precedent in the matter of double-truck cars, but some of the conditions for electric service are new. These are mainly the provision of suitable space and framing members for motor support, provision for sharper curvature than is ever met with in steam practice, and especially important is the problem of throwing a large part of the total weight of the car on two axles, instead of equally dividing it over the four axles of two trucks, as in general steam-railway practice. It is not desirable to multiply the number of motors on any one car; it is desirable—necessary, indeed—that the driven axles should bear such weight as will prevent skidding of the wheels. On wet, especially snowy, rails the frictional resistance between wheel and rail is found at times as low as one-tenth of the total weight on the wheels. That is to say, if an effort of 200 pounds per ton were required to start a car resting its whole weight on the driven wheels, these wheels may slip if the rails be wet. If the rails be dry, the wheels may not slip until a horizontal effort equivalent to 0.35 of their load has been applied; or, expressed in pounds per ton, they may slip when the horizontal effort is 700 pounds per ton.

Experience shows that on a level track, with ordinary running gear, an effort of about 70 pounds per ton is required to start a car—this overcoming the inertia and friction of the

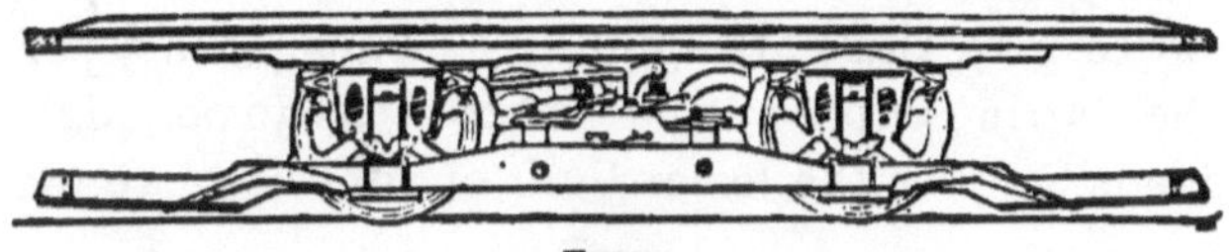

FIG. 52.

FIG. 53.

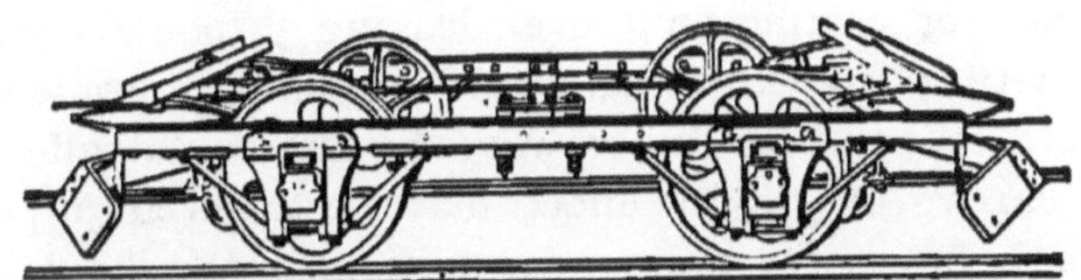

FIG. 54.

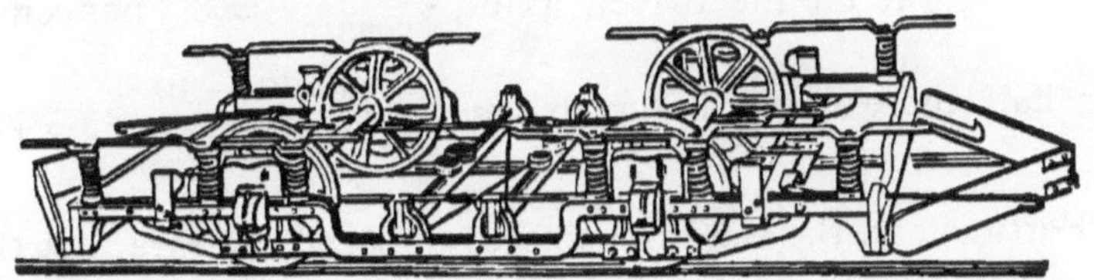

FIG. 55.

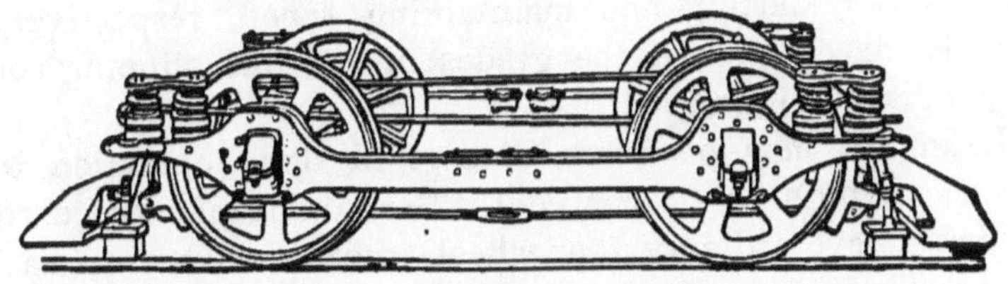

FIG. 56.

TYPICAL ELECTRIC STREET-CAR TRUCKS.

FIG. 52.—STEPHENSON TRUCK. FIG. 53.—ROBINSON RADIAL TRUCK. FIG. 54.—TAYLOR TRUCK. FIG. 55.—PECKHAM CANTILEVER TRUCK. FIG. 56.—MCGUIRE TRUCK.

parts. On grades, add 20 pounds per ton for each 1 per cent. of grade to give the starting effort. Thus, on a 3 per cent. grade the starting effort $= 70 + (20 \times 3) = 130$ pounds. Since 200 pounds may be the lower limit of adhesion, we may calculate that on a grade of $\frac{200 - 70}{20} = 6.5$ per cent. the wheels may slip, though the whole weight of the cars be on the driven axles, as in the ordinary 16-foot single-truck car carrying one motor on each axle. A similar calculation shows that if only half of the total weight be on the driven axle, or axles, there may be slipping at the start, on a 1.5 per cent. grade. The adhesion in this case for the driven axles is only 100 pounds per ton of total weight; of this take 70 pounds for starting on a level, leaving 30 pounds per ton for gravity resistance, which would be met on a grade of 1.5 per cent. When once under way, the horizontal effort per ton, exclusive of gravity effect, may be taken at 30 pounds as a high figure for ordinary conditions. We have, then, *when the car is in motion*, as possible slipping grades, for the whole weight on the driven axles, $\frac{200 - 30}{20} = 8.5$ per cent., for half the weight on the driven axles, $\frac{100 - 30}{20} = 3.5$ per cent.

Fortunately, the conditions here supposed are rare, but the sand box should be filled in order to meet them. If either the start or the run must be made on a curve or on a very rough track (or on both), the figures, 70 pounds and 30 pounds, for starting and maintaining speed, respectively, may be exceeded, and the grades of possible slipping correspondingly diminished.

Examples may be given of cars in daily operation on grades as high as 13 per cent. Sand is frequently used. The cars have ordinary four-wheel trucks, a motor on each axle. There is rarely any commercial demand for operation over any heavier grades.

For single-truck cars we may, then, summarize the case as follows:

For summer service (*i.e.*, when rails are generally in good condition) on grades up to 5.0 per cent., and for winter service on grades up to 3 per cent., the required adhesion may be had from one driven axle. On grades above 5 per cent. for summer and 3 per cent. for winter both axles must be driven; this may be done by using two motors, as is now ordinarily the case, or by some connection of a single motor to both axles, should any method of accomplishing this prove successful.

The case of the long cars may be thus stated. For the lower grades, as above mentioned, for summer and winter respectively (requiring only half of the total weight to be on the driven axles), adhesion may be had (a) by gearing a motor to each of the two axles of one of the two ordinary trucks; (b) or by gearing one motor to one of the axles of each of the two ordinary trucks; (c) or by gearing one motor to both axles of one of the two trucks used, if such gearing can be perfected.

For the heavier grades (requiring all or nearly all the weight to be on driven axles) the requisite adhesion may be had (a) by gearing one motor to each of the axles of the two ordinary trucks, making four motors in all; (b) by gearing one motor to both axles of the two trucks, if such gearing is successful; (c) by using two trucks, modified in such a way that one axle of each shall carry nearly all the weight of one-half of the car, this axle having a motor geared to it; the other axle carrying only enough weight to keep it on the track, and having no motor geared to it; (d) by using only three axles altogether, those under the ends of the car to carry nearly all the weight, and having a motor geared to each—the middle axle serving only as a guide, and carrying no motor. The first method (a) is objectionable by reason of the great number of parts to be maintained and the relatively greater losses in transmission of power to the axles. The second (b) has not been generally demonstrated. The third (c) may utilize 90 per cent. of the total weight, and has been successfully applied on considerable grades, through what is called the "maximum traction truck."

The fourth (d) may utilize about 90 per cent. of the whole weight, and has been successfully applied on considerable grades through what is called the "radial truck." To further explain these last two methods, Figs. 50 and 53 are given, showing the trucks mentioned. Both these designs are of great importance. As compared with ordinary truck construction, they represent a departure which seems essential to the application by present methods of spur-geared motors to long-car service (20 to 35 foot bodies) on grades above 5 per cent. The use of a sprocket chain, connecting the motor axle with an empty axle (as recently designed for some special cases), may indeed serve the same purpose; its success has yet to be demonstrated. The efforts about to be made should be watched with great interest. Further, it is of course possible to use four motors on trucks of ordinary construction. This method has, indeed, been proposed by a large railway company, for covering its winter service on slippery tracks; in summer, one motor truck is to be replaced by an empty truck taken from a summer or open car; the motor truck thus borrowed is to be put under the open car, which would thus be ready for service. The closed car, when needed during the summer, could also be run by its remaining motor truck. In other words, both classes of cars, open and closed, would at any time during the summer be ready for immediate service—the total equipment, counting both classes, being only two motors per car. Further, it is argued that the tractive power and the adhesion against slipping are thus both kept in a reasonable proportion to the demand, which, for both, is greater in winter than in summer.

It was also proposed that the individual weight and capacity of the motors might be diminished as compared with present standards. If each motor were of 10 horse-power capacity, then in summer each car would have a total of 20 horse-power, and in winter a total of 40 horse-power—the supply in both cases being considered sufficient for grades not exceeding 5 per cent., and for a schedule speed of about 8 miles per hour, including stops, the cars in question being about 28 feet long in the body. While this method has some advantages, yet on account of the objection to multiplication of motors already pointed out, it does not seem as desirable as that based on the use of trucks such as the maximum traction type, limiting to two the number of motors per car, winter and summer.

If it be still required that open and closed cars shall be in constant readiness during the summer, two methods are open. If the grades are light, one motor may at the opening of warm weather be transferred to the open cars, each car thus having for summer work one motor of, say, 20 horse-power capacity. If the grades be heavy, requiring even in summer much more than half the weight to be on the driven axles, then a double equipment of two motors per car may be had, each of, say, 15 horse-power capacity. If one motor can be successfully geared to the axles of a truck as by sprocket chains, by gearing or connecting rod, the shifting of one "live" truck from the closed to the open car, as above described, would, for the lower grades, produce the desired result of having both classes of cars in readiness during the summer months. If this condition be required on roads of heavier grade, a complete independent equipment must be provided for open and closed cars, no matter how the motors may be connected to the axles. It is not probable that such an expense would be incurred by many companies; they would in general prefer to run only the closed cars, or to run the open cars with curtains down during summer rains; or, again, to run the open cars only as trailers. This latter method has the disadvantage of attracting the live load away from the motors, thus making more difficult the problem of adhesion.

TROLLEYS.

To obtain a satisfactory "running contact" was four years ago a serious question. To-day, the trolley apparatus as a whole is in a satisfactory condition. A brass grooved wheel about five inches in diameter is centered on rawhide or graphite journals, and mounted at the end of a pole about 12 feet long; this pole trails back from the middle of the car roof at an angle of about 30 degrees with the vertical, when the trolley wire is 18 feet above the ground; it is pivoted, a few inches from the lower end, in a frame attached to the car roof, and springs acting at the lower end press the wheel against the trolley wire; by expansion or contraction of these springs, the pressure is continued as the height of the wire varies; by a cam arrangement, the lever arm through which the springs act against the pole may be changed in such a way as to keep the pressure against the wire nearly constant, whatever the degree of compression or expansion of the operating springs. This general description covers a number of satisfactory forms varying in detail. The pole may be of wood or steel; the steel may be straight, tubular, in sections of different diameter, or drawn tapering in one piece; the current may pass through the journal of the wheel or be taken through brushes; thence it may go through the

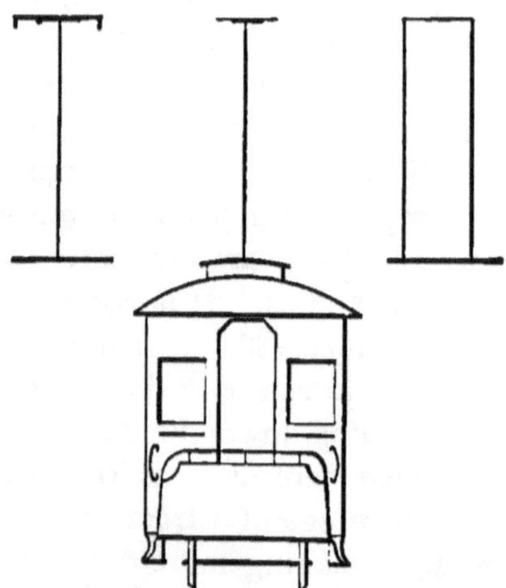

FIG. 57.—UNDERRUNNING ROD TROLLEYS.

pole, insulated at its base to the wire leading to the motor; or it may pass from the wheel to an insulated wire passing through the pole.

Two variations from this general form are important. In one, the wheel just mentioned is replaced by a shoe which slides against the wire. Usually a lining of soft metal is placed in the shoe, being replaced every day or two at an expense of about one cent. This has been used by the Short Electric Company and seems to have given satisfaction. In the other the wheel is replaced by a straight bar of considerable length, placed at right angles to the pole and to the longer axis of the car. To prevent the possibility of trouble at crossings and switches, while doing away with overhead frogs, this bar or cylindrical rod should be curved down at its ends, or the "T" form should be replaced by an inverted "U." (See Fig. 57.)

This "T" form has lately been brought forward by Siemens & Halske. It was used in 1887 by the Sprague Electric Railway and Motor Company, and sacrificed, perhaps needlessly, to the desire to please in the matter of looks. While the action of the trolley wheel is on the whole satisfactory, it would certainly be a great gain in line construction to drop the use of frogs, cross-overs, etc.

A few typical forms are shown in Figs. 58 to 63.

CAR WIRING.

As to the car wiring, it may be thus briefly described. There are independent circuits between the trolley and the rails, one for lamps, one for motors, one for the lightning arrester, and there may be one for heaters. In the lamp circuit there may be placed in series 16 candle-power lamps, each at normal brilliancy when it has about 100 volts between its terminals. Since the line potential is approximately 500 volts, if more than five lamps be required, they must be of lower voltage than 100, or another circuit must be made.

The motor circuit is itself in duplicate when the motors are in multiple, and may indeed be said to be constituted of two circuits. One must of course vary its wiring according to the system of regulation employed. Fig. 64 shows the standard wiring of the Thomson-Houston Company. Fig. 65 shows that of the Edison Company. In the motor circuit are placed the safety fuses and controlling switches. As to the use of these, consult the rules of practice given elsewhere. Electric heaters have not come into very general use, but may grow in favor. Thus far the practice has been to place two or more heaters in multiple, in the first few hours of service, causing a current of about 6 ampères to flow. When the car has been well warmed, the heaters are thrown into the series arrangement, reducing the current to about 3 ampères. It is said that this quantity of current will keep a 16-foot

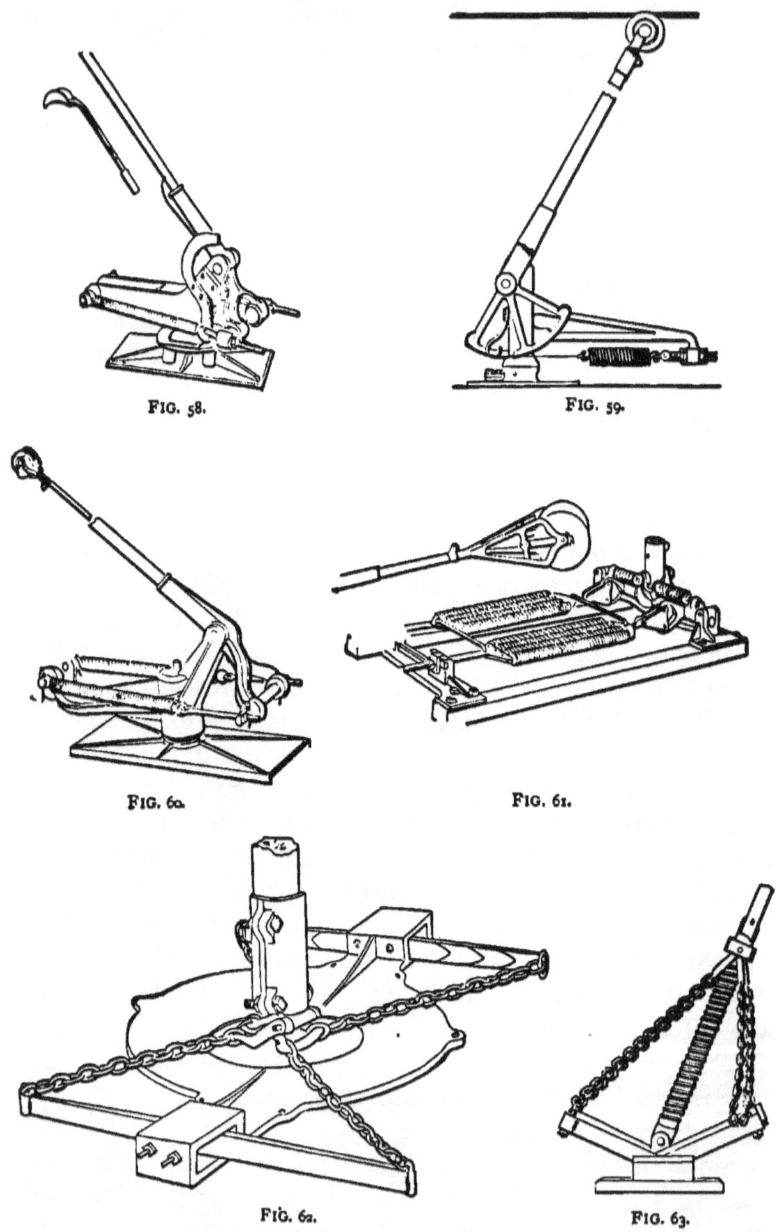

FIG. 58. FIG. 59. FIG. 60. FIG. 61. FIG. 62. FIG. 63.

TROLLEYS AND TROLLEY BASES.

FIG. 58.—SHORT SLIDING TROLLEY. FIG. 59.—BAKER TROLLEY. FIG. 60.—SHORT TROLLEY. FIG. 61.—"BOSTON" TROLLEY. FIG. 62.—LIEB TROLLEY BASE. FIG. 63.—"COMMON SENSE" TROLLEY BASE.

car at a comfortable temperature in very severe weather. Its cost may be roughly estimated on this basis. One ampère-hour will cost from 0.8 cent to 1.5 cents. For convenience, say 1.0 cent. If during an eighteen-hour run the current averages 4 ampères, then the cost per day for the supply of fuel would be 72 cents. The expenditure for fuel in heating by stoves may be taken at about 10 cents.

Cleanliness and convenience are of course great considera-

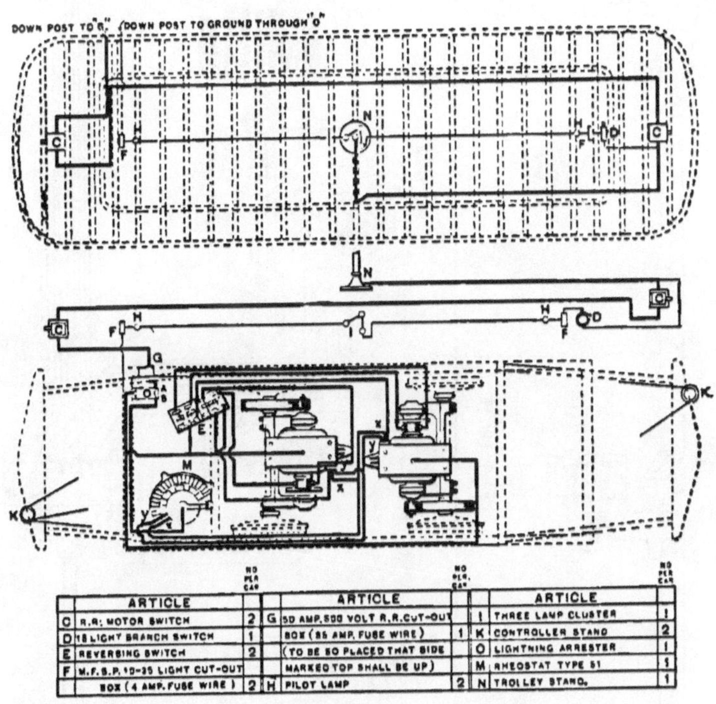

	ARTICLE	NO. PER CAR		ARTICLE	NO. PER CAR		ARTICLE	NO. PER CAR
C	R.R. MOTOR SWITCH	2	G	50 AMP. 500 VOLT R.R. CUT-OUT		I	THREE LAMP CLUSTER	1
D	18 LIGHT BRANCH SWITCH	1		BOX (35 AMP. FUSE WIRE)	1	K	CONTROLLER STAND	2
E	REVERSING SWITCH	2		(TO BE SO PLACED THAT SIDE		O	LIGHTNING ARRESTER	1
F	M.F.S.P. 10-35 LIGHT CUT-OUT			MARKED TOP SHALL BE UP)		M	RHEOSTAT TYPE 51	1
	BOX (4 AMP. FUSE WIRE)	2	H	PILOT LAMP	2	N	TROLLEY STAND.	1

FIG. 64.—THOMSON-HOUSTON CAR WIRING.

tions in favor of the electric heaters. These have been made by imbedding wire or other metallic conductors in clay or a similar substance, the whole being then inclosed in an iron case which may be placed under the seats. The clay, by preventing oxidation, permits a relatively larger current to flow through a conductor of given cross-section, without destroying it.

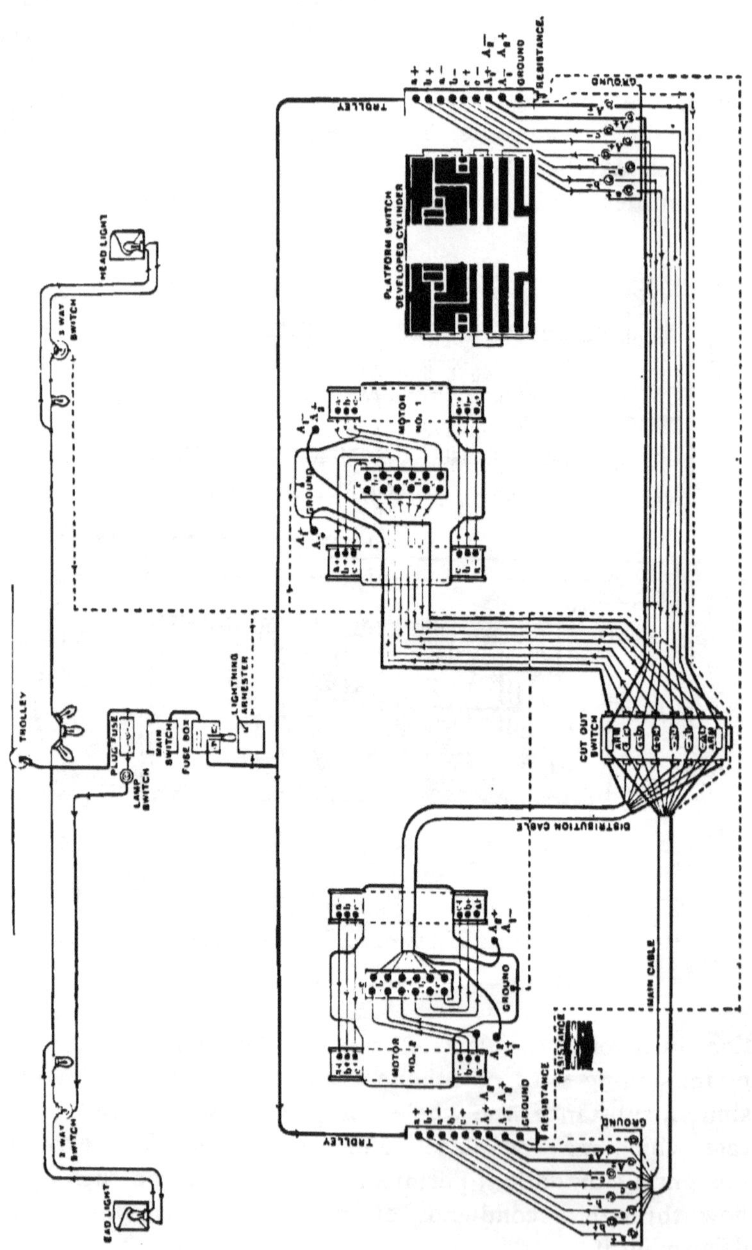

FIG. 65.—EDISON CAR WIRING.

LIGHTNING ARRESTERS.

The device used to protect the machinery from lightning constitutes normally what is often called an open circuit. In its simplest form, the lightning-arrester circuit would show simply a break of say one-eighth of an inch between two points. In Fig. 66 this simple circuit is shown, and in its relation to the other circuits of the car.

It has been demonstrated by experience, that if a high potential be suddenly established at a point, A, Fig. 66, it can more readily force a current across a break, B (of suitable dimensions), than through the complicated windings of wire around metal masses, which together constitute the motors.

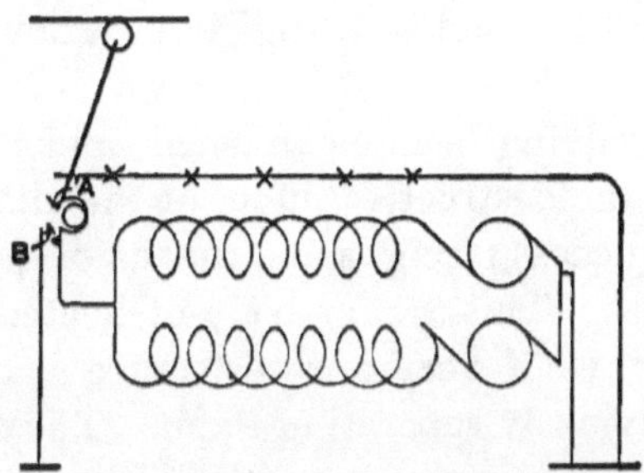

FIG. 66.—LIGHTNING-ARRESTER CIRCUIT.

A steady and moderate potential, such as that supplied from the dynamos, is on the other hand unable to force a current across the air space, B, but sets up a proper current through the motors.

The motor current thus produces in the two cases effective resistances, varying widely as compared with that of the air gap. This is traced to the self-induction of a path containing many convolutions.

Seeking some analogy with more familiar mechanical phenomena, we may liken this to the resistance known as inertia, or the resistance to change of state. Its measure is the measure of the work required to change the velocity of a given mass by a given number of units of velocity in a given time; it is the time element which operates to our advantage in the lightning discharge. The high potential at A can establish an arc at B in shorter time than it can establish a current through the motor windings. The arc having been established, the high potential is at once lowered by the flow to earth, there being nothing now to maintain a higher potential than 500 volts. If between the trolley wire and the ground no other circuit be prepared than that through the motors, then the discharge would force itself through some point in the windings where the insulation resistance between the wire and the body of the motor or between the wire and the general metal of the car (equivalent to earth) is low.

Whether there is a lightning arrester or not, it is best to connect the field windings next to the trolley wire, since the discharge generally, though not always, enters from overhead—injury to the fields being less serious than injury to the armature.

The arc at B having been established, and the motor thus protected from the destructive effect of the discharge, a new trouble arises, necessitating some means of quickly destroying the arc itself. This arc, or body of heated gas between the points at B, is of very low resistance as compared with the cold air previously separating them. The ordinary pressure on the line (500 volts) is therefore able to maintain this arc, and will cause a very great current to flow over this path, practically short-circuiting the motors.

The current, even if not great enough to melt the wires leading to and from B, becomes great enough to materially lower the potential at A, thus checking, perhaps stopping, the motors, and causing a waste of energy. The effort to prevent this secondary action has produced various lightning arresters now on the market. Underlying nearly all these forms may be found one of these two principles. First: to destroy the arc by magnetic action repelling or attracting, and thus weakening it. Second: to lengthen the air gap so much that the arc cannot be maintained across it.

If the latter method be employed, means must be devised for resetting the points with the proper distance between them, or setting new points in position for the passage ot another discharge. In the first method, the pole of a magnet is placed near the air gap, and when the arc is formed the reaction between the lines of force surrounding the magnet and those surrounding the arc strains the latter to the point of disruption. Unless the heat should cause the metal points to "bead," and possibly to bridge over the gap, this device is, of course, constantly ready for action. If the strength of the magnet be properly proportioned, this beading can scarcely occur. As a further precaution, the points may be of carbon, although their disintegration by the heat may increase the gap until the device becomes useless. One of the most successful of the devices in which new points are put in position after every discharge is the Wasson arrester. The arc forms between carbon buttons and the passage of the current

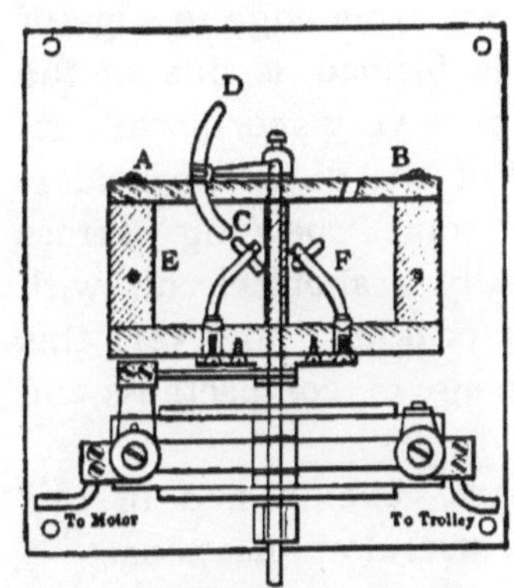

FIG. 67.—WESTINGHOUSE LIGHTNING ARRESTER.

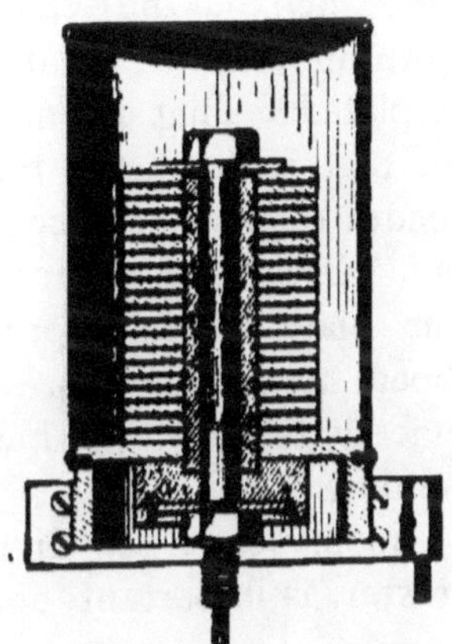
FIG. 68.—WIRT LIGHTNING ARRESTER.

melts successively the fuses, causing a lever to drop into the series of positions which bring the successive couples of carbon buttons very near together. In resetting the lever to its original position, the buttons may be rotated to place fresh portions of their circumferences in the position of nearest approach. This device seems to take care of as many discharges as there are carbon couples which may be conveniently placed in the box, say four or five. In another device the sudden expansion of a body of air, due to the heat of the arc, caused an arm to rotate, as from A to B, Fig. 67, bringing C and D successively into position in the compartments

E and F, respectively. This ingenious apparatus has not been so widely tested as to demonstrate its relative or its absolute merit.

In the Wirt arrester, Fig. 68, there is a departure from the methods just described, in this: that the arc is supposed not to be formed at all. It consists of a number of thin metal plates laid alternately with thin sheets of insulation, the whole arranged in compact cylindrical shape. One end plate is connected to line, the other to ground. In order to complete a circuit through this system of plates, the current must jump from the periphery of the first plate to that of the second, then on to the third, etc. After proper experiment, a certain number and thickness of plates and insulating sheets is found, such that the dynamo current cannot, while the ordinary lightning discharge can, force its way from edge to edge of the plates. That ordinarily no arc is formed, is due to the fact that each of the many air gaps is very short, while its breadth (equal to the circumference of the disk) is considerable. Occasionally there has been some "beading" across from plate to plate, producing finally a short circuit with respect to the motors. Generally speaking, however, this device is reliable, and has the advantage of compactness and economy.

The principle of "dividing the arc," exemplified in this arrester, is important, and has other useful applications.

GENERAL INSTRUCTIONS FOR THE CARE OF MOTORS.

It should be well understood that it is as necessary properly to care for electric motors as that a locomotive should be kept clean and in good working order. Careful attention given to all the parts of any piece of machinery insures to it longer life and from it better service. Electric - railway motors have a very hard duty to perform. Rough and dirty tracks, severe strains, heavy loads, etc., all tend to wrench and shake the motors to pieces, and it is therefore very necessary that every individual part of the apparatus should receive the most careful attention. In order that this matter may be well understood and emphasized, we have arranged a number of rules, which, if followed in their spirit, we believe will greatly aid those to whom the care of motors is given in keeping the apparatus in the best possible condition.

There are so many manufacturers making electrical railway apparatus, and these change so frequently, in greater or less degree, the details of their apparatus, that it is impossible to give to-day any set of rules which may not in considerable part be inapplicable within the next few months. It must always remain necessary that the more minute instructions required for the operation of apparatus shall be obtained directly from the manufacturer of that apparatus.

The rules given we therefore call "General Instructions for the Care of Motors," and they are as follows:

A. Inspection of cars and their preparation for service.

The motors should be thoroughly cleaned and all oil, grease, dust, etc., wiped from them; and all oil wells and grease cups should be filled. The armatures, commutators, and brush holders should receive especial care. An accumulation of dust and oil is a good inducement for short circuits. All parts should be kept as dry and clean as possible; a very little vaseline or paraffine on the commutator, however, if carbon brushes are used, lengthens the life of the brushes and seems to diminish the noise.

All nuts and bolts should be carefully inspected and seen to be tight and in their proper places. Keep the gearing as free from dirt as possible. Many motors now have their gears inclosed in an oil bath, which besides diminishing the noise keeps the gearing in very good condition.

Examine the wiring of the car to see that the connections are correctly and securely made and the insulation of the wires intact. See that the controlling apparatus is in good condition; this is very important, and too much care cannot be taken to see that all controlling mechanisms, switch boxes, rheostats, reversing switches, etc., are in the best possible working order. At least once a week examine and carefully clean the lightning arrester, remove all oil and grease from the boxes, and carefully clean the bearings. Every fortnight examine the armatures, insulation, and fields; repaint and reshellac them if necessary.

The trolleys should also be inspected, and all bearings

cleaned and properly lubricated—especial attention being given to the trolley wheel. There should be good contact between the wheel and "leading-down" wires, otherwise there will be injurious sparking at the wheel. The trolley springs should be adjusted so that the trolley wheel shall be pressed against the wire firmly enough to insure a good working contact when running at full speed.

The brake mechanism in all its parts should be thoroughly examined, for it is exceedingly important that this part of the apparatus be in good condition. Cars should be furnished with good sand if operated on hilly roads.

See that the car is supplied with a screwdriver, monkey wrench, etc., an extra lamp or two, and an oil lamp for use if the power should be cut off.

B. Operation of the cars.

1. When cars are left standing in the car house or on a side track, see that the safety or cut-out switches are placed so that the circuit is open. Generally there is some mark on the switch to show the proper position of the switch lever. The trolley wheel should be removed from the wire and left in such a position as to relieve the trolley springs of their tension.

2. Before placing the trolley wheel on the wire, be sure the circuit is open at the switches, and before closing the switches be sure that the controller handle is at the "off" stop.

Before placing the trolley wheel firmly on the wire, let the side of the wheel just touch the wire; if any flash occurs, except the spark which may be seen when the lamps are on, then something is wrong, and the wheel should be kept off the wire. Probably a switch which should have been open will be found closed.

3. If there is a reversing switch on the car, see that it is set properly before applying the power.

4. Before starting the car, raise the traps and see if the commutators, armatures, gears, etc., are in condition for work. See that the brushes and holders are in good order and properly placed.

5. When ready to start, move the controller handle gradually but firmly. If the car will not move, throw the controller

handle off and look at the several switches between the trolley wire and the motors. Probably some of them will be found open. If the car still refuses to move, throw on the lamp circuit. It may be there is trouble at the power station, and the lamps will tell the story, unless it happens that the dirt on the track prevents a contact between the wheels and the rails. If this is the case, press the switch stick between the rear of one of the wheels and the rail, thus securing a ground. On many roads, an insulated wire is furnished each car for use in just such cases as this.

6. Do not attempt to run the car backward unless the trolley is closely watched by some one who holds the cord in his hand.

7. In throwing power on, move the controller handle step by step, allowing the car to gain headway under one, before advancing to the next step. Too sudden starting strains the machinery and wrenches the gears, etc.

8. In throwing the power off, move the controller handle gradually until nearly at the "off" stop, when it should be turned the rest of the way with a snap; and care should be taken that the power is off before setting the brakes.

9. The brakes should be set gradually, so as not to bring an undue strain upon the gearing.

10. Never run down grade faster than the maximum speed allowed on the level, and always keep perfect control of the car.

11. If the trolley jumps off the wire, a slight slowing of the car may be felt, and at night the lights will go out; stop the car immediately, after throwing the controller handle to the "off" stop.

12. When by accident or otherwise the current from the power station is cut off, throw the controller handle to the "off" stop, throw on the light switch, and watch for the power. When the power is on, it will be well for a portion of the cars only, say those having even numbers, to start immediately, the others waiting a minute or two. If all start at once, a severe strain may be placed upon the dynamos and engines.

13. Never stop a car on a curve, except in case of accident.

This will save the gearing from unnecessary wear and tear, and at the same time relieve the generators and motors from excessive strain and resulting loss of power. Many break-downs and troubles have resulted from unnecessarily stopping on curves. The extraordinary amount of current required to start a loaded car on a curve may endanger the insulation of the motors.

14. Never reverse a car while it is in motion, except to avoid serious accident, and then it must be done very carefully. It is very easy to overdo the matter, blow the fuses, and thus perhaps become helpless to avert some second accident.

15. If there is a reversing switch on the car, be sure that the power is cut off before throwing the switch.

16. In any case of reversal to avoid accident, turn the reversed power on very gradually, as a little will be found sufficient to stop the car quickly, while a sudden application of the power might strip the gears. Break the current as soon as the car stops.

17. Do not reverse a car with the brakes set, for the fuse may be needlessly blown.

18. Run slowly over railroad crossings, curves, switches, rough track, etc. Remember that the cars are heavy and that shaking up the motors should be avoided as much as possible.

19. Run through water very slowly and carefully, and in examining the motors, be sure that no water drips from the clothing or elsewhere upon them. Water on the field magnets will soon cause them to burn out.

20. Be careful that nothing falls from the pockets on the motors, and do not let any metal—for example, an oil can—touch the brass screws on the connection boards, or in any way cross-connect parts of the motor circuit unless the trolley is off.

21. Do not run over sticks, stones, or wires; they may be caught by or knocked against the motors. Remember that the motors are hung very low and are apt to strike such obstructions.

22. If a motor is flashing badly at the commutator, or gives

out a burning odor, or shows weakness in any way, it is best to cut it out. Some companies provide switches to do this easily. If the car has a cut-out switch, insert the key which should be with it in the socket, with pointer up, and turn the pointer around one-quarter of a circle toward the good motor. *Never attempt to move this switch unless the main motor switch is off.*

23. It may be that a fuse is blown, in which case try another, and then two in multiple, which will generally be sufficient for the necessary current required to start the car.

If the double one blows, there is probably a serious ground or short-circuit, which needs careful attention.

24. Unless very familiar with their current-carrying capacity, never put copper or iron wires in place of the lead fuse, for it is the function of this fuse to "blow" whenever the motors are endangered, and if a wire were used, perhaps the current which it would allow to pass would damage the motors seriously.

25. A sure way to stop any electrical trouble in the car is to remove the trolley from the wire.

26. In case of a lightning storm, keep cool, for there is absolutely no danger; a slight noise may be heard, however. If the motors are damaged by lightning, the car will run unsteadily or stop altogether. If lightning damages one motor, cut it out. If both are damaged, pull the trolley down and wait to be pushed back to the car house.

27. Never run down grade with the current on; but the trolley should always be in contact with the wire on such occasions, as it may be necessary to reverse the car.

28. Be careful in the use of sand on the track, as too much will prevent good ground connections. A very small amount will serve to keep the wheels from slipping.

29. The power should be shut off when passing a "trolley break" or "section insulator."

30. In ordinary stopping of a car, always release the brake, but do not let it fly just before coming to a dead stop. The armature will then be able to gather up the lost motion in the gears and shafting, and will be ready for a smooth start.

31. Do not attempt to make up time on grades or rough

tracks; in fact, never, *at the expense of the machinery*, try to make up lost time.

32. Supposing the car to be equipped with Sprague-Edison apparatus (commutated fields), the following rules regarding the use of the switch box will be found helpful: Never dally in the movement of the handle. Move on slowly from step to step, allowing the car to respond to each step and gather headway before proceeding to the next step. In case the switch turns hard, move the pointer to the notch by a succession of slight taps; this will prevent running over the notch.

In case you do run over the notch one-third, when throwing on power, move on to the next notch. In throwing off power, pass the fourth notch quickly. In case (in throwing off) you run over one-third of a space, go back to the last one passed.

The best notches in throwing off are from five to three and from three "off." Throw the pointer from the first notch to "off" with a snap. When leaving the switch, even for a moment, take the handle off and leave it in the car. Always allow the car to gain speed at one notch before giving it another. Never throw the switch on the fifth notch if the car does not start on the preceding ones. Never apply the brakes when the switch is turned on.

When necessary to go beyond the fourth notch, pass slowly to the seventh, never stopping on the fifth or sixth: and always use the seventh notch on grades. Of the lower notches, the third should be used in preference to the first, second, or fourth. Prepare for heavy grades by throwing the switch on to the fifth and the last notch, when the car is going its fastest at the foot of the grade. During wet weather, however, if the car wheels slip while on the grade, work the switch back to the fourth point, or even work it back to the first point and "off" point, until the wheels get a gripe, then work up gradually to the seventh notch.

33. Familiarize yourself with the peculiar noises made by the apparatus, in order that you may detect in this way whether the motors are acting properly.

CHAPTER IV.

THE LINE.

Of an electric railway, the *conducting system*, as a whole, has come to be known simply as "the line."

It may be considered: (A) Merely as a conductor. Taking this view of it, the proper material, its resistance, cross-section, and weight are the points to be determined. Copper is now, and will continue to be, the cheapest conducting metal, so long as there shall be maintained, even approximately, the present relations between the specific resistances and prices of the various metals. (B) As a conductor *in place*, insulated more or less highly, and supplying current in a suitable manner to a number of cars.

Before calculations as to the line resistance can be made, it is evident that the following conditions must be known: (1) The initial pressure, *i.e.*, the voltage at the station; (2) the average or the maximum demand for current; (3) the allowable "drop" on the line at the average or maximum load; (4) the average distance over which the current at the average or the maximum load is to be transmitted.

1. Railway dynamos are now quite generally run at a difference of potential, between the brushes, of 500 volts. This value may fairly be considered as a compromise between considerations of economy in copper on the one hand, and safety to life and facility of insulation on the other. The question of danger to life has been thoroughly and frequently discussed in public hearings before municipal authorities. It seems that nothing need here be said further than this, that as yet no human being has either been killed or seriously injured by the shock due to a 500-volt continuous current; though, of course, many persons while handling electric apparatus have received such shocks. Several horses have been killed by railway wires, while in other cases special efforts to kill a horse in this way have failed.

In the direction of insulation against loss of current, practically nothing remains to be accomplished. A higher pressure than 500 volts could be satisfactorily insulated, so far as loss of current is concerned; but the fear of danger to life, in all systems requiring in any way the use of bare wires in exposed situations, will doubtless keep the service pressure for railways on the city streets where it now is, from 500 to 600 volts.

2. The predetermination of the average or maximum load cannot be made with exactness. Given a certain schedule to be observed by a given number of cars, the power required to perform that service varies with the condition of the track and the running gear, the efficiency of the motors used, the skill of the motorman, the weight of the cars and passengers, the magnitude of the grades and curves, the number of stops, and the relation between the number of cars simultaneously ascending and descending grades. And again, even if all these variables may be given particular values for a given schedule, we must recognize the fact that quite frequently the schedule itself is not maintained. In spite, however, of these uncertainties, it has been possible to adopt rules which are fairly satisfactory guides in the calculation of the copper required. When the service to be performed over given lines is known in a general way, we should then seek to define (a) the maximum number of cars that may be simultaneously on the line; (b) the number that may be ascending known grades and the speeds required; (c) the number descending grades on which gravity is sufficient to propel them; (d) the number on levels and the speed required; (e) the weights (live or dead) to be carried; (f) the horizontal effort in pounds required to propel a given weight, say one ton, over tracks such as may be in question. This last is known as the traction co-efficient. Given the above values, we may easily calculate the whole work to be done, and may express it in terms of horse-power, or finally, in terms of current and E. M. F.

Thus, let us suppose twenty-five cars on the line, of which ten are ascending grades averaging five per cent., and at the rate of five miles per hour; ten are moving down grades, requiring no current; five are on the levels, making ten miles per hour. We will suppose each car and its motors to weigh 10,000 pounds, and the passengers of each car to weigh also 10,000 pounds. The work to be done is of two kinds: first, that against gravity; second, that against the friction of the running gear. Atmospheric resistance at street-car speeds is small, and may be neglected, or considered as combined with that of friction. The work done against gravity is thus calculated: Weight lifted = 10 (cars) × 20,000 pounds = 200,000 pounds.

Height of lift per minute equals the rise per foot on the grade (0.05 feet) multiplied by the number of feet traveled per minute. Now, five miles per hour is equivalent to 440 feet per minute; hence the total rise per minute will be 440 × 0.05 = 22.00 feet.

The work done against gravity per minute will be, then, 200,000 × 22 = 4,400,000 foot-pounds, a rate of 133.3 horse-power.

For the track in question, let us assume that a horizontal effort of 30 pounds per ton will overcome friction. (This is a safe outside figure for all ordinary tram-rail track.) Then, for the 10 cars, we have 100 (tons) × 30 (pounds) × 440 feet per min. = 1,320,000 foot-pounds per min., a rate of 40 horse-power; also for the five cars on the levels, we have 50 × 30 × 880 feet per min. = 1,320,000 foot-pounds per min., a rate of 40 horse-power. Thus, for keeping all the cars in motion, we need, 133 + 40 + 40 = 213 horse-power.

This power, however, is that actually needed at the axles, while that delivered to the motors must include all losses in the motors and gearing. Suppose this to be 25 per cent. of the power delivered to the cars. Therefore, we must supply 213 ÷ 0.75 = 284 horse-power. Since we premise that the pressure under which the current flows is about 500 volts, and since 746 watts = 1 horse-power, we may, with sufficient accuracy, say that 1.5 ampères on the line will be equivalent to 1.0 horse-power. Then, for the total current we shall have 284 × 1.5 = 426 ampères. In this detailed way the power required may be calculated. But it is usually sufficient to allow without such detail 10 to 15 horse-power as the maximum per car for service over considerable grades and in case the total number of cars is in the neighborhood of 20. This is quite sufficient for 16-foot cars. For roads having practically no grades, and for a large number of cars, the allowance may be as low as 7 or 8 horse-power per car.

3. As to the allowable "drop" of potential, it is determined principally by the requirement of a close uniformity of condition, in order that the same motors, moving all along the line, shall be able to give practically uniform results at all points. It has been found that a variation of 10 to 15 per cent. from the station to the point of lowest potential will not injuriously affect the speed demanded of the motors.

In making the following calculations for the copper to be used, we must have clearly in mind just what conditions we wish to produce. There has been a great deal of loose thought and loose calculation in the matter. Thus, have we in view simply a fixed maximum "drop" of say 100 volts? This may seem a definite condition, fixing the copper accurately, yet it is not enough, save in very simple cases, to insure uniformity in the determinations by different engineers, even though they make the same assumptions concerning the value of the rail return. By one of them the line may be wired in such fashion that the maximum drop should occur at many places under many conditions, while by another it may be so wired that the maximum drop could occur only in a few places and under few conditions. The weight of copper in the first case would be considerably less than in the second.

Again, have we in view a certain *average* drop, say fifty volts? This in turn, standing alone, is not definite enough; for we must ask whether we shall consider the average drop to be determined under a fixed maximum load (expressed in ampères or horse-power), or when a fixed number of cars is in ordinary service; and further, shall this average be determined from readings taken on one car while running over all the line or lines, or shall it be determined by readings taken on all the cars at the same instant, or shall the average readings of a round trip on each of many lines be required to fall within the fixed limit? Further, if we aim at an average loss as the determining condition, we must still have in view some limiting maximum drop, otherwise the conditions concerning the average drop might be obtained, while the resulting service might be poor indeed.

If the copper placed on the line has a uniform cross-section along the whole length, then the observance of any given condition as to average drop will carry with it a fixed maximum drop, and *vice versa.* This is because, with a uniform cross-section and, as is usually the case, a tolerably uniform distribution of load along the line, there must necessarily be an approximately regular drop of potential from the station to the farthest point at which service is performed. If, however, the line be broken into many sections, each supplied with current separately, this necessary relation between the maximum and average drop ceases to exist. Thus, let us suppose the governing condition to be that the average pressure on the line when cars are in regular service shall be 400 volts. Let us further suppose the line to be divided into four separately fed sections. On one of these, the average pressure might be 300 volts, while on the other three it might be 433 volts, the average for the whole line still remaining at 400 volts, yet this would certainly not be the condition aimed at by good engineering, since it is not desirable that the pressure should on any section be so low as 300 volts. Generally speaking, indeed, this may be said: that the more complex the wiring, the more necessary is it accurately to define the conditions desired, in order that like results shall be reached by different calculators. This whole matter is of constantly increasing importance, since there is, especially on the larger systems of railways, an increasing tendency toward the subdivision of the lines into separately fed sections.

A very reasonable specification, and one of wide application, may be thus expressed—namely: on any section the average pressure shall not be less than 450 volts, and the minimum pressure shall not be less than 400 volts. This rule will apply as well to the case in which the whole line is electrically continuous as to the case of subdivision, since the whole line in the case supposed would constitute, in the sense in which the word is here used, one section. It should further be specified, that the average and minimum pressure above mentioned should be found under conditions of *maximum load for regular traffic.* Reasonable discretion must be used in determining for any such case what shall be called regular traffic.

It would perhaps be increasing unnecessarily the amount of copper required, to insist that the given pressure should be found in case of maximum *extraordinary* load; as, for instance, when snow covers the tracks, increasing the power required for the propulsion of the cars in service, and usually also causing an extraordinary demand for current to operate snow plows and snow sweepers.

4. Having in view these specifications, we may next consider the distribution of the copper which is to be put up in the line. The following cases arise in practice:

CASE I. (Fig. 69).—A single bare conductor, the trolley wire, usually of uniform diameter throughout its length, is used to convey all the current required for the car service over a given line (or section of line). Only in case of a comparatively small number of cars and a comparatively short line can this method be economically used. The line pressure in such case must always be found continuously lower from

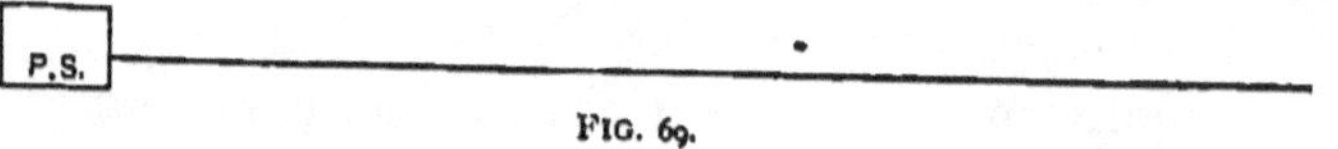

FIG. 69.

the station to the farthest point of service. No two points on such a line could have equal pressure.

This first condition, geometrically illustrated, may be represented by a straight line of uniform thickness, as A B (Fig. 70), in which, let A represent the position of the station. Over any section measured from A toward B must pass, not only the current required by the cars on that section, but also the current for all the sections beyond. If, therefore, the resistance in the section A D be just equal to that in the section F B, the drop in voltage along the first section would be greater than along the second, and still greater than along the third and fourth. Thus, if the tangent of the angle

A E C represents the whole current supplied to the cars, and if we now take A B as representing the total and A E the average resistance over which the total current must flow, then the vertical line A C may be taken as representing the 50 volts total "drop," bearing in mind the geometrical relation that the tangent of the angle, A E C, varies with the ratio $\frac{A\ C}{A\ E}$, analogous to the expression of Ohm's law $C = \frac{E}{R}$.

Following this geometrical method, draw through m' a line to A C, making the tangent of the angle at m' equal to

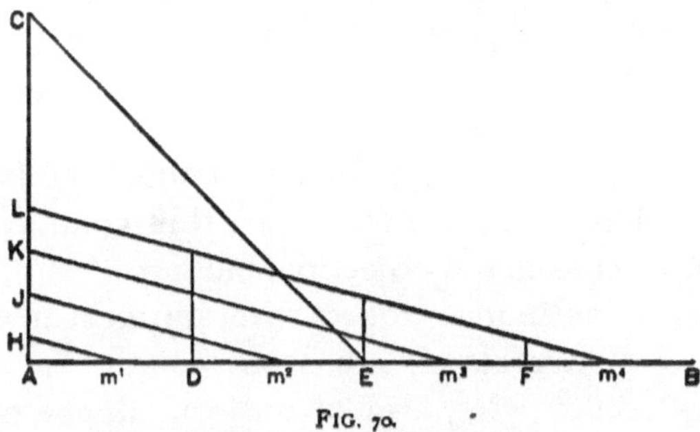

FIG. 70.

one-fourth that at A E C. This tangent will correspond to the current required for the first section flowing over the resistance A D ÷ 2, and the vertical line A H will represent the "drop" over the first section for *its own service*, and will be one-sixteenth of the total "drop," or, in specific value, 3.125 volts. (The fact that the load of each section is presumably distributed along its whole length, instead of being concentrated at one end, renders this illustration inexact, but none the less valuable as setting forth the principle.)

Through m² draw the line to J m², making an angle with A B equal to H m' A. Then A J will represent the number of volts required to force the current for the second section over the resistance A E. Draw other oblique lines at m³ and at m⁴, parallel to the lines through m' and m², and likewise intersecting A C. Then draw vertical lines through D, E, and F, intersecting the oblique lines. The total fall of potential over the first section is now seen to be A H for its own load, H J for the load of the second section, J K for that of the third section, and K L for that of the fourth section—each of these latter quantities being two-sixteenths of the total, or twice as great as the quantity required for the first section itself. This is because the current of the second, third, and fourth sections must be carried over the whole length of A D, while its own current is carried over an average distance of A m', only one-half of A D. Of the total "drop" it appears, then, that $\frac{1}{16} + (\frac{2}{16} \times 3) = \frac{7}{16}$ takes place on the first quarter of the total distance. On the second section, the "drop" in like manner is found to be $\frac{5}{16}$; on the third, $\frac{3}{16}$; on the fourth, $\frac{1}{16}$. The actual pressures in specific values would be at A, 500 volts; at D, $500 - (\frac{7}{16} \times 50) = 478.2$; at E, $478.2 - (\frac{5}{16} \times 50) = 462.6$; at F, $462.6 - (\frac{3}{16} \times 50) = 453.2$; at B, 450.0.

While the "drop" when referred to equal sections of the whole line is thus seen to be irregular, this condition is not generally to be considered objectionable.

CASE II.—A continuous trolley wire, uniform in diameter, is connected at intervals of, say, from 500 to 1,000 feet to a continuous feeder wire, also of uniform diameter. If, in any case, more than one feeder should be connected at intervals to the same section of trolley wire, the case would still be analogous to the simpler one above, since in both there is, over the distance in question, a uniform total cross-section of copper, over which the current passes for the car service. This method will be understood by reference to the diagram,

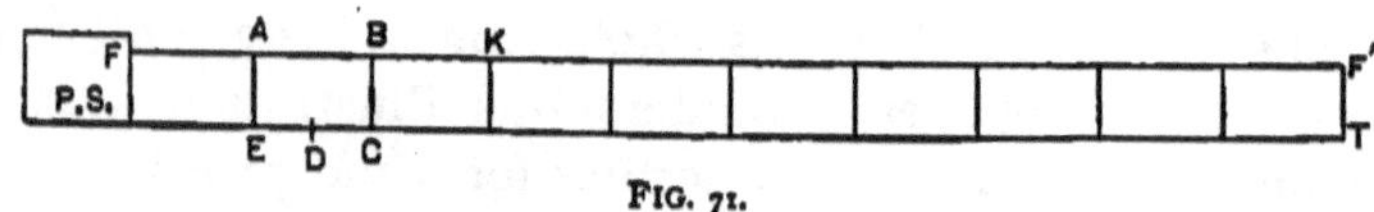

FIG. 71.

Fig. 71. S T represents the trolley wire; F F' the feeder (or feeders) connected to the trolley wire at relatively short intervals, perhaps every 500 feet. In this case, the trolley wire may be of small diameter, since the maximum current over any of its parts is only that required for the service done in about 250 feet—that is, half the interval between sub-feeders, A, B, and K. A car placed at D, midway between A E and B C, would receive a part of its current from A E (the sub-feeder nearer the station) and part along the feeder F F′, through B C and the trolley wire back to D, the proportions being in inverse ratio to the resistances of the two paths divergent at A. Another portion would flow directly from the station along the trolley wire.

In practice, the trolley wire used in this system of comparatively numerous sub-feeders has varied in size from No. 6 B. & S. to No 0. In the system first described, the trolley wire, having to carry the whole current a greater distance, has rarely been less than No. 0, and has been as large as No. 00 B. & S.

CASE III.—A trolley wire, of uniform cross-section, is broken into short lengths insulated from each other, each length being connected at one point to the feeder, one feeder thus supplying the current for a considerable number of trolley-wire sections. This is analogous to the second case just considered, except that the conductivity of the trolley wire itself is of service only over the distance from each sub-feeder to the end of that particular trolley-wire section.

For a car at B, Fig. 72, the whole current required must

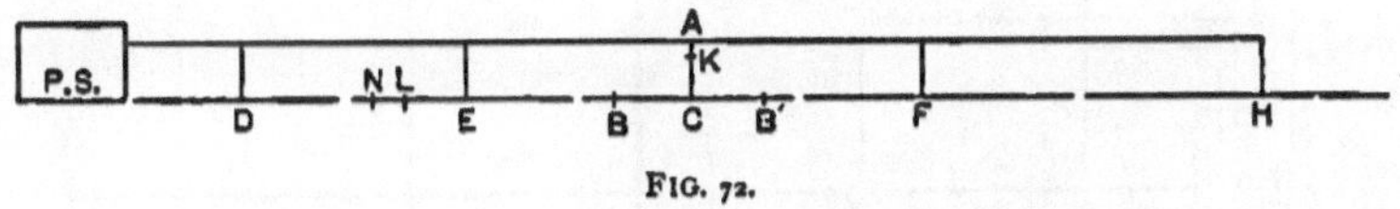

FIG. 72.

pass over the feeder A, thence along a sub-feeder to C, thence to B or B′. Now, although the current cannot come over the trolley-wire sections between the station and B, yet, as in the two preceding cases, the "drop" of line pressure is continuous from section to section of the trolley wire, since the pressure on this wire must be determined by that of the feeders at the points of junction with the trolley-wire sections. The car at B, however, must be working under a lower pressure than the car at B′, if the load from C to B is greater than from C to B′; but a car at E must always be working under a higher pressure than a car at C, and so on, from section to section.

This subdivision into short and independent trolley-wire sections has been required in certain cases, as a safeguard in case of fires along the line. Each of these independent sections may be supplied with a switch, conveniently placed on one of the supporting poles; and the operation of this switch, placed at K in the sub-feeder connecting the trolley wire with the feeder wire proper, enables the trolley wire to be thrown in or out of circuit at will. A fusible plug may readily be placed in the same box containing the switch, and any short circuit of this particular line section may thus be made automatically to cut the section out of service.

The same remarks concerning the feasibility of cutting out these independent trolley sections (either at will, by the switch, or automatically by the fusible plug) that apply in this case (III.) apply in case V., as will be seen when the latter is discussed.

CASE IV.—A continuous trolley wire of uniform diameter is fed at intervals by independent feeders, each running direct from the station. It is plain that each of these feeders may be so calculated as to give, under assumed conditions of load, equal line pressures at the points of junction

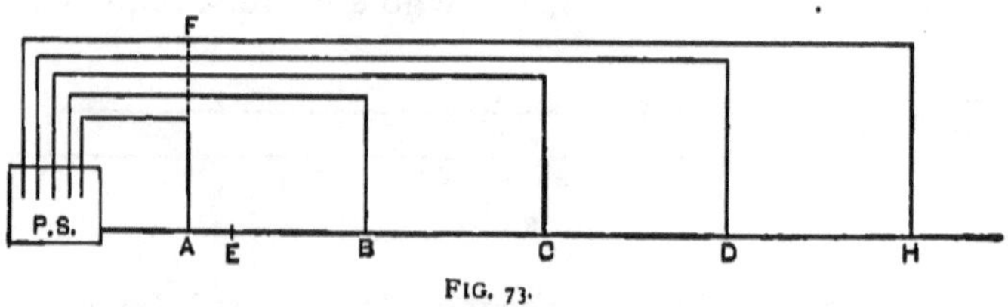

FIG. 73.

(A, B, C, D, and H) with the trolley wire, Fig. 73. Then a car at E would receive its supply of current partly from A and partly from the point B, and these parts of the current supply would be dependent upon the distances of the car from the points A and B, respectively, and in inverse ratio to these distances. It is, of course, not convenient to run a very large number of separate feeder wires. The trolley wire, therefore, is usually fed at a comparatively small number of points, and consequently the distance over which the current is conveyed on the trolley wire alone is much greater than in cases II. and III. The size of the latter has, therefore, usually been greater, the most general practice having been to use a No. 0 copper wire, with a distance between feeding points varying from 1,000 to 10,000 feet. Care should be taken, in using this method, so to limit the length of the intervals that a trolley wire of convenient size shall not be required to carry a current which, over the fixed resistance, will produce an excessive "drop."

A variation of this case consists in connecting two or more of the feeders together at the point of junction of any one of them with the trolley wire. Thus, at F the four feeders may be made practically into one conductor, the four being themselves in multiple with the trolley wire, which is supposed to have at a point very near the station a short connection to the dynamos. The effect of such a connection as that indicated at F will, of course, be to lower the pressure produced at B, C, and E, with the three longer feeders entirely independent each of the other; but such a cross-connection may be advisable if, after beginning operation, the load at any point, as at A, is found to be much heavier than was originally assumed.

CASE V.—The trolley wire is broken into sections insulated from each other (Fig. 74), each section being supplied by one

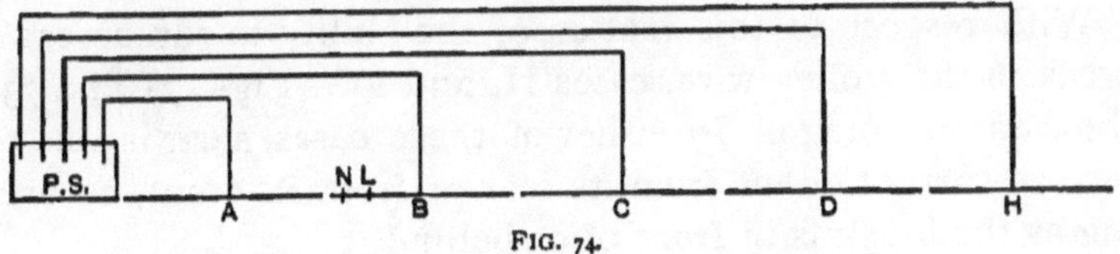

FIG. 74.

or more feeders not connected with the other sections. As in case IV., the potentials at A, B, C, D, and H may be made equal; and, as in case III., the conductivity of the trolley wire for service is lost, except as it serves to convey current from the points of junction with the feeders to the car in any particular part of a particular section. The current for all the cars that may be found in any one section must be carried wholly over the feeder or feeders of that section, the feeders and trolley wire of other sections being electrically independent each of the other. It has been usual to make the trolley wire for this case of about the same size as that for case IV., namely: No. 0 B. & S

The great advantage obtained by this system lies in this, that the effects of several classes of accidents on the line may thus be localized, and each feeder, being supplied with a break switch in the station, may thus be thrown in or out of circuit at will, the other trolley sections not being in any way affected thereby. If, then, the trolley wire be broken at any point, or if there be a short circuit to ground, the movement of the cars over the other sections need not be interrupted.

In any of the cases discussed, it is frequently possible to carry the cars across the break by the headway gained before reaching it. If, however, there be but one feeder connection to any trolley-wire section (not itself connected to the station), and if the break should occur between this feeder connection and any cars in question, as when the break is at L and the car at N, Figs. 72 and 74, the car would be deprived of power to obtain the necessary headway, unless the break should be so near the adjoining trolley section that the car could run by its headway from that adjoining section across the insulating joint between the trolley wires, and also across the accidental break.

With respect to this matter of the ability to run across a break in the trolley wire, cases II. and IV., Figs. 71 and 73, show an advantage. In either of these cases, a car is able to receive current either from its rear or from its front, according as the break is in front of or behind it.

CASE VI.—This consists of a combination of either IV. or V. with III. It is of great value on lines having very heavy traffic. The trolley wire alone, when of convenient size to handle, is (for heavy traffic) not able to supply, without too great loss, the current required by sections (between feeding-in points) of such length as would be needed in order to keep the number of independent feeders within reasonable limits. The remedy is plainly to place between the feeders, at their point of junction, copper in excess of that found in the trolley wire. Such a connecting conductor is called a main. It is connected at two or more points with the trolley wire. This combination must recommend itself in all large systems of electric street railways.

Let us next determine for the cases above described the area and, consequently, the resistance of the conductors that will be necessary to carry the required current to the cars under the conditions given.

We will assume, for simplicity, that we have 10 cars uniformly distributed over a distance of 2 miles; that each one of these cars requires a current of 15 ampères, which, at 450 volts, is nearly equivalent to 10 horse-power.

From Ohm's law, $C = \frac{E}{R}$ (1), or $\frac{E}{C} = \frac{\text{Volts lost}}{C} = R$ (2)

The resistance of any wire is

$$R = \frac{\text{length in feet} \times 10.8}{\text{area in circular mils}}, \qquad (3)$$

10.8 ohms being the resistance of a wire of length one foot and diameter .001 of an inch.

Equating (2) and (3), we have—

$$\frac{\text{Volts lost}}{C} = \frac{\text{length in feet} \times 10.8}{\text{area in c. m.}} \qquad (4)$$

and by transposing we get—

$$\text{c. m.} = \frac{\text{length in feet} \times 10.8 \times C}{\text{Volts lost}} \qquad (5)$$

From this formula, which holds for any metallic circuit, we may readily find the area necessary for any required current, with any desired per cent. of "drop," when the distance of transmission is known.

Let us now, for simplicity, further assume in these calculations that the resistance of the return circuit is *nil.* This puts all the resistance into the overhead line or lines, or, in other words, reduces the area of the wires to one-half of what it would be if the circuit were a double metallic circuit, each side having the same resistance; or, briefly, we may consider only the distance out one way, not the return.

The irregularity of the "drop," when referred to sections of equal length and resistance, has been shown in the geometrical treatment of case I. We can, however, obtain a general expression determining the distribution of "drop" as between any assumed divisions of a continuous wire of uniform size. We may represent the ten cars as regularly placed along the line in Fig. 71.

The distance from the station to No. 1 is 1,056 feet, and that is the distance between successive cars.

Let C be the average current required by each car; for the general case illustrated above, the number of cars by N, the interval by d, the resistance of d by R. The total current from the station will be N C, from the station to the first car the "drop" will be N C R, from the first car to the second (N–1) C R, from second to third car (N–2) C R, etc., the "drop" over the last section being C R. The total "drop," which we may represent by E, is made up of the sum of these quantities or may be expressed:

$$E = C\,R\,(N + (N - 1) + (N - 2) + \quad + \quad) \qquad (6).$$

The sum of the series in parentheses we know from algebraic demonstrations to be represented by the expression,

$$\frac{N^2 + N}{2}$$

We then have $E = C\,R\,\frac{N^2 + N}{2}$ (7); for the resistance of any section, $R = \frac{2\,E}{(N^2 + N)C}$ (8).

Now let us consider any section whatever, as that which is *m* sections distant from the station. The current flowing over this section will be represented by C (N — m). The "volts lost" will be C (N — m) R. In equation (8) we have the value of R, true for all sections. Substitute this value in the expression for "volts lost" and we have

$$\text{Volts lost} = \frac{(N - m) \times C \times 2\,E}{(N^2 + N) \times C} = \frac{(N - m) \times 2\,E}{N^2 + N}$$

Now substitute this value for "volts lost" in equation (5), and for "length in feet" substitute "d," and for current (N—m) x C; we then have

$$\text{c. m.} = \frac{d \times (N - m) \times C \times 10.8 \times (N^2 - N)}{(N - m) \times 2\,E}$$

$$= \frac{d \times 10.8 \times C\,(N^2 + N)}{2\,E} \qquad (9)$$

The supposition of uniform size of conductor has been introduced above, hence this equation, true for any section, gives in fact the cross-section necessary throughout the line.

Now suppose E = 100, N = 10, d = 1,056, we then have

$$\text{c. m.} = 11 \times 264 \times 3 \times 10.8 = 94{,}087.6$$

It is plain that formula (9) may readily be converted into one having more direct reference to the whole length of line. Let D represent that length. Then $d = \frac{D}{N}$; substituting this value for d in (9) we have

$$\text{c. m.} = 5.4 \frac{D \times (N + 1) C}{E} \qquad (10)$$

The quantity (N + 1) C is simply the total current required for one more car than the number originally in view. E, it is to be remembered, represents the allowable "drop" at the end of the line. In this shape the formula is of very convenient application.

The determination of the area of the conductor made above, as applicable to case I., is evidently applicable also to cases II. and III. The differences in actual construction would be as follows: In case I., the whole of the area, 94,087.6 c. m., would be found in a single bare trolley wire. In case II. this area would be found partly in the bare trolley wire and partly in a parallel feeder. Thus suppose a No. 6 trolley wire to be used, of about 40,000 c. m., then the feeder would have to be about 55,000 c. m. in area. To insulate a pound of copper with the material used ordinarily in weather-proof wires costs about 7 cents. The total additional cost of the feeder, over the bare wire, is found by taking this cost per pound and adding it to the cost of placing the feeder in position and connecting it to the trolley wire. These two items may be roughly taken at $100 per mile.

In case III. we must have a feeder of about 94,000 c. m., since the continuity of the trolley wire is broken. The increased cost over either of the other two methods is evident.

It has been assumed above, for convenience, that the resistance of the rail and earth circuit is zero.

Because the resistance, under favorable conditions, is extremely low it has been the habit of some engineers thus to rate it as zero, placing all the allowable resistance in the overhead wires and reducing them to one-half the cross-section that would be required, if the outgoing and the return branches of the circuit were supposed to be of equal resistance, as in the case of the double trolley system. The total weight of copper overhead would thus be made only *one-fourth* of that needed for the double trolley system, where the length and cross-section are both *doubled*, as compared with the single trolley or ground-return system. The facts do not seem to justify entire neglect of the resistance of the ground return. Especially does this seem true when we consider the changes of moisture in the earth, the rusting of the track, the gathering of dust and dirt upon it (causing an appreciable resistance between the wheels and the rails) and the resistance of joints from rail to rail.

The specific resistance of iron is well known, being about six times as great as that of copper. The cross-section and the length of the rails may also be readily determined, hence the actual resistance of the rail return (neglecting joints) may be accurately calculated. It may be taken fully into account, supposing only that the joints be so bonded together that the rails may be considered as practically continuous.

It is in assigning a value to the *earth* as a conductor that the greatest uncertainty exists. Since the earth becomes a conductor through contact with the rail, it is evident that the manner of laying the track and the nature of the paving laid along the track must materially affect the quantity of current which actually leaves the rails. Further, the nature of the soil itself, both in the surface stratum and in the lower strata; the number of iron pipes (as for gas, etc.) placed in the ground; the proximity of large bodies of water—all these variables will affect this quantity. Realizing that the soil is usually to be found permanently moist at a distance of several feet below the surface, electrical engineers very early adopted the practice of connecting the rails with plates or rods, which were sunk to a convenient depth. There was some gain in conductivity, but later it was thought best by some, on account of telephone disturbances, to aim rather at the restriction of the current to the rail than at its dispersion through the earth. To utilize perfectly all the metal in the rails, specifications were accordingly drawn up looking to their thorough bonding both longitudinally and laterally.

To this bonding it has by some been thought wise to add a continuous copper conductor, supplementary to the rails and the earth. The effect of this, however, is generally small, except in so far as it performs the function of a bond-wire, rendering more certain the otherwise difficult maintenance of a good circuit around the joints; and further, in that it will cause the potential of the rails and the earth to be so nearly identical as to prevent the occurrence of the slight shocks sometimes given to horses while resting some of their iron-shod feet on the ground, others being on the rail.

In view of all the uncertainty as to the quantities involved and all the variety of practice in treating them, it need scarcely be added that there has been considerable variety in the results. Fortunately, any error of under-calculation is readily corrected. The first service of the cars over the line may be purposely brought to the conditions of maximum load. The speed then, or, more accurately, the readings of a voltmeter on a car, will furnish immediate data as to the sufficiency of the copper, or will detect local trouble in the rail connections. The erection of an additional feeder, or some intelligent work on the bonds, will soon set matters right. If, on the other hand, the pressure on the line shows somewhat higher than was calculated, no harm is done and the coal consumption has been permanently reduced.

While the assumption of any definite value for the rail and earth resistance (called for convenience the return circuit), or of any fixed ratio between this value and that of the resistance overhead, must generally be in error, it will not be unwise to present a rule which has given fairly uniform results.

Subject to modification, whenever it is possible by actual measurement to determine the rail-earth resistance, we may use without large error the following modification of formula (10):

$$\text{c. m.} = 6.5 \frac{D \times (N + 1) C}{E} \qquad (11)$$

This calls for about 20 per cent. more copper than is required by the assumption of zero resistance in the return circuit. Should there be considerable differences in the quantities of current required by uniformly distributed cars, sufficiently accurate results will yet be obtained by the same formula (11), if the *average* current C be properly determined. Should there be considerable "bunching" of cars, the required copper may be calculated separately by the general formula (5) (increasing it by 20 per cent., as above in (11)), the "length in feet" being taken from the station to the middle point of the section (supposed of moderate length) on which "bunching" is found; and the required areas may then be summed. If it be desired to change the size of the feeder, after placing any length S A from the station to A, the area up to A may be determined as above, a fixed "drop" being then allowed (less than the total allowable) and the point A may then be taken as a new datum point of initial pressure. By diminishing the size of the feeder over the part nearer the station, we may maintain the *total "drop"* at 100 volts, while increasing the *average "drop"* beyond the value it would have if a wire of uniform diameter were placed along the whole line or section.

Failure to understand this relation has not infrequently caused copper to be placed where it could *not* do the greatest good. Thus, suppose a uniform wire gives a line pressure of 450 volts at A and 400 volts at B. It is found that, there being heavy grades at B, the motors become unduly heated. It is desired to increase the pressure, thus decreasing the current required for a given amount of work. A length of wire, S A = A B, is bought and (1) is placed between the station and A, and connected at intervals to the trolley wire; or (2) it is connected only at S and at A; or (3) it is connected at A, thence at intervals along A B to B; or (4) it is connected simply at A and B. For *raising the pressure at B*, the fourth is the most efficacious method; and if increase of pressure is needed at B, and is not important elsewhere, the fourth method should be followed. The formulæ already given will show the distribution of pressures.

In treating cases IV., V., and VI., there are required two determinations of different characters. First, to determine the diameter of the feeders, the function of which is to carry a certain current over a certain distance at a certain percentage of loss, the whole current to pass out of the feeder at one point, the end. Such a calculation, made by formula (5), is very simple. It should be borne in mind, however, as explained above, that some assumed value for the earth circuit must be used. The general formula (5) is

$$\text{c. m.} = \frac{\text{length in feet} \times 10.8 \times C}{\text{volts lost}}$$

If we were to use a double metallic circuit, this distance in feet would be the distance *out* plus the distance *in* (return). For the single trolley system, however, we have in mind the distance measured only in one direction. We increase the value of the constant 10.8 by 20 per cent., as above, and then have

$$\text{c. m.} = \frac{\text{length in feet} \times 13 \times C}{\text{volts lost}} \tag{12}$$

The value of C in this formula depends upon the number and service of the cars which it is proposed to supply by the feeder in question. The volts lost must, of course, be less than the total "drop," since a part of this total must take place on the trolley wire, or on trolley wire and main, as in case VI. The terminals (feeding-in points) of the feeders become, with respect to the trolley wire, points of initial pressure; and the second calculation to be made consists in the determination of the length of the trolley section between feeding points, if the trolley wire (of fixed diameter) be used alone, or the determination of the diameter of the mains for a fixed distance between feeding points. This latter determination would be entirely analogous to that made above for cases I., II., and III. The formula to be used is that for earth-return (11). The same formula, slightly transposing its terms, becomes

$$D = \text{c. m.} \frac{\text{volts lost } (= E)}{6.5 \times (N + 1)\, C} \tag{13}$$

in which C is the current per car, or per short section into which the line may be supposed to be divided, the volts lost being the difference between the maximum allowable and the "drop" on the feeder alone, from station to feeding points. If the pressures at these points be maintained practically equal, and if the work between them be distributed uniformly, the point of lowest pressure (maximum "drop") will be found on the trolley line midway between feeder ends.

By the system of independent feeders the pressure may readily be made higher at some point distant from the station than at some point near it. And if the work at any distant point be very trying, the establishment of such differences may be wise.

Since No. 0 has been largely used for the trolley wire, it may be well to deduce from formula (13) the distance over which it alone may supply current to a given number of cars. Let us assume, as heretofore, that 400 volts is the minimum allowable line pressure; assume, further, that we maintain 450 volts at the feeder ends. Then the total "drop" over the trolley must not exceed 50 volts. A mean between the Birmingham and Brown & Sharpe gauges gives 110,000 for No. 0. Suppose, further, that we have about 10 cars to the mile, or 150 ampères total current. Then

$$D = 110{,}000 \frac{50}{6.5 \times 150} = 5{,}641 \text{ feet.}$$

For any other car service the distance is readily found.

In large systems, having lines crossing and recrossing each other, there will arise some rather perplexing questions as to potential distribution. The simple formulæ (11), (12), and (13) will serve all practical purposes; since, after all, the whole matter consists in a sometimes complicated application of Ohm's law.

Except in the matter of the rail-and-earth circuit, all that has thus far been said applies equally well to the single and double trolley systems. The latter has now been almost wholly superseded by the former. There are three reasons for the survival of the single trolley method as the fittest: first, greater simplicity of overhead turn-outs and frogs, in so far as the mechanical operation of the trolley is concerned, and this is the controlling reason; second, greater facility in insulating the out-going from the in-going side of the circuit, for when the rail return is used these sides are about 18 feet apart, while with the double trolley wire they are from 8 to 18 inches apart; third, greater economy of copper in large systems. This advantage is not as great as has appeared from the comparisons already given in discussing the copper calculations. An offset must be made by considering the cost of bonding the rails thoroughly, as compared with the cost of supplying and erecting the copper return, which would serve instead of the rails and earth. In case of very light service, the first cost may be less for the double than for the single trolley system. A fourth advantage is usually claimed, in that fewer wires are actually required to be erected, thus diminishing the objections made to the whole trolley system merely on the score of "looks." To this, the advocates of the double trolley (for there are a few) answer that in either system all feeder wires may be buried; that the comparison would then rest as between bare wires alone; that the single trolley wire requires one or two guard wires, stretched parallel to each "live" wire, and these guard wires require span-wires for their support—this being done to prevent foreign wires from falling across the "live" wire and to the ground, where they may or may not make such contact as will cause them to be entirely burned out; that the double trolley system does not require guard wires, since in case of any foreign wire crossing the two live wires it would at once be destroyed and the trouble at the railway station would end.

There has been so little extension of this double trolley system that it cannot now be stated whether or not the public authorities would generally allow this difference of construction as between the two systems. It does not seem probable. In Cincinnati, Ohio, where the Cincinnati Street Railway Company has produced the most notable example of double trolley service, guard wires were, however, not erected.

The advantages of the double trolley system are two: First, it causes little interference with the telephone circuits that use ground return in the neighborhood of the railway lines. This has been the cause of much litigation, urged by the telephone interests, endeavoring to force the use of that system of electric railways which would least interfere with the established telephone service. Thus far the courts have ruled, for the most part, against such requirements.

The case of the telephone labors under this disadvantage:

that it strives to destroy the most practical system of railway circuits in order to continue a relatively poor system of telephone circuits. The remedy for the evils brought upon the telephone service by the disturbing earth currents of the railway service is to be found in the use of complete metallic circuits for the one or the other or both. To apply the remedy to the telephone circuits is to produce the only installation which may be called first-class, with or without consideration of railways. The currents required for telephones are exceedingly small; the forces by which they may be disturbed are correspondingly small. To insist that when passing through the earth these currents shall not be perceptibly disturbed by other currents, is to insist upon a practical monopoly of the earth as part of an electric circuit. The best English and European telephone practice, uninfluenced by any trouble from railway currents, tends decidedly toward complete metallic circuits. To apply, on the other hand, the same remedy to the railways, is to impose serious difficulties in the way of the practical success of the operation of cars over the complexities of switches, turn-outs, cross-overs, and the like. It will afford a pleasing exercise of ingenuity to plan overhead apparatus suitable for the following conditions: The double-tracked road of one company has single-tracked branches, one to the right, one to the left, at a given street crossing, while continuing its double-tracked course beyond the point of junction. The double-tracked road of another company crosses that of the first at this point of junction, running parallel to the branches. The two companies refuse interchange of current and want to use the double trolley system.

We cannot do better, in setting forth more fully the merits of this controversy, than to give the opinions of the Superior Court of Cincinnati, and the Supreme Court in the State of Ohio. These are found in Appendix A.

The second advantage to be noted is this: that effective insulation of the motor windings may be more readily secured than in the case of the rail-return circuit. In the earlier stages of the art, this was indeed an important advantage; for on the single trolley roads no accident was more common than the "grounding" of armature or field coils. Previous practice in winding for comparatively high potentials had rarely to deal with the case in which the metal of the machine was in fact part of the circuit. To produce a short circuit through the body of the dynamo, it was generally necessary that the insulation of the wires should break in at least two points of different potential. When, however, one brush of the motor was connected, as was often the case, directly to the motor frame (this latter being hung on the axle, whence the current passed to the wheels), it needed a break at but a single point to cause trouble. In the violent fluctuations of current strength and magnetic density incident to car service, there were often produced very high electromotive forces of induction, and these, if not the normal pressure of 500 volts, often destroyed the insulation of armature or field. In some designs this trouble has been practically met by attempting to insulate the frame of the motor from the car truck. This will be seen to impose considerable mechanical difficulty. Generally, so much improvement has been made in the details of armature and field winding, that the matter has ceased to be big with misfortune, as was once the case.

There are now in the United States not more than ten double trolley lines, only one of which, that at Cincinnati, is of what may be called a modern construction—that is, built within the last two years. Of single trolley lines there are several hundreds.

The third method of power supply, involving an extensive conducting system, is that through underground conduits. Thus far, all efforts, and they have been many and ingenious, successfully to insulate the conductors in a conduit, have failed, save when topographical conditions were very favorable. Some of these efforts will be described elsewhere. The principles governing the calculation of copper, as already set forth, will of course apply to conductors underground as well as overhead.

It should be stated that in using the rail-and-earth return it is desirable to avoid setting up a high current density in the earth at points along the line or at the station. When the line current is very great there may occur localized earth currents of such magnitude as to seriously interfere with underground circuits or even to produce considerable electric effects on metallic pipes or cables. As this condition is unlikely to exist except in roads with very heavy traffic, where railroads are necessarily inadequate, it would seem best to use in such cases numerous ground plates along the line and especially at the station, placing them below the level of water and gas pipes and in moist earth. This will both raise the general conductivity and avoid high local current densities in the earth.

We have now treated the subject-matter of this chapter in its more general aspects, and in Appendix B will be found a set of instructions for the erection of an overhead line.

There are now on the market many varieties of insulators, frogs, pole clamps, etc. Simply for the sake of consistency, throughout the instructions, some special types are referred to. Modifications may readily be made in such of the language as applies only to particular forms or dimensions. It is intended simply to illustrate the general mechanical rules which should be observed.

We go into considerable detail in these specifications because many railway companies now prefer to do this work directly rather than through contractors.

For much of what is valuable in them we are indebted to Mr. W. E. Baker, of Boston, who prepared similar instructions, which we have slightly modified for this publication.

CHAPTER V.

TRACK—CAR HOUSES—SNOW MACHINES.

TRACK.

FOR half a century steam-railway engineers have studied track construction. Great advances have been made in the last ten years, and change is still considerable; this means that the present railroads are not perfect, are not final; yet, withal, the steam-railway practice is the best guide in a general way for street-railway men using electric cars to follow. True, the conditions as to pavement and wagon traffic are widely different in the two cases, and render the street-railway problem for a given weight of car or locomotive much the more difficult of the two. But it remains that the best track, considered simply as to the company's own traffic, is in both cases the approved "T" rail construction found generally, with varying detail, on our steam railways.

To-day the problem of track-construction—especially that of joint construction—is perhaps the most important one pressing upon street-railway men for their attention.

It is beyond the scope of this work to give a treatise on track construction; we desire, however, to present a few examples of good practice—at least, of that which is good now. Much must yet be learned in this matter.

Broadly speaking, light track construction must give way to heavier construction. Meanwhile we show in Fig. 75 the construction at present largely used by the West End Railway Company of Boston; in Fig. 76 the track used by the Pittsburg, Allegheny and Manchester Traction Company, and in Fig. 77, that toward which the West End Company of Boston now leans. This last is noteworthy, in that the chairs are electrically welded to the rail, the only bolts used being those at the fish plates. To reach these from time to time a joint box has been advised, which is shown in Fig. 78. This permits ready access to the nuts on the outside of the rails, so that they may be inspected and set up frequently and with little trouble. One side of the box serves as a fish plate, the bottom as a chair.

Through the kindness of Mr. F. H. Monks, General Man-

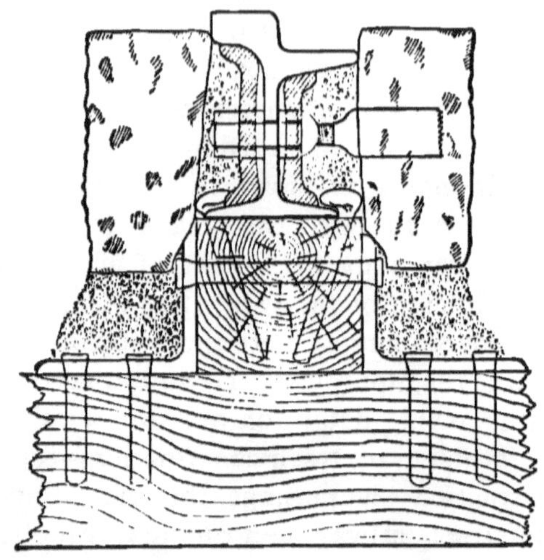

FIG. 75.—TRACK CONSTRUCTION OF WEST END STREET RAILWAY COMPANY.

ager of the West End Street Railway Company, we also show, in Figs. 79 and 80, a large number of rail sections used in various parts of the country. The lighter, especially the flat sections, have seldom been laid originally for electric-car service. There is, however, almost no type of track so

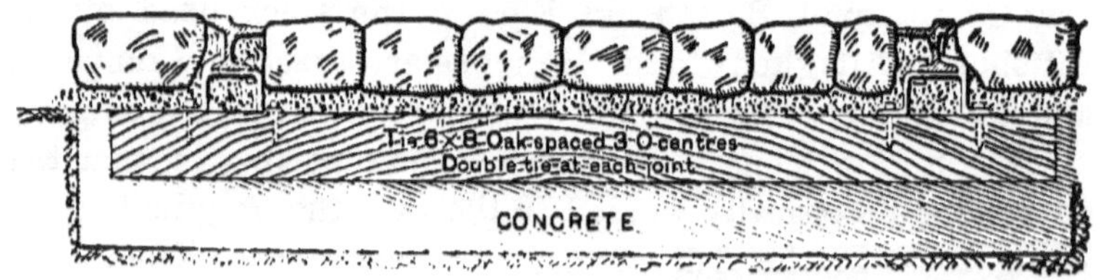

FIG. 76.—TRACK CONSTRUCTION OF P., A. & M. TRACTION COMPANY.

poor but that some bold operator has been willing to use it as a destroyer of electric apparatus and of dividends.

We will close this very short treatment of this subject by the following quotation from a very excellent and succinct letter recently addressed by Mr. F. H. Monks, General Manager of the West End Street Railway Company of Boston, Mass., to the *Street Railway Journal:*

"The coming construction for electric roads will contain but few jimcracks to break, wear out, and rust away. Strength and durability are surely the qualities which should be sought. Since the problem is so nearly akin to that which steam roads have solved, why not follow closely on their lines, so far as is possible, adding only such parts, forms. and appliances as may be necessary to conform to the difference in the conditions under which the two are laid and operated? The writer suggests the following:

"*Ties.* Sound oak or chestnut, six and one-half feet long, six inches face, five inches thick, placed three feet on centers, and two ties set eight inches apart under each rail joint.

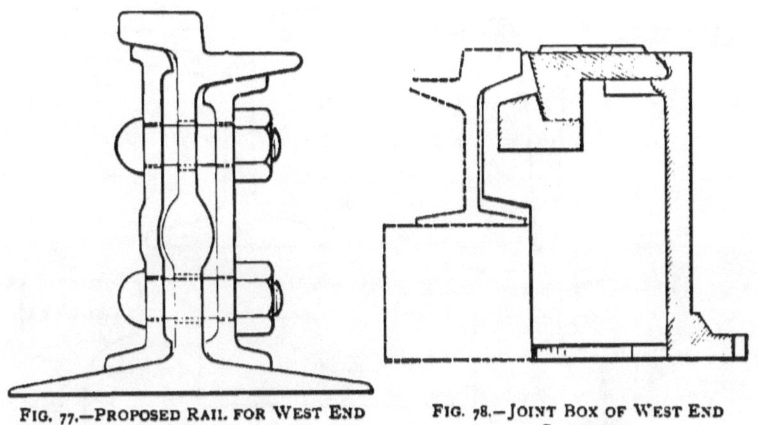

FIG. 77.—PROPOSED RAIL FOR WEST END COMPANY.

FIG. 78.—JOINT BOX OF WEST END COMPANY

"*Gravel.* Sharp, clean, free from loam and clay. Tamp thoroughly under and between ties, particularly in case of those carrying rail joints.

"*Rail.* A girder of sufficient total height to allow two inches of gravel between top of tie and bottom of paving stone; head of rail two and one-eighth inches, flat flange two and three-quarter inches, web one-half inch to nine-sixteenths of an inch, lower flange five and one-half inches to six inches. Rail to be laid directly on ties and spiked thereto by railroad hook spikes four and one-half inches long. Rail to weigh about one hundred pounds per yard, and to be drilled for copper connections, tie bars, and splice bars.

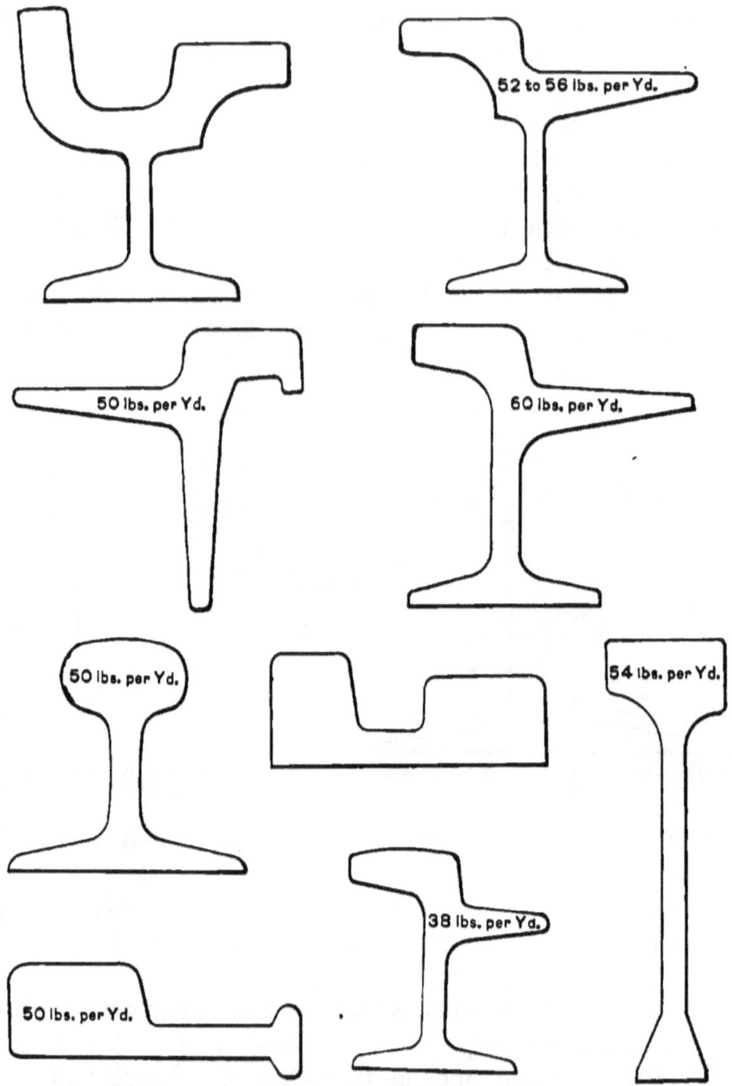

FIG. 79.—TYPES OF AMERICAN RAILS FOR ELECTRIC TRAMWAYS

" *Tie bars.* Flat iron bars six feet apart, round and threaded ends, to pass through web of rails and set tight by proper nuts.

"*Splice bars.* Thirty-inch steel channel bars, well fitted to rail, and so drilled for six bolts as to allow for expansion and

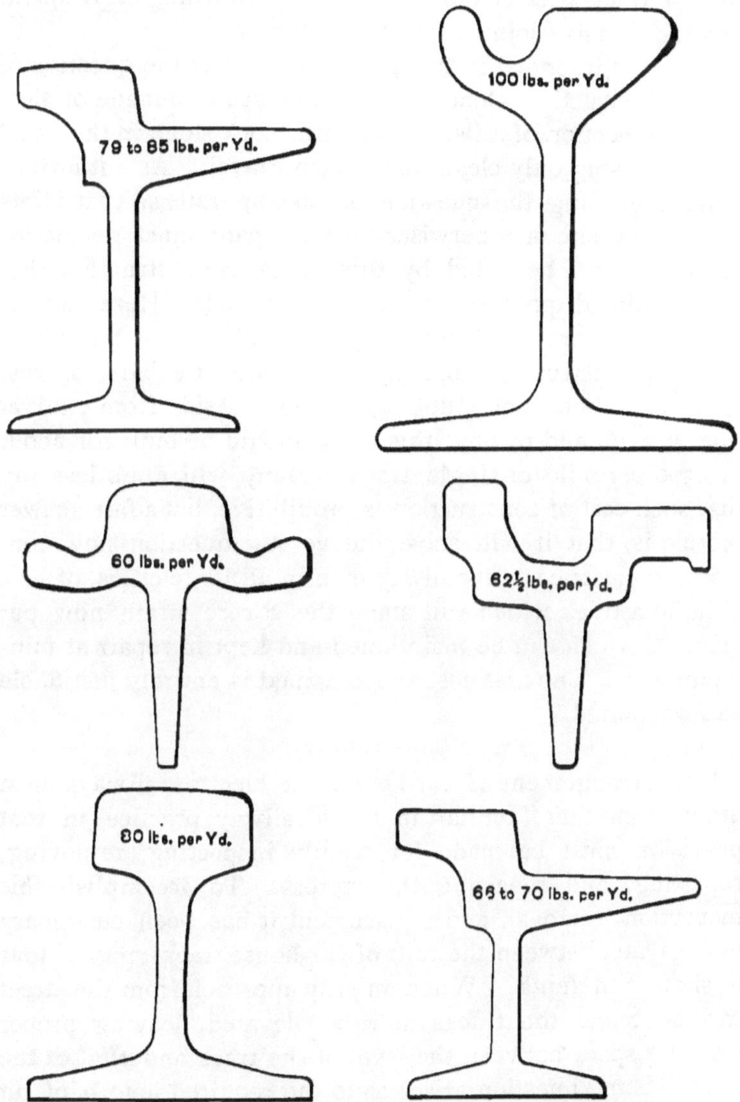

FIG. 80.—TYPES OF AMERICAN RAILS FOR ELECTRIC TRAMWAYS.

contraction of rail. Best quality bolts and nuts to be used. Head of bolt to be continually struck with twelve-pound hammer when nut is being set up.

"*Joint boxes.* Heavy cast-iron boxes with removable covers, corrugated tops, to be placed on and spiked to ties, and set outside track at each rail joint, thus permitting of frequent and inexpensive joint inspection and repair.

"As it is intended in this paper to consider the question of track construction alone, no mention need be made of electrical connection of rails. Pave with block stone in the usual manner, using only clean, dry, sharp gravel. As authorities differ regarding the question of placing rails so that joints will be opposite or otherwise, each company must decide for itself. It will be noted by this mode of construction that steam-railroad practice is closely followed. Here are no chairs, stringers, or other supports. There are but seven parts in construction, namely, ties, rails, tie bars, spikes, splice bars, bolts and nuts, and boxes. Aside from excavation, gravel, and paving, this track should be built for about $10,500 per mile of single track. Many will doubtless say that such cost of construction is prohibitive, but a fair answer thereto is, that it will solve the vexing question now confronting electric street-railway men in all large cities, of how to build a track which will stand the severe strain now put upon it, which can be maintained and kept in repair at minimum cost. The first cost above named is entirely justifiable and warranted."

CAR HOUSES.

The arrangement of car houses for electric railways must differ from that familiar in horse-railway practice in that provision must be made for readily inspecting, removing, replacing, and repairing the motors. To accomplish this inspection, removal, and replacement it has been customary to excavate, between the rails of car-house tracks, pits of four to six feet in depth. When an easy approach from the street can be made the tracks may be elevated, leaving proper working space between the level of the track and that of the floor. Some question arises as to the required length of pit room as compared with the total length of car-house trackage. This ratio must depend upon the number of injuries to mechanism, which in turn depends on the quality of the machinery and the quality of the service maintenance.

It may be roughly stated that pit room should be supplied for at least one car out of ten in service. If the excavation and protection of the pit space be not costly it will be well considerably to increase this proportion. We show in Fig. 81 a very convenient machine used by the East Cleveland Street Railway Company, Cleveland, Ohio, for the expeditious handling of motors in the pits. An hydraulic hoisting apparatus having a vertical range of a foot or more is mounted in a low truck which runs on light rails laid in the pit. The cut shows a motor on the platform of the hoist ready to be

FIG. 81.—HYDRAULIC HOIST FOR HANDLING MOTORS.

lowered or raised. It is clear that a motor or any of its heavier parts may thus be handled with much less time and labor, and with more safety, than by the "slings" or "main strength and awkwardness" not infrequently employed. It is in the pit that the closest acquaintance with the motor itself can be formed. We know of one large company which requires division superintendents, inspectors, and motor-men to serve a period in the pit, until the electrical department of the company shall certify to the fitness of the employee for the discharge of his duty.

In order to store cars at night quickly, and to get particular cars out quickly when ready, most companies of considerable size have used transfer tables—usually running near the front of the car house. Those that have been employed for the ordinary horse cars may, of course, be used for electric cars.

We show in Fig. 82 a transfer table used by the West End Company of Boston. It is propelled by a cable, which, in turn, is operated by a stationary motor, the controlling apparatus for which is placed at one end of the transfer table. Another feature of this installation consists in the fact that a car may cross the pit to the leading-out track just opposite that on which the car may be standing. The tracks are made continuous across the transfer pit by short inclined rails covering the difference of level between the floor proper and the bottom of the pit, whereon are laid also the rails on which the car crosses the pit. The crossings of these rails

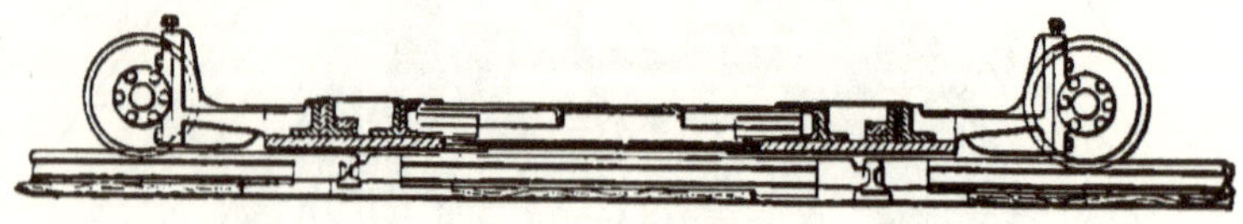

FIG. 82.—SECTION OF TRANSFER TABLE.

with the rails on which the table runs are of course quite numerous, but the great facility of car movement thus gained is worth much to a road working its cars on a close schedule.

In order to avoid the frequent climbing upon the car top to inspect the trolley, the East Cleveland Street Railway Company has arranged a raised platform by the side of an entering track to the car house in such a manner that the inspector is, when on the platform, at the proper level to inspect the trolley.

Such a construction costs but little money, and it has been already a source of much satisfaction to the management. Fig. 83 * has been drawn from a photograph of this arrangement.

Near the car house—preferably under the same roof—should be placed the repair shop. If, as in a larger system, there be several car houses, the principal repair shop should

* View furnished through the kindness of Mr. C. W. Wason, General Manager East Cleveland Company.

be near one of the largest car houses, or, if that is not convenient, near the station. The machinery of such a shop may be run conveniently by a stationary electric motor. A company operating cars even as few in number as ten will find it not unprofitable to become the owners of a few good

FIG. 83.—PLATFORM FOR READY ACCESS TO TOP OF CARS.

machine tools. A lathe and a drill press are, of course, the most generally useful. We cannot pretend to give here any valuable advice as to just how far a railway company should go in the manufacture of repair parts. Local facilities and the personal qualities of the management must determine that question. We can only suggest that when either tools or men in the repair shop are found to be idle half their time, question should be raised as to the wisdom of keeping them in service. It may well be cheaper to carry a liberal stock of spare parts. In the repair shop, perhaps even more than in other parts of the plant, is the value of order and neatness displayed. Have a separate bin for each kind of supplies; keep the floor well swept; especially have an eye to metal filings; keep the winders and their supply of wire well away from litter of any kind; keep armatures in racks, not on the floor, where they may easily be bruised; have plenty of light, either natural or artificial; keep a record of every armature and field repair, giving to each its number, date, and character of injury; if you manufacture supply parts look well to their cost, that you may be sure as to the

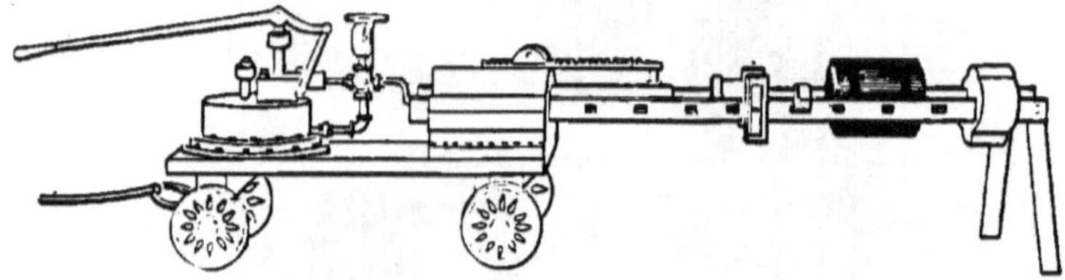

FIG. 84.—HYDRAULIC JACK FOR REPAIR-SHOP USE

wisdom of making them at home or purchasing from dealers; if you rewind armatures or fields manage to have at least a 500-volt current—if not something higher from an adjacent lighting circuit—with which to test the parts in the shop before replacing in the motor, remember that your winders should have good will as well as skill—their capacity to do harm easily is great; repair injured parts as soon as possible after discovery of the injury; have a care to get good mechanics—not electricians—for the work of the shop. A man who has been "handy" with tools all his life will devise economical ways of doing the work. If the repair shop is of considerable magnitude he will design a few special tools that will quickly pay for themselves. This should not be construed as meaning that the number of special devices needed is considerable.

As a suggestion for effort in the right direction we show

in Fig. 84 an hydraulic press, valuable for pulling off or pressing on commutators, pinions, or gears, or almost any other use involving a straight pull or push.

This machine was made for its own use by the East Cleveland Street Railway Company.

SNOW MACHINES.

To remove snow wholly from city streets seems beyond the power or the will of most municipalities. Certainly, in many cities of the United States such removal would be a task beyond reason. The work of the electric-railway com-

FIG. 85.—SNOW PLOW FOR ELECTRIC ROAD.

pany is, then, to displace the snow from its tracks; for this purpose "plows" and "sweepers" have been used. Both were already familiar in the service of horse railways. For perfect cleaning of the track both seem necessary.

It has been thought by the management of one of the largest electric-railway corporations that the sweeper does its work too well—in this, that other vehicles, traveling on runners, are so much interfered with by the cleanly-swept rails, and in turn so much interfere with cars that it is best to leave an inch or two of snow on the track, as when cleared by plows. Unless the snow is very much compacted, or has become a sort of frozen slush, it is then possible to obtain fair contact between wheel and rail, thus keeping up the car service. It is intended to prevent packing and freezing by prompt and constant use of the snow plows. Holding these views, the corporation in question has supplied itself with plows only, and these of simple type, such as are used in connection with horses. The plow is driven by two 15 or 20 horse-power motors, one for each axle, the motors being on a platform protected by a cab, and driving through the medium of sprocket chains. It is shown in Fig. 85. Such a

FIG. 86.—SNOW SWEEPER FOR ELECTRIC ROAD.

machine is of course less expensive to maintain than a sweeper. Other companies, on the contrary, use sweepers only. If the question of sleighs be of small importance, and the snows rather light, it would seem that, as between the two machines, if only one can be had, the sweeper would be best. We show in Fig. 86 a snow sweeper run by three 15 horse-power electric motors, two for the axles and one for the brushes, or revolving brooms.

The sweeper brooms are both driven from a single counter-shaft placed on the deck, this shaft being on the same angle as the broom shafts, and consequently parallel to both of them The sprocket wheels are on each end of the counter-shaft, one for the front broom and one for the rear broom. The object of using but one counter-shaft on the deck of the sweeper is that it gives more room to the motor-men. The sprocket chain which drives the brooms raises and lowers on a radius, so that the chain is always kept tight and the possibility of its ever dropping from the sprocket wheels is prevented.

A powerful machine that claims to combine the functions

FIG. 87.—COMBINED SNOW PLOW AND SWEEPER.

of plow and sweeper is shown in Fig. 87. The flier, or rotating brush and plow, has a number of steel blades disposed much as are the buckets of a steamboat paddle wheel. It also carries steel brushes, one brush parallel to and touching each steel blade. The brushes extend beyond the blades about two inches, thus clearing away the snow or ice which has not been cut away by the blades. Each flier is driven at a high speed by a motor of 15 to 25 horse-power. By vertical adjustment, this machine will leave either a perfectly clean road-bed or a light covering of snow, the rails being clean in both cases.

The combination is expensive, compared to the simple plows and sweepers, and whether its value is in proportion to its expense remains for the present winter to determine.

Salt should not be used for removal of snow unless nothing else is at hand. Its use is likely to cause rapid corrosion of the rail bonds. The lesson most distinctly taught during the few winters whose snows have fallen on electric railways is this: "Do not let the snow get ahead of you." A watchman should inform the proper authority of any considerable fall occurring after service hours, steam in at least one engine should be kept up for immediate use, and, generally, everything should be in readiness at all times for putting the snow gang at work within one hour, or less, from the start. Such a course is hard to follow, perhaps, but it pays.

CHAPTER VI.

THE STATION.

ONE of the first questions that arises in the construction of an electric railway is, naturally, the position and character of the power station. It is not at all a simple matter, for there enter into it considerations of a rather complicated nature, all having a direct bearing on the economy of the future system. It is needless to say that the line should first be roughly located, in order to arrive at an intelligent understanding of the conditions that must be fulfilled. This done, the subject of the proper location of the station can be taken up.

So far as its position with reference to the line is concerned, it is perhaps sufficient to say that, other things being equal, it should be as nearly central as possible. In the chapter on line construction enough is said to impress upon the mind the conditions imposed by the distribution of the copper, but it should be remembered that the line is, comparatively speaking, a permanent investment, and from the relatively small interest charges and repairs it does not involve an important part of the expenses.

In any given station, whatever its position, a considerable item of the running expenses will be labor, and this is virtually independent of position. The factor that should go farthest in determining the proper location is the availability of fuel. If it is possible, the power station should be so placed that coal cars can be run up to its very door and the fuel shoveled directly into the coal bins. In some few favored places an arrangement even nearer the ideal has been found possible, as, for example, in Scranton, Pa., where the street railway power station is located just at the foot of a gigantic pile of culm, refuse from the neighboring coal mines, so that the fireman could almost climb up the side of the pile and kick the fuel under the boilers. Under such circumstances, especially as the culm is secured at a nominal price, the cost of fuel is almost negligible. Ordinarily, however, one must depend on the railroads for the transportation of coal, and hence it is most desirable to place the station close beside a railroad track, even if the site be rendered somewhat more expensive by so doing. If the coal has to be carted at all, a few hundred yards more or less distance makes very little difference, as the expense is largely in the handling.

Aside from fuel, water is the next prime necessity to be considered: and it goes without saying that if a site can be found close both to railroad tracks and water of suitable quality for use in the boilers the station should be placed there, even if at a considerable distance from the point that would be indicated by consideration of the line alone. In small stations, where condensing engines are not to be employed, the matter of water supply becomes somewhat less important, but still deserves careful examination. If water for condensation is required it is almost imperative to get within pumping distance of a plentiful supply.

In locating a railway power station, then, the first thing to be considered is nearness to the supply of fuel, and, after that, central position with reference to the line and availability of water.

In finally determining the best location for a station nothing but actual expense estimates will enable a decision to be formed. If the station is placed far away from the center of the system additional copper will be needed in the line to compensate for its increased distance, but to offset this there may be the gain that comes from cheaper fuel and water. In two places, one convenient to the supply of fuel, the other in the center of the system, there will probably also be differences in the cost of real estate.

If to obtain cheap fuel it becomes necessary to move the station so far from the center of the system that the interest on the increased amount of copper necessary is greater than the probable annual saving in fuel, it is sufficiently obvious that it will not pay to put the station near the fuel supply. The relative cost of real estate in the two places must also be taken into consideration; it often happens, however, that a position at some point near a railway track, where cheap fuel can be obtained, is also a location where real estate is not unreasonably expensive; while the center of the system is quite likely to be near the center of the town, where land is decidedly costly. In choosing between two possible sites for a station the interest on real estate at the two points should be considered, as well as the interest on the investment in copper and the saving of coal. A few rough estimates will show the more economical location. In the majority of cases it is possible to get near coal and water without getting very far away from a reasonably central position on the line, but now and then the question becomes more complicated and recourse must be taken to the estimates just mentioned.

It may here be well to take some notice of the use of water power. It is not under all circumstances that the water-wheel can successfully compete with the steam engine, and the question should be determined on its merits in each particular case where it arises. Estimates should be made of the cost of installation of the wheels and the necessary water ways to feed them; the cost of water rights should be found, and the approximate amount of extra loss entailed by the generally more inconvenient location of the power station.

With these data can be compared the probable cost of coal necessary for producing the equivalent horse-power, and the difference in investment if engines are to be employed. The constancy of the water power must be carefully considered, with reference both to its failure at a critical moment and the regularity of supply necessary to give the wheels sufficient margin of power for good regulation.

It must be borne in mind that, particularly in a small station where the variations in load are excessive, water is by no means easy to regulate; and the greatest care should be exercised in the installation for the purpose of securing uniformity of speed, even under the large changes of output required. This irregular output, that is a prominent characteristic of most electric-railway plants, in contradistinction from every other sort of installation, is a thing which should be most watchfully considered in designing future stations.

It is hard, even by diagrams, to give any adequate idea of the changes of load to which a small railway plant may be subjected; they are extraordinarily great and exceedingly sudden. Fig. 88, which is a fac-simile of a record for ten minutes on a recording ammeter, may give some faint idea of the condition of things. It will be seen that at one point the output jumped from zero to 150 horse-power and back inside of a single minute, and during the latter five minutes shown in the diagram there were no less than twenty-five sudden *variations* of 50 to 100 horse-power, each taking place

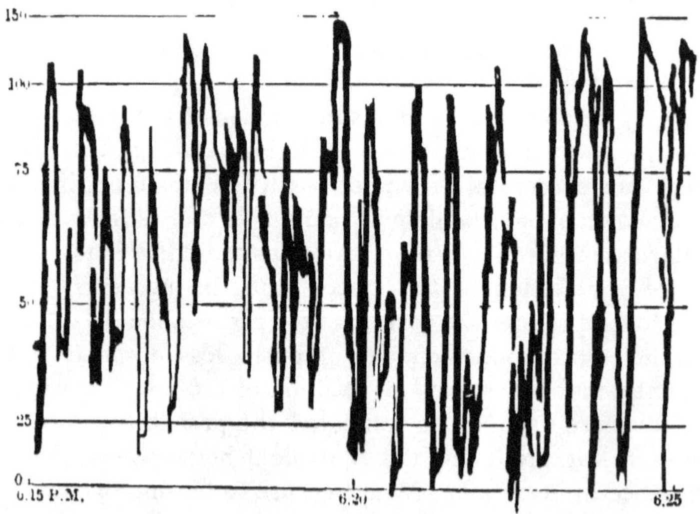

FIG. 88.—DIAGRAM OF VARIATION OF CURRENT IN AN ELECTRIC RAILWAY SYSTEM.

within a few seconds. The road from which this record was obtained is about four miles in length, and was operating seven cars at the time of the test.

The effect of such a state of affairs is twofold: First, the regulating power of the machinery is taxed to its utmost, and even the best governed high-speed engines do not respond quickly enough to keep the voltage on the line constant under such circumstances, while water-wheels are totally unable to keep pace with changes so sudden. Second, owing both to poor governing and great variations in actual mechanical strain, the machinery, whatever it may be, is all subjected to severe and unusual tests of its strength and stability.

It therefore becomes necessary that in a railway power station the foundations, engines, shafting and dynamos should be of the best mechanical construction. The use of heavy fly-wheels on the engine or the shaft of the water-wheel is to be recommended, especially in small stations; and all engines designed for this class of work should be exceptionally strong and solid in construction, and most securely bolted to an unusually firm foundation. The same is true of all shafting that is to be employed, and the dynamos should be fixed in position as solidly as possible. On extended street railway systems, running a considerable number of cars over a comparatively level track, the variations in load are of course much less, and consequently such extraordinary care in station construction becomes unnecessary. But in these, as in all similar cases, it is best to err on the side of safety.

Not only should this condition of variable load that we have mentioned affect the mechanical design of the station, but, to a certain extent, it should also modify the general arrangement of the plant with reference to the location of the various portions of the machinery. Especially is this true in small stations where only one engineer is employed.

It is most desirable, under such circumstances, so to place the engines, switchboard and dynamos that they can be readily seen and their performance watched from a single point. If, for example, lightning enters a station, and one of the dynamos begins to blaze at the commutator, the engineer ought to be able to reach the switches without rushing the length of the dynamo room and around various portions of the machinery. If a fuse blows, it ought to be in such a position as to be readily noticed without going around a corner to look for it. This condition of easy accessibility is not difficult to fulfill, but is sometimes neglected. Wherever the entire equipment is subjected to often repeated and severe strains particular care should be observed in getting all the safety devices and switches where they can be easily reached. In large stations where several men are constantly employed the necessity does not become so imperative.

With respect to the general design of a railway power station marked differences must necessarily exist between small and large installations; of course this is true of electric-light stations as well, but to a far less extent, for the same reasons that we have endeavored to impress on the reader in the preceding paragraphs.

The differences that should exist between plants of various sizes will be pointed out later, by giving the sketches of three designs for widely divergent station capacities. As a general principle, it is safe to say that the subdivision of power should be carried rather further in railway plants than is ordinary in the construction of an electric-light installation, for the reason that the various component parts generally being subjected to far greater strains, and worked at a point much further from their normal capacity, both security against accidents and general economy demand rather smaller units of power than in cases where the load is, relatively speaking, uniform and somewhere near the full capacity of the machinery. A fundamental law of economy in installations of any kind is to work at all times as near the full capacity of the machines employed as considerations of safety will permit. Dynamos and engines are both designed to carry their normal rated loads economically and continuously, and while in most power stations for railway work the great variations require a somewhat larger factor of safety between the normal and maximum load than in the case of electric-light work, the same broad idea must govern the arrangement of the plans for either.

In beginning the task of designing a railway power station, the first problem is to decide on the amount of power to be required, both for present needs and future exigencies. Neglect of looking forward to the possibilities of a few years hence has caused inconvenience in many an electrical installation. On the other hand, construction for the distant period when traffic may be several times its present amount is certain to produce an inefficient station.

The best policy is to build at first fully up to the immediate capacity anticipated, leaving the design of the station in such form that additional engines and dynamos can be installed whenever needed. If a fairly good estimate is made for the initial requirements the plant can be increased at a rate quite sufficient to keep pace with any probable demands. It is worth noticing, too, that if the number of cars on a given line is doubled the maximum capacity of the plant is not increased in anything like the same ratio, unless in cases where very large numbers of cars are involved. If, for example, an electric road starts with twenty cars and provides ample equipment for them, the addition of twenty more will certainly not require doubling the capacity of the engines and dynamos; for as the number of cars becomes greater the average output demanded comes more nearly to approximate the maximum load, without very much increasing the latter.

Where only three or four cars are in use it is quite possible that all the motors may be making severe demands upon the station at the same time; for example, two cars may be on heavy grades at the moment that another is starting, but with forty or fifty cars the varying outputs of the several motors tend to balance each other, and it is safe to assume that, even with a ten-car line, at no time will all the motors be worked simultaneously up to their full capacity.

So, in deciding on the proper capacity for the plant to be constructed, a very important factor is not only the size of the road, but its size with reference to the maximum output that the conditions of service will require.

Not only do these conditions go far to determine the proper capacity of the various prime movers employed, but, in the case of steam engines at least, they afford good reasons for selecting one or another type of machine. As has already been mentioned, in the chapter on prime movers, various forms of valve gear for steam engines permit of very different ranges of cut-off; for example, some of the automatic high-speed machines in frequent use to-day allow the steam, when necessary, to follow the piston throughout nearly the full stroke, while, as a rule, the Corliss valve gear—than which there is nothing more perfect as regards the results obtained when working under favorable conditions—usually permits a range of cut-off corresponding to only about two-fifths of a whole stroke.

It therefore goes without saying that when extreme variations in load are to be expected, as is the case in comparatively small roads, the use of the Corliss engine would entail a cylinder big enough to supply the maximum output—often several times the average output—working at a cut-off of about two-fifths the stroke. This, as has been mentioned in the chapter on prime movers before referred to, virtually compels working, on the average, at a cut-off too short for anything like good efficiency. To certain types of compound engines the same considerations apply, although it should be said that, as a rule, these machines, particularly when used condensing, stand underloading rather better than simple engines. In the case of very small roads, the choice is limited to high-speed engines with a wide range of cut-off; of these the types on the market are almost too numerous to mention, and most of them give excellent performances. As the size of the road increases and the ratio between maximum and mean loads approaches unity, a point will be reached where it will become possible to employ economically large compound engines, with Corliss or equivalent valve gear. Condensation, even in small engines of all classes, should be practiced wherever water is available, unless fuel is extremely cheap.

With very large stations, triple, and even quadruple, expansion engines are coming into use and do admirable work. It is hardly probable in any practical electric road, however, that so good economy can be obtained with these complex machines as is usual when they are employed for other purposes where the load is far more regular.

Some striking examples of this have been brought to the authors' attention; in one case a triple-expansion engine of an excellent modern type was put into use in the power station of a comparatively small road, with the idea that great economy of fuel would be obtained; the actual result was very disappointing, for by careful tests it appeared that the electrical horse-power hour, instead of being obtained by the use of about two and one-half pounds of coal, or less, actually required between six and seven pounds—a result no better than could have been reached by using a far less expensive and complicated simple, non-condensing, high-speed engine, of the kind generally employed for such service. It was one of the cases where neglect of properly considering the question of load led to unhappy results.

If we were, then, to attempt to obtain a general idea of the kind of engine suitable for any given station, we must be able to predict, roughly at least, the ratio between maximum and mean loads. In a general way, whenever the former is three times, or thereabouts, the latter, nothing will give better results than a plain, solid, high-speed engine with a heavy fly-wheel; with the maximum load about double the mean load, the various forms of Corliss and similar engines come into play with excellent results; and where the load ratio approaches even more nearly to unity and the road is of considerable size, requiring 500 horse-power or over, it is probable that the best results will be obtained by using triple-expansion engines. These are only approximate figures, but they at least give an idea of the circumstances that should cause the selection of one sort of engine rather than another.

Perhaps the best insight into the methods to be followed in laying out a power station for an electric road may be obtained by considering plans for several plants of particular sizes and the special circumstances that lead to their adoption in each case. We shall select, then, three specimen cases, and investigate them in considerable detail. First, we shall take up a five-car road of average character; second, one with twenty-five cars or thereabouts, and finally, a large city system with one hundred cars in regular service. These three cases may fairly serve as examples by which to show the various conditions that have to be met in designing a permanent and efficient power station.

Let us suppose, in the first place, that the problem before us is to install in a town of ten or fifteen thousand inhabitants a small electric road, not, as will be the case in larger places, for the purpose of enabling the business streets of the city to be reached from the suburbs, but to facilitate transit from one part of the town to another. The track will probably be less than five miles in length and usually a single track with turnouts.

The grades, of course, are liable to be of almost any amount, depending on the configuration of the country; but, as a rule, no continuous grades of more than six or seven per cent. are likely to be encountered, although there may be short pitches of slightly heavier gradient—nothing, however, of sufficient length to require particular consideration. On most small roads of this sort the service is rather infrequent, the cars being run on from fifteen to thirty minutes' headway, and five cars will generally be enough to give ample accommodations. Such a road is typical of a large portion of those now in existence and of very many of those to be built in the future; for the electric road, from its comparatively small cost of installation, can be utilized in a vast number of places where otherwise there would be no street railroads at all, or at most miserable horse affairs, giving very poor service.

Starting, then, with the assumption of our five cars operating over not more than five miles of track and over grades of only moderate severity, the question that must be considered is the character and size of the power plant required. In such situations the cars employed are, for the most part, sixteen or eighteen foot bodies on six-foot wheel bases; these have sufficient capacity for the work, can be employed to draw trailers if necessary, and are of reasonable weight. The same considerations that in Chapter IV. were used to determine the distribution of copper will serve fairly well for determining the capacity of the power station.

It may be useful here to give a brief table showing approximately the power required to drive a sixteen-foot car weighing, with its equipment and a moderate load of passengers, eight tons, up grades from one to ten per cent., at the uniform rate of eight miles per hour, which is not very far from being a fair average. On very light grades the determining factor of the power required is the condition of the track, and on very well laid track the so-called "coefficient of traction" should be fifteen or sixteen pounds per ton; on ordinary street-car track it is more frequently twenty, and rises from that figure to twenty-five, thirty, and in some cases even to forty pounds per ton; twenty pounds is quite nearly correct for the average conditions, and the table is founded on that assumption.

Per cent. grade.	Power at wheels.	Per cent. grade.	Power at wheels.
0	3.5	6	22.5
1	6.5	7	25.5
2	9.5	8	28.5
3	13.0	9	32
4	16	10	35
5	19		

This table gives the mechanical power required at the car axle, to the nearest half horse-power.

The average commercial efficiency of the motors is here taken at the figure that is to-day true for most of those in use, of about sixty per cent. To a very considerable extent grades compensate themselves, so that their effect on the *average* power required is not by any means so great as upon the *variations* in the power. Heavy grades mean a widely fluctuating maximum load, but they increase the average daily output by only a comparatively moderate amount.

With the ordinary car equipment of two fifteen horse-power motors, and the usual speeds, from eight to twelve miles per hour, experience has shown that five to six electrical horse-power per car is necessary on nearly level track. Ordinary grades upon the road will not, in general, increase this amount greatly; they will, however, cause more or less extended periods of exceedingly heavy output, rising to twenty-five or thirty horse-power for each car on the grade.

On large roads, where these times of abnormal output are infrequent, one can safely count on the figure just given for the power required per car; under such circumstances about ten indicated horse-power, at the station, per car will therefore prove quite sufficient. On a road like the one we are contemplating, with, possibly, severe grades and only five cars in operation, it may easily happen, for example, that a couple of the cars may be simultaneously upon the grade and a third starting under somewhat unfavorable circumstances. The amount of current ordinarily taken in starting a car is momentarily more than fifty ampères, which at the ordinary voltage corresponds to about 25,000 watts; we therefore might have 80 or 90 horse-power demanded for a minute or two, and a longer call for power of 50 to 75 horse-power, depending of course on the length of the grades which are to be surmounted.

In towns where there are small roads such as we are considering, it frequently happens, too, that inordinate demands for power are made on the occasion of a somewhat infrequent theatre night, a political demonstration, a base-ball game at some point far out on the line, and other public gatherings. This usually means bunching three or four heavily loaded cars, sometimes all the cars on the line, at one point, and starting them out within a minute or two of each other: for a small road is dependent for its revenue largely on its willingness and ability to accommodate just such unusual demands.

It would therefore probably happen that on our five-car road there would be times when the output would have to reach nearly or quite 100 horse-power for a few minutes at a time, although it would be strange if the *average* electrical horse-power required throughout the day should exceed 30. The reader will therefore readily understand that, under the circumstances supposed, a far greater margin of power is required than in case of a larger system.

In a comparatively small place it is fortunately not difficult to obtain a favorable location for a power station, and the expense of securing a site convenient to fuel is generally not great. Our station should be located as centrally as feasible, and close alongside a railway track or wharf, where coal can be easily obtained. It is highly desirable, even in so small an installation, to have a spare engine and dynamo, although it frequently is impracticable; it may be laid down, however, as a rule of fundamental importance, that a road should never be trusted to a single dynamo for continuous running, and even if the total output required be small, two dynamos should be employed: for nothing damages the reputation of a street-railway system more permanently than a breakdown that requires the suspension of traffic for a day or so, and such breakdowns are sooner or later bound to occur where only a single dynamo is used. Railway generators are peculiarly susceptible to injury from lightning, and for this, if no other reason, the above precaution is necessary. For our specimen five-car road, a proper outfit of dynamos would be two of about 40,000 watts rated capacity each; this means an ability to supply 100 electrical horse-power, or more if necessary, and sufficient capacity in either dynamo to keep up a tolerable service on the road if its mate should unfortunately be disabled.

As regards the type of engine that should be employed there is little room for dispute, for the only thing that would answer the purpose properly is a high-speed simple engine, belted direct to the generators, and having as much weight in frame and fly-wheel and as wide a range of cut-off as is practicable. If it can be run condensing, so much the better, as in most cases of employing steam engines. Its capacity should be, approximately, 80 indicated horse-power at one-quarter stroke cut-off and 90 or 100 pounds steam pressure.

It will be observed that the rated capacity of the engine of this plant is much less than that of the dynamos; and this is the correct arrangement, for, as will be shown in the chapter on efficiency, a dynamo can be run at half its nominal output without serious loss of efficiency, while under similar circumstances an engine not only loses from its friction becoming a large portion of the total output, but the machine *per se* becomes less economical of steam. Occasionally compound engines of as small size as that mentioned are used, but they are not to be recommended for such a small and widely varying output, as the increase in economy of fuel is not sufficient to counterbalance the increased first cost and the slighter greater complication and consequent danger of trivial accidents, which, though easily repaired, will, where a single engine is employed, cripple the entire road.

The boilers for the plant should be quite capable of supplying steam for 100 horse-power easily, and rather more if pushed a little. Two boilers should be employed, to permit giving one of them careful attention. A single boiler is undesirable for the same reason as a single dynamo. It is even better to have two boilers—either of them sufficiently large to handle the plant if necessary—and employ them alternately; this principle, which, although more expensive in first cost, is cheaper in the long run, will be dwelt upon in speaking of the arrangement to be employed in large stations.

We have, therefore, for this first specimen station, an equipment consisting of two 40,000-watt dynamos, one 80 horse-power high-speed, simple engine belted directly to them, and two boilers of about 50 nominal horse-power each.

As to the type of boiler to be employed, it is largely a matter of taste and convenience; probably, for general use, nothing is much more satisfactory than the plain tubular or return tube types, well set and provided with ample furnace capacity.

Excellent results are obtained from the water-tube form of boiler, of which there are now a number of meritorious patterns. We should hesitate, however, to advocate their use to the exclusion of the regular tubular boiler.

In any case it must be remembered that the intrinsic differences in boilers are vastly less than the individual differences introduced by general care and skillful firing. Boilers should be periodically cleaned and overhauled in the most thorough manner; hence the necessity for a spare boiler, which will permit this to be properly done without suspending operations, unless in part. It should be remembered always, then, that a plain boiler with a skillful fireman will give better results than the most expensive and elaborate form of patented boiler with a poor fireman.

A very convenient arrangement for such a small station is shown in Fig. 89. Everything is under a single roof and on one floor, with a stack at one side; the floor space is divided into two nearly equal portions, one devoted to the boilers and coal bins, the other to the engine and dynamos. The engine is placed at the end of the dynamo room close to the boilers, and belted directly to the two generators with slightly different lengths of belt, so as to render both machines more accessible.

The switchboard is placed as shown, on the wall close to

both engines and dynamos, so that the engineer, sitting in any convenient spot, can watch the operation of the entire plant and be almost within reaching distance of the apparatus. Of course circumstances will necessarily vary the plans of the station, but the example given is convenient in its arrangement.

As regards material, the station should preferably be a substantial brick structure, if necessary, however, a frame

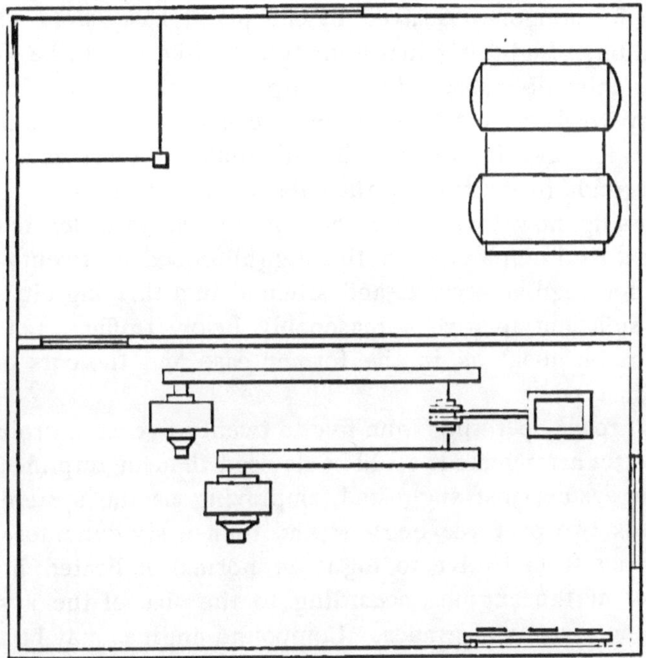

FIG. 89.—ARRANGEMENT OF POWER STATION FOR A FIVE-CAR ROAD.

building answers the purpose fairly well. If, as is sometimes the case in either construction, an iron roof is used, it should be carefully sheathed, either by roofing felt or by a wooden ceiling at the eaves; if this precaution be not taken, there is likely to be constant trouble from moisture condensing on the machines under variations of temperature, a contingency which is carefully to be avoided. For a permanent station, a brick building and substantial chimney should uniformly be erected.

The foundations are a matter of prime importance; both engines and dynamos should be given a most solid bed constructed of rubble and cement, or brickwork, as convenience dictates, but far more substantial than would be employed in supporting ordinary machinery. The dynamos ought to be carefully insulated, the wooden base on which they are usually placed being generally sufficient for this, provided proper foundations are employed. The interior fittings and switchboard ought to be the subject of careful, thorough construction: for no money is saved by cheap and hasty work about a station. Lightning arresters and the like should be given non-combustible bases of a size large enough to avoid danger of any woodwork catching fire in case of accident. The usual insurance rules for electric installations will serve as a sufficient guide for putting up the subsidiary apparatus.

Passing now to the next case to be discussed, let us consider a road employing in the neighborhood of twenty-five cars for regular service, and situated in a thriving city of a size sufficient to give a reasonably heavy traffic. Let the grades be about as in the former case and the cars be of similar type.

For roads operating from five to twenty-five cars, probably no better arrangement could be devised than an amplification of the system just suggested, employing as the system increases two or three engines and four or six dynamos, and allowing from twelve to eighteen normal indicated horse-power at the engines, according to the size of the system and the severity of grades. Compound engines may be used with great advantage for machines of 100 horse-power and upward, especially if it is possible to obtain water for condensation. As for a twenty-five-car road, it is on debatable ground, where a question necessarily arises between the slight economy to be gained by direct belting and the great convenience of being able to operate any and all the dynamos from either of the engines. At about the same size of plant, too, the low-speed engine begins to come into play.

For a road of the size we are considering, an allowance of ten or twelve indicated horse-power per car would nearly always be sufficient, and a proper equipment for twenty-five cars will consist of four 60-kilowatt dynamos. When long cars or snow sweepers are used, each should be reckoned as equal to two ordinary cars. It is probably preferable to have recourse to a countershaft, and this should be situated at one end of the dynamo room and on very substantial foundations. The dynamos, arranged as shown in Fig. 90, should each be driven from a pulley provided with a friction clutch, enabling any of the machines to be employed. The shaft itself should be

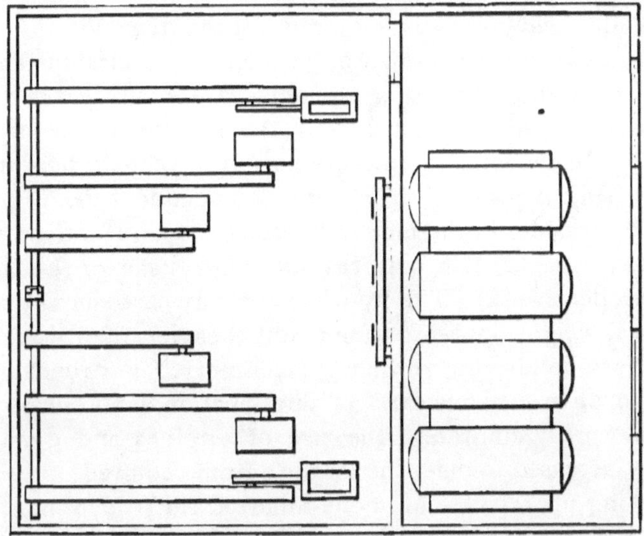

FIG. 90.—ARRANGEMENT OF POWER STATION FOR A 25-CAR ROAD.

divided into two sections, connected to each other and to the engines by friction clutches.

With an installation of this size two engines should always be employed, belted direct to driving pulleys at the ends of the line shaft; these engines should be of similar or identical pattern, and may be either simple or compound, as the conditions of fuel expense may dictate. Corliss or similar low-speed engines may very well be used under these circumstances, as, except in certain cases, the fluctuations of load are not likely to be great enough to put such an engine at any considerable disadvantage. With slow-speed engines, either simple or compound, a rated capacity of 150 horsepower is desirable for each. If high-speed engines having a

12

wide range of cut-off are employed, 125 horse-power nominal capacity for each will probably be sufficient. The boilers should be three or four in number, aggregating about 300 horse-power. Fig. 90 shows the general arrangement of such a station as has just been described. In operating it, a little judgment will enable excellent results to be attained. During a large part of the day one engine operating two, or possibly three, of the dynamos will handle the load easily and evenly; in the morning and the evening, and at times when especially heavy work is required, all the dynamos and both engines can be put into use. Under these circumstances, repairs on the engines or dynamos are easily carried out without interfering seriously with the regular service of the road. If compound engines are used, they should be worked condensing if possible. The cross-compound type is probably preferable, in the matter of ease of repairs and accessibility, to the tandem patterns, although some of the latter do excellent work. The fly-wheels in any case should be of unusual weight—at least 50 per cent. heavier than would be employed in driving ordinary machinery. For roads operating over twenty-five cars, an amplification of the plans just given works admirably, the size of engines and dynamos being increased to meet the larger output required.

Taking up the case of a one-hundred-car road, something of the same line of equipment may be followed, but the dynamo units should of course be larger. For one hundred cars six 150-kilowatt machines are desirable, allowing a sufficient surplus of power to reserve one dynamo as spare equipment. The same arrangement of the line shaft may be advantageously followed. It is probably advisable, however, to divide it into three sections instead of two, providing, as before, friction clutches for each dynamo pulley and each driving pulley. Usually two or three engines may be used, preferably three, as this number enables a better adjustment of the load and permits one of the engines to be thrown out of use for repairs without causing serious inconvenience. For so large a plant, triple-expansion condensing engines are very strongly to be recommended; three of 400 nominal horse-power each would answer the purpose admirably.

Under these conditions the mean load can be kept reasonably near the full output of the machines in use all the time.

Except for a few hours each day, two engines will do the work and do it well, and one dynamo could be, and should be, reserved to be thrown in only when it is absolutely necessary. Slow-speed engines, with Corliss or equivalent valve gear, are in their element in an installation of this size; and all the great advantage of their unusual efficiency can be

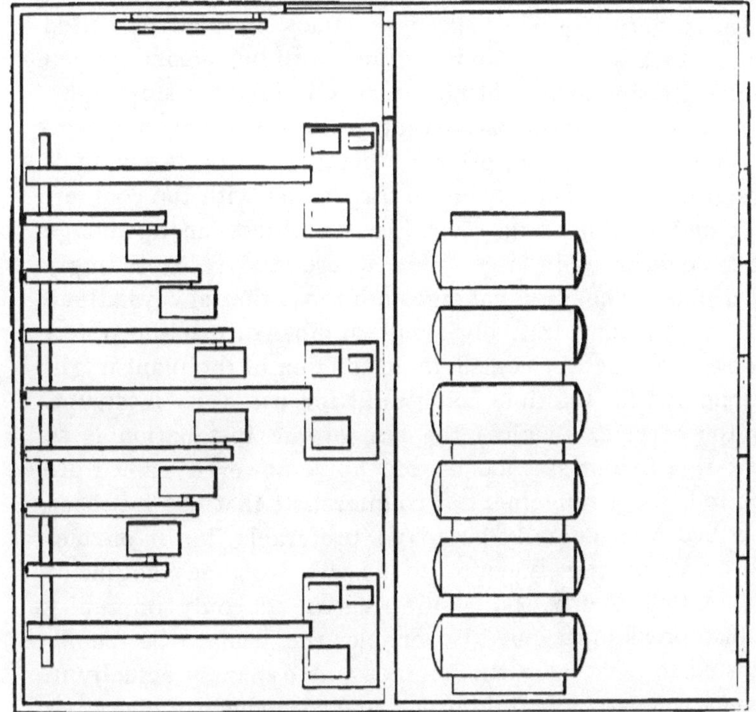

FIG. 91.—ARRANGEMENT OF POWER STATION FOR A 100-CAR ROAD.

enjoyed. The boiler capacity, aggregating say 1.200 horse-power, should always be divided into five or six units, so that the boilers can be thoroughly overhauled, one at a time, without causing any special inconvenience. A good arrangement for a one-hundred-car station is shown in Fig. 91.

In a road of this size it is decidedly advantageous to employ some double-truck long cars, with two motors of 20 horse-power or thereabouts; and these should be counted as the equivalent of two standard cars in making up the total equipment. On small roads such cars are not advisable, as they are very heavy and would have to be run very lightly loaded a large part of the time; but as the size of the system and the traffic increase, their use becomes desirable, as in every case where high speed is to be attempted.

In unusually large electric-railway installations, operating several hundred cars, it is often advisable to employ an arrangement radically different from the one just mentioned —that is, very large low-speed dynamos, coupled or belted directly to triple-expansion engines, forming a combination plant quite similar to that now generally used for ship lighting, but on an enormously larger scale. The advantages to be found in this description of plant are, first, a saving of perhaps five per cent., owing to dispensing with the countershaft and belting; and, second, compactness, an advantage not to be despised in large cities, where land is exceedingly expensive. The disadvantages are those that always attend the use of a comparatively small number of machines; for, in case of accident, a considerable fraction of the plant might be rendered for the time being unfit for use, thus interrupting the service. Unless the size of the installation is so great that five or six 300 or 400 horse-power dynamos are required, the arrangement of countershaft that has just been described will generally be found preferable, for it enables any or all of the dynamos to be run from any engine or combination of engines. This secures not only immunity against breakdown, but also enables the load to be readily adjusted so as to keep the engines and dynamos actually in use nearly up to their full output, and consequently operating at their best efficiency.

A number of our present electric roads have employed for service aggregating as high as one hundred cars merely an exaggeration of the plans just laid down for the roads of the smallest size, some few designing engineers having carried the subdivision of power to a very needless extreme. If the conditions of high average load are fulfilled, such systems as those just mentioned may work very economically; but it is believed that with equal care the designs we have shown will produce somewhat better results in economy of fuel, continuity of service, and small repair bills. As regards the subdivision of power, one general rule may safely be laid down, deviation from which is very likely to lead to high running expenses or disastrous uncertainty of service; it is this: *The number of power units should be just such that the disabling of one of them will not interfere with the successful operation of the system.* It will very seldom happen that more than a single unit, composed of an engine and dynamo, or engine and two dynamos, will be disabled at the same time. Owing to the fact that accidents are rather more likely to happen to dynamos than to engines, the engine units may be larger than the dynamo units. In very small plants it becomes necessary sometimes to take one's chances of being crippled by an accident, but every station should be so designed that an engine or dynamo can be stopped for repairs at any time, without seriously interfering with the car service.

Since writing the above, there has appeared a very valuable statistical article by Mr. J. S. Badger * on the operation of electric roads; and from this we cull some data regarding station operation that are of the utmost value as emphasizing the general principles of station design here laid down. The very best performance as yet recorded for an electric-railway power station is furnished by the East Cleveland road—No. 4 of Mr. Badger's paper.

The station equipment consists of horizontal return tubular boilers with Murphy furnaces, furnishing steam at 100 pounds pressure to three 200 and three 125 horse-power Armington & Sims simple, high-speed engines, belted directly to Edison generators. Of these there are six No. 32 and six No. 16, the former being of 80 kilowatts and the latter 40 kilowatts rated capacity.

The road is in general level, with, however, one five per cent. grade 400 feet in length; seventy motor cars and seventy trailers are on the average in daily use. One electrical horse-power is obtained for every five pounds of slack or four pounds of nut coal consumed, the evaporation being 7½ pounds of water per pound of slack. The cars run

* *The Electrical World*, October 31, 1891.

on an average 91.1 miles per day each, and the consumption of fuel per car mile is 4.3 pounds of slack or a corresponding proportional quantity of nut. This is a very excellent example of correct proportioning of station capacity to work, although even a better performance would have been attained had compound condensing engines been used.

It will be noted that the aggregate capacity of the dynamos at the full rated load was 980 horse-power, a little less than half the rated capacity of the motors, thus following closely the proportion we have suggested elsewhere. Further, the engines were of a trifle less capacity than the dynamos, 975 horse-power, and at full output the dynamos would give 12½ electrical horse-power for each motor car. Trailers were steadily used, a practice the advantage of which has often been pointed out, so that the actual electrical horse-power provided per car at the rated output of the dynamos would probably be about eight—at all events, somewhat less than 10; and this also is in accordance with the estimates we have given.

In the actual operation of the station, the coal required was but five pounds of slack for every electrical horse-power hour, which corresponds very well with the evaporation given, showing a consumption of nearly 35 pounds of water per indicated horse-power, a thoroughly consistent figure, and one not by any means unusually great for high-speed engines running under variable load.

Had compound condensing engines been substituted for the simple machines, preserving practically the same proportion of engines to dynamos, even this excellent performance could have been improved; since five pounds of slack, at an evaporation of 7½, was required per electrical horse-power hour—certainly much more than would have been needed with large condensing engines.

Were any evidence needed to show that the good result was due to proper proportioning of the station, it can be readily found from the details of Mr. Badger's road No. 2. In this case the engine power aggregated 675 horse-power, while the dynamos were five in number, of 80 kilowatts each, total about 535 horse-power; the engines were all high-speed machines of standard make, running on a steam pressure of 80 pounds; the grades of the road were decidedly heavier than in the case previously mentioned, the steepest being 9½ per cent. and about 400 feet in total length. Sixteen motor cars only were in regular use, and although the company had trail cars, they were seldom used. The full dynamo capacity was a trifle over 35 horse-power per car, and the fuel used was nut and slack, of which the consumption was no less than 11 pounds per car mile as against 4.3 on the East Cleveland road. Although the conditions mentioned were somewhat more severe in road No. 2, so far as economy is concerned, the direful effect of the preposterously large engine and dynamo capacity is only too evident in the enormous consumption of fuel.

It will be noticed, in planning the model stations, that in every case the whole installation has been shown on a single floor, and that the ground floor; and whenever it is practicable this is the best arrangement, because thereby immunity from vibration and ready access to all parts of the plant are secured. Sometimes, though rarely, in city stations, ground space becomes so expensive that it must be economized so far as possible; it then becomes necessary either to use the direct-coupled dynamos just mentioned, or to place the engines on the ground floor and belt upward to the dynamos, either directly or through the medium of a line shaft. This simple procedure is likely to cause a great deal of trouble from hot bearings. A number of such plants are running, usually for electric-light apparatus, and in many cases they will be found in operation with the caps removed from the bearings and convenient means at hand for cooling the journals with water. In any such case the dynamos should be provided with a substantial support directly under them, and should not be trusted to the supporting floor beams. It is far better to use the direct coupled engine and dynamos where circumstances compel contracted quarters for the station.

A word may well be said here on the subject of belting. It is advisable to use leather belting of the very best quality, special attention being paid in selecting it to closeness of grain and pliability. There are now in the market a considerable number of patented beltings of various descriptions—canvas belting, perforated belting, and link belts of different kinds; these various sorts are frequently used, and, as a rule, give satisfaction; but we have yet to learn that they are productive of better general results than first quality plain leather belting kept in good order

With regard to the minor arrangements of a station, there are many details which can hardly find a proper place in this volume, unless it be sufficient merely to suggest them. The switchboard in any station should be placed where it is very easily accessible and in plain view of the dynamos; the other apparatus should be easily visible from the regulating devices, and these latter should be grouped together for convenience and placed where they can be reached in a moment if anything happens. Plenty of room should be allowed for switches and cut-outs, and both should be of the most substantial description. Whether the cut-outs are magnetic or merely fuses, they should be looked after at frequent intervals and kept in good order, and never should be set to act at the nominal rated capacity of the dynamos. It is an unnecessary refinement of cautiousness and will result in stoppage of the cars without any good results, for any well-constructed dynamo will carry for a few minutes nearly 50 per cent. over its rated capacity without doing any damage whatsoever.

In case, for example, of a short circuit on the line, it is far better to overload the plant by 25 or 30 per cent. and keep the cars running, than to save the fuses at the expense of stoppage of traffic; for fuse wire is cheap, but the loss of public confidence that results from frequent cessation of current on a line is a very serious matter, particularly when the railway company has to ask any favors of the municipality. The dynamo itself gives plenty of warning of an overload before it is in any way injured; and a judicious and watchful engineer will very often save blocking the cars by keeping his hand on the switch and his eye on the ammeter. Of course it should be remembered that with a large station a given short circuit on the line may cause no serious results, while in a smaller station it might necessitate cutting out the dynamos to save the armatures.

Every station should be equipped with a good set of testing instruments, consisting of a standard ammeter, voltmeter, one or more magnetos, and a portable testing set for the measurement of resistance. It is desirable that the ammeter and voltmeter should be portable, so that they can be used on the cars if necessary; and in large stations special instruments should be provided for this purpose. In addition, at least one engine indicator should be provided—better still, two—and for stations where there are compound or triple-expansion engines, enough to use simultaneously on each cylinder of a given engine. The expense is not great, and the instrument is more than likely to save many times its cost in fuel. A standard steam gauge for testing the gauges in the boiler room is a useful addition to the equipment, also some form of portable tachometer for getting the speed of engines and dynamos.

It is worth while stating something here about lightning arresters. Enough has been said in the chapter on motors to give some idea of the importance of providing such apparatus, but everything said about the danger to motors is thrice true of dynamos. A railway generator, being a shunt or compound wound machine of high voltage, is peculiarly sensitive to the attacks of lightning; for there is direct communication between the line and the armature, and a lightning stroke entering the station is very likely to do mischief.

The direct damage done to the machine by the lightning itself is usually small, and a powerful stroke is not at all necessary to very disastrous results. The action is usually as follows: the lightning enters the station and reaches the machine in sufficient amount to puncture the insulation of the armature; the current follows the spark, and the machine promptly destroys itself. The extent of the damage may go all the way from a few wires melted off to an almost complete destruction of the armature windings and the commutator.

It therefore becomes necessary in protecting railway machines against lightning to ward off if possible every particle of the discharge. Ordinary lightning arresters, the function of which is simply to provide an alternative course to earth, are practically useless under these circumstances; it is not sufficient that the lightning may choose between going through the machine and to ground through the arresters; it must if possible be *compelled* to take the latter course. It is a well-known fact that both arc dynamos and railway motors, in fact, all series-wound machines, are much less liable to damage from lightning than shunt or compound wound machines; this is owing to the fact that before any lightning can enter the armature from the line it must pass through the series magnetizing coils of the machine, which possess quite high self-induction and choke back the entering discharge in a more or less efficient manner.

Bearing this in mind, the use of powerful impedance coils situated between the lightning arresters and dynamos, as

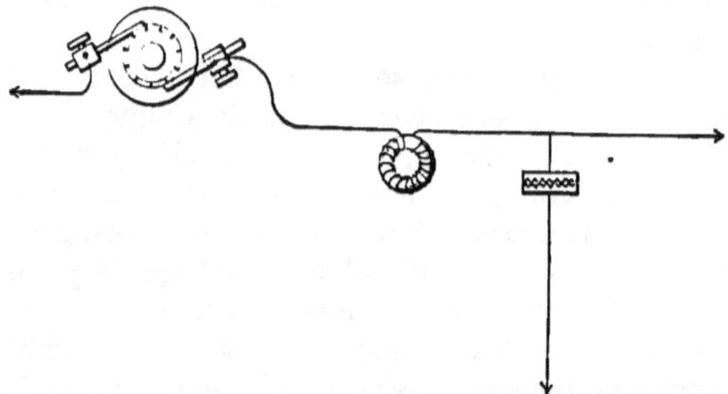

FIG. 92.—ARRANGEMENT OF IMPEDANCE COIL WITH LIGHTNING ARRESTER.

shown in Fig. 92, is strongly to be recommended; by offering a powerful resistance to the passage of a sudden discharge through them in the direction of the armature, they are likely to drive the lightning to ground and save the machines. Without such assistance it is quite possible for the lightning to jump to ground through the lightning arrester, and also partially to discharge into the dynamo and destroy it. It is quite impossible to give explicit advice as to the necessary power of the impedance coil required, but one having approximately the magnetic dimensions of the field magnets of the motors will probably answer the purpose.

As lightning usually enters a station from the line rather than from the ground, it may be a useful precaution so to connect the series coils on compound-wound dynamos that they shall be between the armature and the line, and by their considerable impedance help protect the former.*

On account of lightning, if for no other reason, a railway dynamo ought to be carefully insulated from the ground; for under these circumstances the armature runs less risk, inasmuch as the lightning discharge is then not so likely to jump to and through the armature core. Aside from this, the frame ought not to be grounded, on the score of safety to the men who are handling the dynamo. As to the protection of the station itself from lightning, it may be well to suggest that in case an iron roof is used, as is not infrequent, grounding it at one or more points converts it into a most efficient lightning rod. (See, further, Appendix E).

In the boiler room accessory apparatus should not be forgotten. In large stations, mechanical stokers are a highly desirable addition to the plant, as they enable an economical use of grades of coal that will much reduce the fuel bill. Separators and feed-water heaters should be regularly employed, unless the arrangement of the boilers with reference to the engines is such as to render the former unnecessary. For feeding the boilers, duplicate or triplicate means should be provided, so that no easily supposable series of accidents can prevent the proper feeding of water. If condensing engines are used, care should be taken to have the supply of water for condensation ample; and in every case the engineer should be watchful both of the quantity and quality of the water for the boilers. Where good, clear river or pond water can be obtained, nothing better is to be desired; where this is not available driven wells may answer the purpose. If one is compelled to depend on the city water supply, the bill is apt to be unpleasantly large, particularly if condensation is to be attempted, and arrangements should be made for procuring an independent supply of water.

In the dynamo room there are divers devices that may be profitably introduced. One of these is a permanent traveling crane, supplied with differential tackle of sufficient power to

* This proceeding is useless where the series coils are shunted by a resistance.

handle the component parts of any of the dynamos used. With this apparatus the replacing of an armature or field becomes a very simple and easy matter, while without it much time is consumed and considerable danger of injuring the machinery is incurred. It is advisable to fit up in connection with the dynamo room a little workshop where minor repairs can be executed; for an engineer who is handy with tools can oftentimes save his company a good deal of money, and he should certainly be provided with a sufficient kit of machinists' tools to encourage him in their use. Large stations usually should have a repair shop connected with them, and many companies even go so far as to rewind armatures—sometimes making an excellent job and saving a very considerable amount of money thereby.

In constructing a station of any size whatever, room should be left for an increase of plant—not to be put in, however, until it is needed. A few feet extra length in line shafting will easily permit the introduction of a couple of extra dynamos in a station of some size, and in a very small station ground space is usually cheap enough to permit leaving room for an additional engine and pair of dynamos when they shall become necessary.

As regards the general care of dynamos, little special instruction need here be given. As railway dynamos run for the most part at five or six hundred volts, they are somewhat more liable to short circuits, both in armature and commutator, than are ordinary incandescent or arc machines. As a result of this, they should be kept scrupulously clean; the commutators should be cleaned with great care, especially if carbon brushes are used; the brushes should be kept nicely trimmed, and, in fact, extra precautions should be taken throughout the station.

The variations in load on railway machines are apt, as we have many times before stated, to be very violent and very sudden, so that there may be more sparking at the commutator than is usual in other dynamos. A sharp lookout should be kept for this, and the brushes shifted whenever necessary.

The carbon brush has come into very considerable use,

and a word regarding its employment is appropriate.* Where carbon brushes are to be used a little extra work should be spent on the commutator, in keeping it perfectly clean, otherwise it is likely to heat from short circuits caused by carbon dust, and frequently exhibits the symptom of minute sparks flashing from under the brushes from segment to segment. When carbon brushes are working at their best they simply impart a grayish tinge to the entire commutator, without causing either heating or flashing. If there is a noticeably unequal blackening of the commutator it is likely that trouble will follow, and the commutator should be rubbed clean at once.

Dynamos of different makes and sizes cannot always be run in parallel on a railway circuit, for the reason that they may be magnetically unlike, and will not then respond with equal promptness to magnetic changes; therefore, when the load varies, as it may often and violently, the machines will not respond equally, and the load will be unevenly distributed, causing unnecessary strains on one or more of the machines. It is likewise very undesirable to run in parallel dynamos driven by engines of different makes and speeds, for when a sudden load is thrown upon the system the engine governors will not respond with equal promptitude, throwing the system out of balance and causing certain machines to do far more than their share of the work. Engine makers will often point with pride to the fact that their machines are so well governed that they will not vary in speed more than one or two revolutions per minute, when running empty and when running loaded. This boast is sometimes true, but nevertheless the speed of these very engines at the time of throwing on the load may *momentarily* vary 10 or even 20 per cent.

A slow-running engine necessarily governs with less promptitude than a high-speed engine, and in most cases the

* Probably the handiest way of trimming a carbon brush is to strap a piece of sandpaper, face out, around the commutator after the current has been shut off for the night, and drop the brushes into the brush-holders, letting the machine run slowly until the sandpaper has smoothed off the end of the brushes into the proper form. A supply of brushes thus prepared can be put in whenever necessary, and will run smoothly and quietly from the very moment they touch the commutator. Take care not to let the carbon dust get to the commutator connections, as it may make trouble.

practiced eye can detect at the moment of sudden load a very marked slowing down of the fly-wheel, unless it be of unusual weight. It is for this reason that heavy fly-wheels are so strongly recommended, for they enable the engine to tide over the period of sudden load with greater ease than would otherwise be the case; it must not be forgotten, however, that no engine governor in general use to-day begins to act until the speed has actually begun to fall, so the fly-wheel is of particular service only in helping the engine over very sudden changes.

It is worth remembering that from the high voltage of railway dynamos and the consequent violence of any short circuits that occur, the machines are subject to serious accidents under circumstances which would lead to but trivial results in low-voltage stations. It is therefore advisable to cut out the machine as soon as it shows signs of trouble that cannot be immediately removed by caring for the commutator a little. The insulation should be frequently tested after shutting down for the night, and every precaution taken to avoid the beginning of difficulties, for when once begun they grow with extraordinary rapidity.

In case of a short circuit on the line the engineer should stand by the machines, watching them and the ammeter, and ready to cut them out if it becomes necessary; it is a good, safe rule, however, to hold on until the fuses go, in order to avoid shutting down the line. With copper brushes a short circuit on the line was frequently followed by a very considerable injury to the commutator, but now that the carbon brush has come into general use the machines will stand a pretty severe ground on the trolley wire without danger.

Finally, the engineer of a power station should co-operate with the superintendent of the road in keeping watch of the performance of the entire system. The engines should be indicated at least once a day, and the ammeter and voltmeter readings should be taken simultaneously with the indications. The amount and time of abnormally large currents, the occurrence and severity of short circuits in particular, ought to be noted in an engineer's log-book, as well as the effect of any unusual service occurring on the line, such as the employment of a snow plow, the effect of trailers, if they are used, and the result of any particularly heavy service. If possible, the amount of coal burned each day should be noted, and also a rough check kept of the amount of water used. The former amount can sometimes very conveniently be ascertained, if not every day, at least occasionally, by taking a time when the coal pile can be easily cleared out and measuring in barrels the coal used.

If, as is usually the case, particularly in stations of any size, feed-pumps are used for supplying the boilers, the weight of water delivered per stroke of pump can be very readily determined; and by attaching to the pump a simple stroke counter like the registering mechanism of a gas meter, the total amount of water pumped per day may be at once found; then, by bringing the water at the close of the day's work in each boiler back to the point where it was at starting up in the morning, the total amount of water evaporated per day can be roughly ascertained. By taking the coal consumption and water consumption in connection with the indications of the engine and the simultaneous volt and ammeter readings, the whole performance of the station can be periodically determined without much trouble, and sources of loss detected and estimated.

An engineer's log-book in which can be entered the data mentioned is highly desirable, and in Appendix C is given a convenient form of page for noting these various facts. It is filled in so as to show in a general way its application. The report can be and should be made as full as practicable, and in case of special tests it is only necessary to fill the log-book more completely than is shown in the example, taking indicator and ammeter readings at more frequent intervals. Now and then, at least, voltmeter and ammeter readings, at intervals of a few minutes throughout the time of running, should be made; and, in fact, the engineer of the power station ought to take pride in knowing all the details of the performance of the machines, for not only are these of importance in helping him to operate the station successfully, but they possess a money value to the company that employs him.

In closing this somewhat brief consideration of railway power stations, we can do no better than to impress again on the mind of the reader the necessity for solid and substantial work throughout, the importance of using only the most carefully made engines of extra weight and strength, and the prime necessity of so arranging the installation that it can be kept loaded as nearly as possible to its full capacity throughout the hours of running. And finally, the last place to economize is in the engineer's wages; for a cheap engineer can do enough mischief in one day, through sheer ignorance, to eat up the dividends for an entire year. It requires an unusually intelligent and skillful man to handle a railway power station properly, and such an one must be obtained, even at a cost considerably above that familiar to those whose experience has been in other lines.

HINTS FOR THE CARE OF RAILWAY POWER STATIONS.

It may be well to append here, for the benefit of those who have to do with the running of stations, a brief series of hints on the special care that must be given to station apparatus; although what has previously been said is quite sufficient for those who are already somewhat familiar with the operation of dynamos such as are used for lighting.

BOILERS.

To begin at the boiler room, it is almost unnecessary to state that scale is the fireman's worst enemy. Boiler scale is the natural and logical result of insufficient attention to the boilers. All water holds in solution a certain amount of mineral matter, varying from the merest traces—next to none at all—to quite a perceptible fraction of an ounce per gallon, in mineral waters. Hard water, so called, is that containing an unusual proportion of mineral matter, most frequently salts of lime.

During the process of evaporation that goes on in the boiler all this material is left behind, and unless the boiler is periodically cleaned out, a deposit, calcareous in nature and sometimes almost as hard as stone, will be formed wherever the water touches the iron shell or tubes of the boiler.

In addition, there is likely to be a certain amount of suspended matter in water—particularly river water—which is added to the general collection of foreign matter and is likely to coat the interior of the boiler with mud. We have seen boiler scale a quarter of an inch thick and nearly as hard as flint taken from the boilers used in an electric-light station; and, in fact, it is only too common to find scale allowed to accumulate through carelessness, although seldom to so great an extent.

The result is not only to diminish very much the evaporative power of the boiler, but to expose the iron, unshielded by circulation of water from the effects of direct heating, in such a way as to rapidly deteriorate it and very likely to cause a boiler explosion. The remedy for scale is to periodically clean out the boilers thoroughly and carefully, and to use from time to time suitable scale preventers, which usually act mechanically by hindering the scale from sticking to the iron. A large number of these are in use, and perhaps the simplest of any is crude petroleum, a small amount of which is placed in the bottom of the boiler after it has been thoroughly cleaned and before the water is turned on; as the boiler fills, its interior is coated with a film of oil which does not allow the hard deposit of scale to cling.

No general rule can be laid down as to the frequency of boiler cleaning necessary, for it depends entirely on the quality of the water used. A very little experience will enable an engineer, under any particular circumstances, to tell in what quantity deposit is forming and to apply the proper remedies at once. As mentioned previously, it is best in installing any station so to arrange the boiler units that one or more can be thrown out of use and thoroughly overhauled without interfering seriously with the working capacity of the plant.

It should be the engineer's business to see that the boiler-room accessories are kept in proper condition, that the safety valve is all right, the steam gauge registering correctly, the joints tight, and the pumps in working order. Duplicate means should invariably be provided for feeding the boilers, for one of the most serious accidents that can happen is the derangement of the feeding apparatus so that water cannot be properly fed into the boilers. If injectors of the Körting or any other type are employed, a steam pump also should invariably be supplied; and even if but seldom used, it should be tested at frequent intervals to see that it is kept in perfect working order against the time, which must inevitably come, when the injector gets clogged and refuses to do its duty. All steam pipes of any length should be jacketed with some one of the many non-conducting coatings now in use—magnesia, asbestos, mineral wool, or the like.

ENGINES.

With respect to the engines, one general rule may be laid down: watch their performance and overhaul them at the first sign of trouble, for trouble goes on from bad to worse with the greatest speed.

A well-made, well-set and well-cared-for engine is as reliable a piece of machinery as the ingenuity of man has yet devised, but if ill-treated, even the best engine will go on strike with extraordinary persistence. As previously mentioned, the foundations for engines to be used in railway power stations should be extra heavy, and the engine itself thoroughly well balanced. There should be very little or no vibration when the machine is running, either in the engine itself or in the steam pipes. If the engine is regularly indicated, as it should be, the valves can be put in their proper order very readily, and any abnormal features about the engine noted at once. Any thumping, rattling, or unusual irregularities in the running of an engine should be investigated at the earliest possible moment and the proper remedies applied. We cannot go here into the details of running an engine, but can only give these general cautions—rendered all the more necessary in railway power stations on account of the severe strains to which the machinery is sometimes subjected.

DYNAMOS.

Some general hints have already been given as to the proper care of this most important part of a power-station equipment, but it may be worth while to go somewhat further into detail, and to give an idea of the troubles likely to be encountered and the ways of reaching them.

The list of faults to which a dynamo may be subject varies somewhat with the size and character of the machine; and what applies to small motors, or to arc-light dynamos, is not likely to apply with equal force to railway generators. These latter are usually very well made and efficient machines, as, indeed, they have to be to perform the severe service exacted from them.

As regards their general arrangement, they should be on firm foundations, readily accessible, and provided with means for loosening, tightening, and aligning the belts. First of all, they should be kept clean and dry, and usually, if well cared for, will give very little trouble. However, every dynamo is liable sooner or later, from one cause or another, to operate in a somewhat unsatisfactory manner. Whatever the cause of such abnormal performance may be, it must be hunted up and the proper remedy applied at once.

Speaking in a general way, the troubles most likely to be met are sparking, heating of the commutator, armature, or field magnets, heating of bearings, and failure to generate current at all; this last is unusual, and the cause may be either very serious or very trifling. Of course, several of these "bugs" often develop at once, as, for example, a prolonged overload is likely to cause sparking, and heating of the commutator, armature and bearings.

Taking, however, the causes mentioned in their order, we may note that sparking at the commutator may be produced by a rather wide variety of causes. The commonest one—so common as to hardly require comment—is a combination of overload and wrong position of the brushes; the former, a glance at the ammeter will tell, and in addition it becomes manifest by overheating of the armature, severe strains, and sometimes even slipping of the belt; the latter cause can be remedied by moving the rocker arm to the point of minimum sparking.

In the earlier railway dynamos it was quite common to find a sudden variation of load requiring a considerable change in the position of the brushes; in the later types, however, there is little or no change of lead with load, and whether the current is one ampère or several hundred, the brushes can be left very nearly in the same position. If you chance to be operating a dynamo where the non-sparking position of the brushes does shift, eternal vigilance is the only way to avoid sparking at the commutator.

The cause of sparking just mentioned presupposes that the machine is in good condition, but it is not altogether uncommon to find sparking produced by faults either in the commutator or in the armature. If the former is eccentric, irregularly worn, or has one or more bars loose or set irregularly with respect to the others, sparking is sure to ensue; for the tension on the brushes is irregular, they are thrown into vibration, and the result makes itself evident at once. An examination of the commutator will readily detect these defects. A rough commutator needs no description; and if it is eccentric, the fact can be readily detected, either by a visible wabbling or by holding a stick lightly against its surface, when any irregularity in motion will make itself apparent to the sense of touch. Loose or irregular bars are found with equal ease.

In these days of carbon brushes, worn commutators are not so common as they once were; if from using copper brushes and keeping them in the same position too long the commutator should become untrue, it should be smoothed very cautiously with a fine file, or very fine sand-paper—never emery-paper—great care being taken to brush the minute particles of metal out of the insulation between the commutator segments before starting up the machine. In very bad cases, a thin cut taken off the commutator in a lathe will remedy the difficulty; but this is considerable trouble, and on some large machines a tool holder is arranged to be fitted alongside the commutator, so as to turn it down very easily by simply running at low speed and feeding the tool by hand. Similar in its results to a rough commutator is a rough brush. A little examination will show when the brush is irregularly worn, so as to make poor contact, and if so, it should be trimmed or ground down.

Sparking due to faults in the armature may be very severe, but is almost invariably local in its character; that is, affecting only a few of the commutator segments. This peculiarity is very readily noticeable, and such an effect may be due either to a short-circuited coil in the armature or a broken coil, either one of which will produce violent sparking at the commutator bar or bars most immediately concerned with that particular coil. When a coil is short-circuited it heats very violently, may burn out entirely, and is quite likely to make itself obvious by a smell of overheated insulation. On running the machine slowly the current may show pulsations, and on stopping, running the fingers over the armature will generally locate the short-circuited coil immediately by the heating, particularly if one is guided to it by the visible burning at the edges of the corresponding commutator segments.

Short circuits may occasionally be produced by stray particles of metal getting between the segments of the commutator or among the armature connections. Under these circumstances the trouble can be found by inspection and easily removed. More often the short circuit is in the coils themselves, and may usually be looked for at the head of the armature.

Perhaps the most elusive of all faults in armatures is what may best be described as a flying short circuit—that is, a short circuit between two or more of the armature coils of such a character that it does not appear when the machine is at rest, but becomes noticeable as soon as the revolution of the armature, by centrifugal force or magnetic drag, presses the offending wires together. Under these conditions the machine may fail to excite, as the shunt winding is completely short-circuited; there will be little or no heating, because of the small excitation and little current flowing in the coils; and oftentimes even a careful measurement of the resistance of the armature coils all around the commutator will fail to disclose the trouble, for the simple reason that when the armature stops the short circuit no longer exists. Such a fault is perhaps best found by separately exciting the fields and then running the machine rather slowly, when all the characteristic signs of short-circuited coils will appear and can be located by the consequent heating.

If there is a broken circuit in the armature, as sometimes happens, there will be very serious flashing at the commutator—localized as before—but there will be no special heating of a single coil. The defect may be sought at the commutator end of the armature, as breaks in the wire are most frequent where the connections are made with the commutator segments. In either of the cases we have been discussing it is best not to try any makeshifts, but to cut the machine out until it can be properly repaired. A temporary remedy sometimes applied in such a case is to cut off the defective coil from the commutator bars with which it is joined, and then temporarily to connect the disconnected bars to those next succeeding; but it is not advisable to do this unless driven to it.

In addition to the causes of sparking already enumerated, there is a tendency for sparks to flash around the commutator over several segments. This is most often observed where carbon brushes are used, and is usually a result of carbon dust getting into the insulation between the segments and over the surface of the commutator generally. It is often accompanied by heating of the commutator, and never occurs where proper care is taken.

Heating in either the field-magnet coil or the armature of a dynamo is generally due to one of two causes—overloading or short circuits; overloading can be told by the condition of the ammeter, and most often results in railway power stations from a ground on the line, although the fuses will usually take care of the machines. The heating due to short circuits in the armature we have already mentioned. In addition there may be—particularly when the machine is first set up—moisture in the armature coils, producing a general case of short circuit through the insulation. Where this is present the armature usually feels moist, and may even steam. It is a rather unusual condition, and can be best remedied by drying the armature very gently for a considerable time.

Passing a current through it is the easiest way to do this,

although the amount used should not exceed the regular current for which the armature is intended.

Heating in field-magnet coils may arise from forcing the voltage, thus sending through the coils a greater current than that for which they were designed; or from a genuine short circuit in the coils, not a usual occurrence, and easily discovered by measuring the resistance of the two or four coils with which the machine is wound and comparing them each to each. If the difference in resistance rises to more than a few per cent., there is probably a short circuit. Moisture may get into the field coils just as in the armature, and is expelled in the same way.

Heating of the commutator may be due to general overload, or to short circuits between the segments, due usually to particles of metal or carbon from the brushes. The first case may be detected by the ammeter, the second by the localized heating and the sparking that ensues. Cleaning very carefully is the only suitable remedy.

One of the commonest troubles encountered in an electric station of any kind is the heating of the bearings in one or more of the dynamos; this is generally due either to lack of oil or to excessive pressure coming from an overload. With the ordinary types of railway generators, in which the bearings are self-oiling, there is little likelihood of the first cause producing any serious results, unless the oil well leaks or is extraordinarily neglected; and a glance at it will tell whether the trouble is there to be sought. Occasionally the armature shaft may be sprung, or the bearings may be out of line, although neither of these conditions is common. In the former case, the armature turns hard and is likely to stick at a particular point; in the latter, the shaft still turns with difficulty, but with nearly equal difficulty all the way round, and the revolution becomes much easier if the bearings are slightly loosened from their foundations. There is no help for a sprung shaft, while the bearings, if out of place, can be aligned. Neither of these difficulties is as common, however, as a hot bearing, due either to the pressure of the shoulder of the pulley sidewise against the bearing or too great belt tension.

In either case the bearing on the pulley end will be heated more than the other. If the trouble is due to lateral thrust—as can be easily told by trying to push the armature back and forward in its bearings while the machine is running—the belt pull should be aligned. Lateral play of perhaps an inch is allowed on almost all dynamos, so that with ordinary care there need be no trouble from this source.

If the heating is simply a case of too great belt tension, which can generally be told by inspection, the only thing to be done is to ease the belt; the trouble is unlikely to occur at all if the bearings are looked after and kept in proper condition. The greater the area of contact on the pulley can be made, either by using a larger pulley on the machine or a smaller one on the driving shaft, or by using wider pulleys and belting, the more easily the same power can be transmitted, with less pressure.

It is not a good plan to try to cool dynamo bearings with water, for water is not a pleasant thing to have in the vicinity of an armature; better cut out the machine until it can be fixed, or, if it be absolutely necessary to run, the very careful application of ice may sometimes relieve the difficulty.

Occasionally a dynamo will positively refuse to do its work; and perhaps the most frequent causes are a flying short circuit in the armature, such as has just been mentioned, or a break or bad connection in the field coils. If the former is the case, separate excitation by one of the other machines will soon locate the trouble; if the latter, measuring the resistance of the field coils will generally disclose the difficulty. Sometimes in setting up the machine an accidental bad connection may pass unnoticed until an attempt is made to run. If a dynamo has been assembled recently, there may be a bad magnetic joint between the yoke and the magnet cores, when the machine will fail to excite properly. This will be found by inspection, and it is remedied without difficulty by doing the work over again more carefully.

Setting up or taking down a dynamo is a rather difficult matter, and in a permanent station it is a very good idea to have a simple traveling crane for the purpose of doing such work as may be required. Where this does not exist, temporary supports can be erected and a differential tackle will do the work. If the traveling crane is at hand, the removal of an armature becomes a simple matter, for its weight can be taken by the tackle and it can then be slipped out of position by a single movement of the crane; the yoke or other attachments which may be in the way being previously removed. Such a course is only necessary in very large armatures, for small ones can be easily handled by a gang of men, the shaft always being so blocked up as not to bring the armature down on the pole pieces. In shifting an armature with the crane or temporary tackle the ropes and chains should never be allowed to touch the windings; the grip should be entirely on the shaft.

MINOR APPARATUS.

As regards the subsidiary apparatus required to run a station, every switch, cut-out and lightning arrester should be kept in thoroughly good working order, with no bad contacts, and no dirt allowed to accumulate. Lightning arresters should be set with the plates or points across which the lightning is expected to jump very close together, from a sixteenth to an eighth of an inch, and should be kept free from dust, which might otherwise cause a short circuit. Generally only those forms should be used in which after one lightning discharge the arrester is automatically reset and ready for another.

The measuring instruments of the station should be compared now and then with each other and with standard instruments, and the engineer or superintendent of the station ought to take pains to understand the function of each piece of subsidiary apparatus, of whatever kind, with which he has to do. Specific information in these matters can best be obtained from the company which furnishes the particular forms of apparatus employed, as there are frequently minor, though important, details peculiar to each especial make.

CHAPTER VII.

THE EFFICIENCY OF ELECTRIC TRACTION.

WHATEVER may be the advantages of electric traction, whatever its convenience as a means of rapid transit, it is on its efficiency that its ultimate importance must depend. We must realize at the start that the electric motor is not a prime mover, a fundamental source of energy; it only furnishes a very perfect and elegant means of utilizing electrical energy, already generated by some prime mover, at the point where it may be most convenient to employ it, whether that point be fixed, as in the case of stationary motors, or moving, as in the case of street railways. Advantages may be and are gained by employing motors, sufficient to offset considerable losses in the necessary transmission and transformation of electrical energy, but if these losses rise above a certain amount the system must inevitably be a commercial failure.

Let us look, then, deliberately at the series of transmissions and transformations necessary in electric traction, and form as close estimates as possible of the losses, their magnitudes, and the most practicable means for reducing them to a more satisfactory figure. The first transformation of energy is from the pressure of steam generated in the boilers to the rotary motion produced by the engine and employed in driving the dynamos.

Then the mechanical energy obtained is first transferred, through the medium of shafting or belting, to the dynamo, where it is again transformed and appears as electrical current on the line. In this convenient shape it is transferred, with little loss, to the point on the line where the motor or motors may happen to be. There it undergoes another transformation in the motor back to mechanical energy, which is then transferred, through the medium of gearing of one sort or another, from the armature shaft to the car wheels.

Fortunately, the losses at several points in this somewhat complicated system of transmutations are comparatively small; and for practical purposes we may consider the losses to be substantially as follows: first, the losses in the engine and attachments; second, those in the dynamo: third, those on the line; fourth, those in the motor, fifth, those in the gearing. Luckily, not all these are serious. In the art of electric traction as to-day practiced, their relative magnitudes are about as follows: the most formidable are the first and the last; they are of about the same magnitude, varying enough in different cases to render it quite impossible to say offhand which is the larger. Then come the losses in the dynamo and motor, generally smaller than either of the former, and that in the motor being somewhat the larger. Finally, the loss on the line, which in many cases is the least of all. Reduction of gearing has in some cases made that source of loss relatively small, but too often at the expense of motor efficiency.

Let us take up these several causes of inefficiency in order:

ENGINE EFFICIENCY.

Of the total indicated horse-power developed by ordinary engines running at full load, nearly 10 per cent. is consumed in friction of the moving parts. Occasionally the loss may be less than 5 per cent.; not infrequently it rises to 15 per cent. Its amount is nearly a constant, not a constant per cent. It may increase or diminish with the load, but generally remains somewhere nearly uniform.

It is probable that every engine has a particular load for which the friction is a minimum, owing to the fact that the strains to which the machine is subjected produce slight flexures that vary the resistance at the bearing surfaces; for the purpose in hand a safe estimate for the friction is 10 per cent. of the rated capacity of the machine, and this, by the convention usually adopted among engine builders, is its output for boiler pressure of from 80 to 100 pounds and a cut-off of about one-fourth stroke. A few concrete cases may be of interest to the reader.

TYPE OF ENGINE.	SPEED.	I. H. P.	FRICTION IN H. P.
Buckeye....................	280	23	5
Westinghouse	300	84	7
Armington & Sims.........	290	80	9
Corliss.....................	88	120	10
Compound condensing......	—	347	44

These are merely examples to show the general magnitude of the friction. The differences in friction between two engines of the same make and the same engine at different times are as great or greater than any difference that is likely to exist between two similar types of engine of like size. Relatively, friction is less in large engines.

In railway power stations it is the almost invariable rule, at present, that the driving engines are used not at their full output, but at something considerably below it. The exceedingly irregular demands for power on small roads make the average load throughout the day decidedly less than the full output of the engine. The result of this state of things is that the friction, not a serious loss when working at full load, becomes an important cause of inefficiency. Take, for example, an engine rated at 100 horse-power, driving a dynamo of similar capacity: the loss by friction in the engine will be about 10 horse-power, irrespective of load; and used as such engines generally would be on roads of comparatively small size, the average load seldom would be more than one-half the figure above named. Consequently, 20 per cent. of the indicated horse-power would be wasted in worse than useless friction. If the average load should fall to 30 horse-power, a condition by no means infrequent on small roads, about 30 per cent. of the indicated horse-power is lost, and so on.

In two cases that have fallen under the writers' observation the efficiencies of the engines, owing to underloading, were reduced to 70 and 68 per cent., respectively; while even at the greatest output noted during the day's run the efficiencies were only 85 and 84 per cent. On another road, under very different conditions, the efficiency of the engine at mean load was 88 per cent.; at maximum load for the day, 92 per cent.

The differences between these three cases are exceedingly instructive. In the first-mentioned one the engine was a 125-horse-power Corliss, driving two 30,000-watt dynamos through the medium of a counter-shaft. The mean load was about 40 horse-power and the maximum load 80 horse-power —six cars being in regular service on the line.

In the second case two Armington & Sims engines were employed, belted directly to two 60,000-watt generators running in parallel. Seven cars were used on the line. The average indicated horse-power was 51.4 and the maximum indicated horse-power 108. In the third case the engine in question was a Ball machine of 150 rated horse-power, driving two 50,000-watt dynamos. Fourteen cars were in regular use. The mean indicated horse-power was 86.4 and the maximum horse-power 124. The first two cases are examples of how *not* to design a railroad power plant. The third is a specimen of fairly good engineering practice, although the dynamos might have been somewhat larger to good advantage.

The more nearly the average load on the engine approximates to its full rated capacity, the greater the efficiency that can be secured. A properly constructed steam engine can be worked for short periods considerably above its nominal horse-power without injury; and it is more desirable to do this than to allow its average load to become low.

For example, the engine of the third road mentioned was worked occasionally as high as 215 horse-power for a few moments, to supply a sudden and unusual demand for power on the line. It was quite capable of doing this without damage, and it was only by this means that the unusual efficiency mentioned could have been secured. To work an engine, then, most efficiently, so far as loss by friction is concerned, it is necessary to keep it loaded as nearly up to its rated capacity as possible. On small roads, where the maximum electrical output may be three or four times the average, this can only be done by occasionally working the engine up to its extreme capacity—in other words, by employing an engine with a wide range of cut-off, so that the steam may follow almost up to full stroke if necessary. In addition to losses by friction in the engine there are also losses in shafting, if it is employed.

It almost goes without saying that the best engine efficiency will be obtained by belting or coupling the dynamos directly to the engines; and for this purpose, unless the machines are very large, a high-speed engine is preferable, since it not only regulates better under extreme variations of load, but usually allows a wide range of cut-off.

And this brings us to another very important consideration in efficiency. We have already seen that underloading an engine produces a needlessly large loss in friction and reduces the efficiency of the system; but it also does worse than this—it lessens the efficiency of the engine directly and considerably. If the steam be permitted to expand to many times its original volume in the cylinder, there will ensue an amount of condensation that very materially increases the amount of steam necessary to produce a given effort. In other words, if the steam is cut off too early in the stroke it is employed at a great disadvantage, and the amount of coal that must be burned under the boilers to furnish the required number of indicated horse-power in the engine will be greatly increased. On the other hand, if the cut-off be too late, as would happen if an engine were regularly overloaded, the steam passes into the exhaust before it can be employed to the best advantage; and there is a corresponding increase in the amount of fuel required.

For any given pressure of steam there is a particular ratio of expansion that will use the steam in the most economical way, and consequently will require the minimum amount of coal per horse-power hour. A considerable amount of investigation has been spent in finding this point; and the best formula that has yet been produced is probably the following, due to Emery:

$$\frac{1}{r} = \frac{1}{1 + \frac{p}{22}}$$

Where r is the ratio of expansion, that is, the ratio of the volume of steam admitted to its final volume, and p is the absolute pressure in pounds per square inch, that is, the pressure of the steam plus the atmospheric pressure.

For the boiler pressures generally employed, 90 to 100 pounds to the square inch, the most economical point of cut-off is between one-fourth and one-fifth of the stroke. This is about the point generally taken by engine builders in designating the nominal horse-power of their engines, as has been previously explained. There is, then, a double reason for working an engine up to its full rated horse-power. First, it reduces the friction to as insignificant a factor as possible; and, second, it employs the steam to the best advantage. As a matter of fact, overloading is less injurious in the latter respect than underloading; so that we have an additional reason for not employing engines larger than are necessary to do the work required.

Compound or triple-expansion engines, in which the total expansion of steam is carried through several cylinders instead of being confined to one, do not suffer as severely from variations in the expansion ratio as simple engines; so that they are not only more economical *per se*, but are better adapted for work in which the variations of load are likely to be large. It is not our purpose to go again here into the details of the question of engine economy, but it may be stated that compound engines require for a given output about two-thirds as much coal as simple engines, and triple-expansion condensing engines only about two-fifths as much—the machine in each case being worked at its full normal load.

For maximum engine efficiency and economy, the facts would, then, indicate the use of the compound or triple-expansion condensing engine, loaded as nearly as possible to its full rated capacity under ordinary circumstances, and allowed now and then to work up to nearly its extreme power, in case of an emergency. If simple engines are used, the same requirements as to load follow. Engineers are often called upon to decide between the use of a high-speed engine with slide or piston valves and the low-speed engine with Corliss or equivalent gear. For most kinds of service the latter has the advantage in economy; in street-railway work, however, the conditions—except in quite large stations—are reversed; since, in the first place, the Corliss engine has a smaller range of cut-off than the high-speed types, and consequently cannot be regularly worked at so near its full power, if the variations in load are to be large. Secondly, when worked considerably below its capacity it suffers more severely from cylinder condensation, on account of the large exposed surfaces. In cases where the maximum load does not exceed twice the average load the Corliss valve gear will

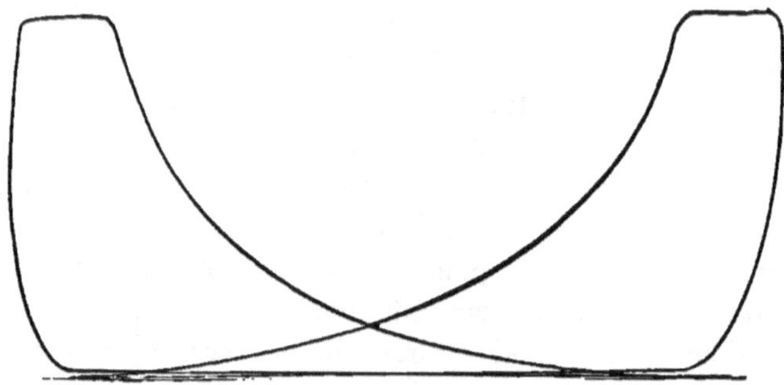

FIG. 93.—AVERAGE LOAD CARD. CORLISS ENGINE.

give excellent results.* We cannot emphasize this question of underloaded engines more forcibly than by showing the annexed indicator cards, taken from the first two engines referred to at the beginning of this chapter.

Figs. 93 and 94 are respectively the diagrams for mean and

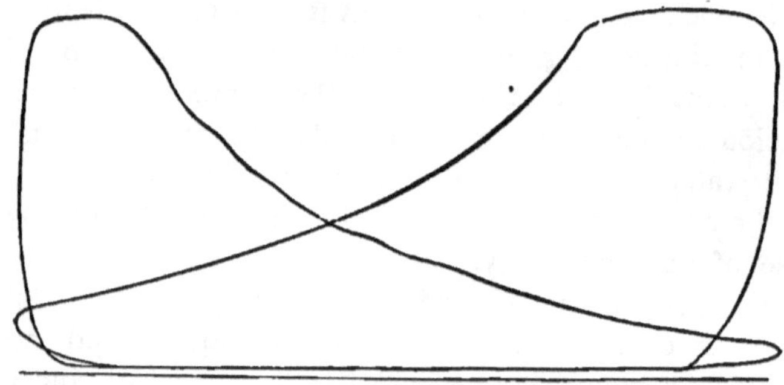

FIG. 94.—MAXIMUM LOAD CARD. CORLISS ENGINE.

maximum loads of the Corliss engine in the power station at Lafayette, Ind.† Figs. 95 and 96 are the same for one of the

---

* We speak of Corliss engines in this treatise as embodying the important principle of independent admission and exhaust valves. All that is said of them applies with equal force to some other engines possessing this same characteristic. Some of these have a "positive" valve gear (*i.e.*, with all the valves opened and closed by the engine itself) and can be run at a higher speed than the regular Corliss type.

† *The Electrical World*, June 22, 1889.

engines in the Syracuse, N. Y., power station, tested by Mr. Hulett.* A glance will show the advantages that would exist in a case where the mean load is at an economical cut-off instead of a somewhat inefficient one, as at Syracuse, or the even worse state of things at Lafayette.

Nevertheless, the card given by the Corliss engine is noticeably better in configuration than the other; and were

FIG. 95.—AVERAGE LOAD CARD. HIGH-SPEED ENGINE.

it not for the exceedingly short cut-off and consequent heavy cylinder condensation, the superiority of the valve gearing would have asserted itself in a diminished amount of fuel. As the case stood, however, the Syracuse engine gave the indicated horse-power hour on 5.3 pounds of coal, as against 7.3 in the other.

To sum up, then, the golden rule for efficiency in power stations is to use the engine at its most economical point of cut-off when delivering its average output; and this will be in nearly every case at the full nominal capacity of the en-

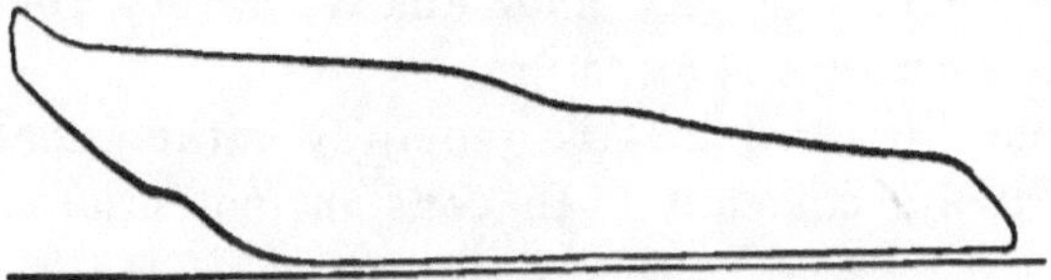

FIG. 96.—MAXIMUM LOAD CARD. HIGH-SPEED ENGINE.

gine. Employ compound or triple-expansion condensing engines wherever the cost of fuel renders it desirable. If simple engines are used, employ heavily built high-speed engines belted directly to the generators, unless the conditions are such that the variations in load are not much more than 50 per cent. of the average, when an engine of the Corliss or similar type equipped with a heavy fly-wheel will econo-

* *The Electrical World*, August 16, 1890.

mize fuel. With a thoroughly well-designed power station, the efficiency of the engine, so far as frictional losses are concerned, may be raised to between 85 and 95 per cent., with direct belting to the dynamos or good counter-shafting; while under less favorable circumstances it is likely to fall as low as 70 per cent.*

DYNAMO EFFICIENCY.

Much that has been said regarding the evil effects of underloading engines applies also to dynamos, but there are important differences between the two cases. There is nothing in the dynamo corresponding to the inefficient utilization of fuel that we find in an underloaded engine, and whereas the losses by friction in an engine remain nearly constant, quite irrespective of load, in the dynamo there are certain losses that decrease very rapidly as the load diminishes. Taking the dynamo as we find it, it may be worth while to glance for a moment at the various causes that tend to diminish its efficiency.

There is, of course, a certain amount of friction in the bearings, and besides this there are electrical and magnetic losses due to the reversals of magnetization in the armature core (hysteresis), losses by eddy currents in the metallic parts of the machine, energy spent in keeping up the magnetism of the field magnets, and, finally, a very considerable loss due to heating the armature.

The true frictional loss is generally rather small and remains virtually constant. In constant-potential machines,

* It may be well here to note the practical efficiency of the steam engine as a mechanism for transforming the heat energy of coal into mechanical power. As has been explained in Chapter II., the possible efficiency is limited by the practical range of temperature attainable, but *within this limitation* the best engines are tolerably efficient. With compound or triple-expansion condensing engines the actual efficiency may be from one-half to two-thirds of that theoretically possible between the given temperature limits. With most engines three or four tenths of the possible efficiency would be an estimate more nearly in accordance with fact. To take a concrete case, the 150 H. P. pumping engine at Pawtucket, R. I., cards from which are shown in Figs. 23 and 24, Chapter II., during a careful test gave the following results: Total heat received by the engine per hour, 2,427,198 units; indicated work, 143.49 H. P.—that is, 368,769 heat units; actual efficiency, 15 per cent. Upper limit of temperature, 356° F.; lower limit, 100° F.; $\frac{T_1 - T_2}{T_1} = .31$. Therefore the engine in question gave very nearly half the possible efficiency. The coal consumed was 1.54 pounds per indicated H. P. hour, the steam used 13.64 pounds.

losses from hysteresis and eddy currents and the loss in the field magnets are also practically constant; but the energy spent in heating the armature increases very rapidly as the current increases. It is, in fact, equal to the resistance of the armature multiplied by the square of the current. Thus, with an armature resistance of one-tenth of an ohm and a current of fifty ampères flowing, the loss in the armature will be 250 watts; while with 100 ampères it will rise to 1,000 watts. Consequently, when the load on the dynamo is diminished, its efficiency does not fall anywhere nearly so rapidly as the efficiency of the driving engine falls, because, while the losses in the latter remain constant, those in the armature are much diminished by the lessened heating of the armature.

With machines having a comparatively high armature resistance the total efficiency may fall off but slightly as the current diminishes. We frequently hear the electrical efficiency of a dynamo mentioned, but the phrase is apt to mislead; for in practice we do not deal with the electrical, but with the commercial, efficiency, and this includes not only losses in the armature and field coils, but also all the others to which we have alluded.

With the best forms of constant-potential dynamo now constructed, the commercial efficiency, that is, the ratio of the power supplied at the pulley to the electrical energy developed on the line, may rise slightly above 90 per cent., but more often falls two or three per cent. below that figure.

Thus a dynamo giving 60,000 watts (about 80 horse-power) at full load will require nearly 90 horse-power to be supplied at the pulley and nearly 100 indicated horse-power at the engine. As we have just seen, on lighter loads the commercial efficiency of the dynamo will diminish, though not very rapidly, until very light loads are reached; in other words, until the loss in the armature becomes quite small in comparison with the losses from friction, hysteresis, and eddy currents. Fig. 97 shows the efficiency curves of several types of machine at various loads. The horizontal scale shows the proportionate load on the machines, the vertical scale the efficiency obtained at those loads. The curve numbered I is for an English incandescent machine with magnets of the inverted horseshoe type, intended for an output of 45,000 watts. Curve II is from a Wenstrom incandescent dynamo of about the same size, and Curve III is from a Thomson-Houston railway generator of 60,000 watts capacity. It will be seen that from half load to full load the efficiency varies but little, while below half load it falls off quite rapidly. It is very apparent, then, that to secure the best efficiency in a

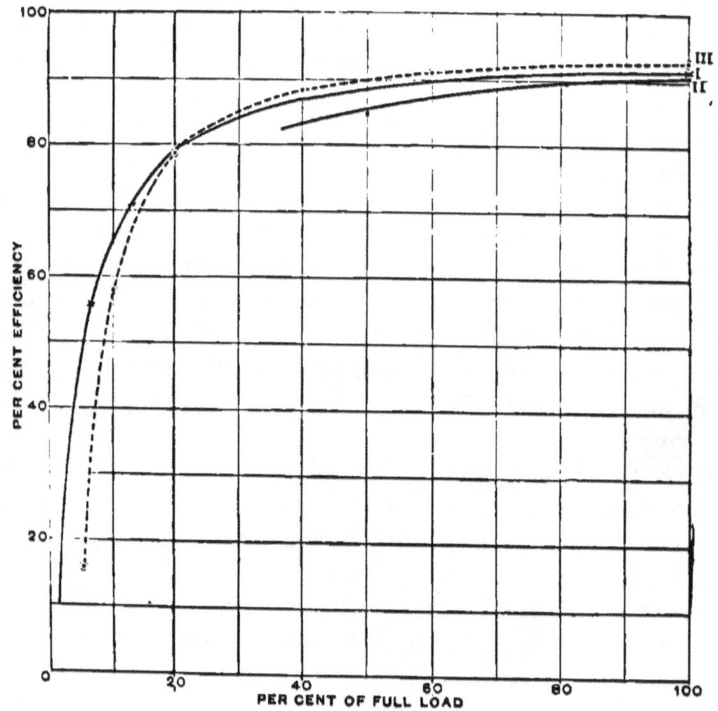

FIG. 97.—CURVES OF DYNAMO EFFICIENCY AT VARYING LOAD.

railway power station it is necessary to operate the dynamos at somewhere nearly their full output, in which case the capacity of the machine should nearly correspond to the maximum continued output required.

If the variations in load are likely to be great it may become necessary to work the dynamos above their intended capacity, for a few minutes at a time, as occasion demands; but while this overloading is quite unnecessary on large roads, it is not to be recommended even on small ones, unless in cases where the overload is not likely to continue for more than a few moments. A safe rule is to provide a dynamo of a rated capacity equal to about half the aggregate rated capacity of the motors. This will be ample, unless in the case of very small roads with heavy grades. The average applied power required is not far from 6 to 8 horse-power for each 16-foot car, supposing the latter to be equipped, as usual, with two 15-horse-power motors; it is unlikely to rise above 10 horse-power. This subject, however, has been already fully treated in the chapter on station design.

With a dynamo equipment thus designed the average output would be generally a little over one-half the full capacity of the machines; and their efficiency should consequently be high, something like 85 or 90 per cent. Under very favorable circumstances, when the number of cars in operation is so large that the maximum current required at any time upon the line does not greatly exceed the average current, this efficiency may be somewhat increased; though it cannot be expected to reach, or to rise above, 90 per cent., except in rare instances. Taking, now, these figures with regard to dynamo efficiency in connection with what we have previously stated about engines, we may be able to form a fairly accurate estimate of the probable station efficiency—that is, the ratio between the indicated horse-power at the engine and the electrical horse-power available upon the line.

There is certainly room for considerable improvement in the usual practice, particularly in the earlier stations. The power station of the Lafayette, Ind., street railway, tested by one of the authors, gave a station efficiency of but 40 per cent., owing to the fact that the engine was worked at less than a third its full output and the dynamos at less than a fifth. The power station at Syracuse, N. Y., tested by Mr. Hulett,* is an example of decidedly better practice. Its engines are belted directly to the generators, and were run on an average at about a third of their full output; the dynamos, however, of twice the capacity of those used at Lafayette, were under considerably more than twice the average load;

* *The Electrical World*, August 16, 1890.

and consequently, since at these small outputs the dynamo efficiency varies rapidly with the load, the Syracuse dynamos must have given altogether better commercial efficiency. In addition, by avoiding the use of a counter-shaft, the power wasted in friction was relatively less than in the case first mentioned; so that the total average efficiency of the station rose to 62.8 per cent.

Another road, instanced by Mr. G. W. Mansfield,* gave

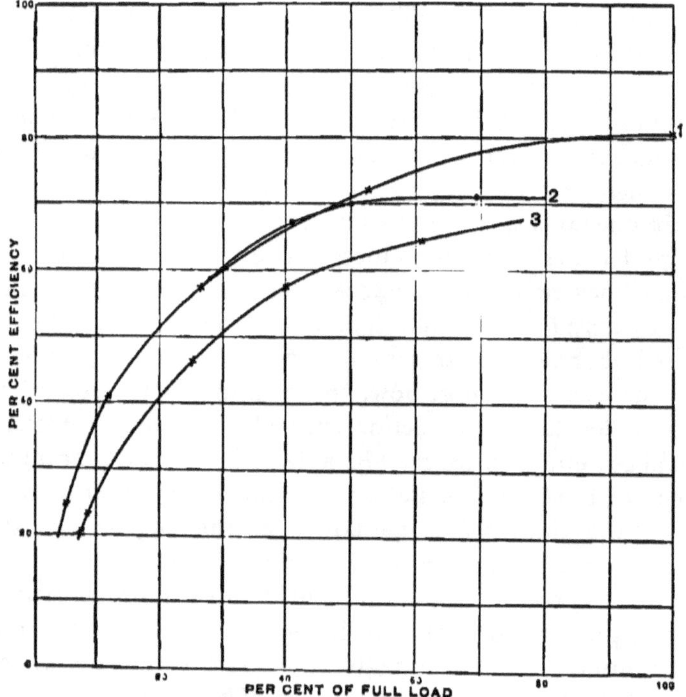

Fig. 98.—Curves of Plant Efficiency at Varying Load.

an average station efficiency of 54.6 per cent. The number of complete power-station tests published is comparatively small, and the number of accurate ones smaller still. The three above mentioned bear internal evidence of approximate accuracy, which is more than can be said of some others that have been reported.

With our present dynamos and engines, a station efficiency

* *The Electrical World*, August 17, 1889.

of 75 per cent. is about the best that can be hoped for on ordinary roads; and this can only be obtained by the greatest care in designing the station. This statement is amply illustrated by Fig. 98, which shows the commercial efficiency of the combined engine and dynamo plant in three cases.

The first curve, 1, represents the efficiency, at various loads, of a Goolden incandescent dynamo coupled direct to a Willans engine; the dynamo alone had a commercial efficiency at full load of 91.2 per cent., and the engine showed about the amount of friction usual in machines of its size. It will be seen that the highest commercial efficiency reached was 81 per cent., while at half load it fell to 72.2 per cent. Curve 2 shows the commercial efficiency of an Armington & Sims engine coupled directly to an Edison dynamo; the half-load efficiency is seen to be about 70 per cent., and the maximum load efficiency was 71.4 per cent. Curve 3 shows the efficiency of the Lafayette, Ind., power station—a Corliss engine driving three Edison dynamos from a counter-shaft, two of them railway dynamos of 30,000 watts capacity each, and the other a smaller power dynamo. The maximum commercial efficiency of the combination was 65 per cent., and the half-load efficiency less than 60 per cent.

This third case very plainly shows the effect of the counter-shaft, and the use of two railway dynamos instead of one of double the capacity.

The obvious moral is to belt or couple the dynamos directly to the engine, whenever possible. Cases often arise where it becomes desirable to drive more than two dynamos from a single large engine, but the advantage of convenience thus secured is obtained at the expense of a certain small amount of efficiency.*

Passing now from the consideration of the power station where the electrical energy is generated, let us take up the problem of its efficient transmission to the point where it is needed.

---

* Recent developments of the *steam turbine* may affect engineering practice considerably. A late test of a 100-kilowatt combined turbine and dynamo gave at full load the *electrical horse-power hour* for 27.6 pounds of steam, and at *half load* for 29 pounds This is a better result than has ever been obtained from any except compound engines, and renders possible the economical use of direct-coupled dynamo and engine plants of a much smaller size and lower first cost than have been generally contemplated.

The losses encountered in transmitting electrical energy over a system of conductors are two in number. First in importance is the loss due to the resistance of the line. This is equal, in watts, to the resistance multiplied by the square of the current, and can readily be determined and allowed for. Second, and of but very little importance under ordinary circumstances, is the loss due to leakage—usually, but not always, imperceptible. In overhead systems of distribution as now constructed the insulation resistance, by actual measurement, can be readily brought to over a megohm per mile of line. Therefore, even on very extensive systems, the insulation will be amply high to prevent any appreciable loss of power. It should be noted that doubling the length of a line halves the insulation resistance, so that a line of the insulation above mentioned and ten miles long would have a total insulation resistance of 100,000 ohms.

EFFICIENCY OF THE LINE.

The loss due to line resistance may evidently be made as small as may be desired by increasing the cross-section of the conductors. Of course it is not economical to do this beyond a certain point, for the cost of copper would then more than offset the cost of the small extra amount of power demanded by a line of higher resistance. At great distances from the power station the line is comparatively inefficient; near the power station there is almost no loss, and as lines are constructed to-day the average line efficiency is generally from 90 to 95 per cent. For example, on the Lafayette street railway previously mentioned the average loss on the line is about 7 per cent., on the Syracuse railway about 9 per cent., and in other cases even smaller. Where the system is a compact one with plenty of feeders, 5 per cent. will readily cover the losses in the line. If a wide district is reached from a single power station, twice this figure is a fair allowance. In certain cases even greater loss may prove economical in the long run; as by locating the power station at a point where coal is cheap and water for condensation is available, the operating expenses may be lessened in spite of the slightly increased output. The details of this matter are, however, not suitable for treatment here, as they have been discussed at length in the chapter on the line.

Taking into account, then, the losses that occur in the transmission of the electrical energy from the point of generation to the motors on the cars, and combining with it the station efficiency, we find that in the most favorable case a trifle over 70 per cent. of the indicated horse-power at the engine may be applied to the motors. In the roads now operated the proportion is much more likely to be between 55 and 60 per cent.

It is instructive thus to summarize the losses as each successive step in utilizing the power of the engines at the cars is passed, for it helps one to recognize the real difficulties that have been met in electric traction and the success that has been obtained in spite of them. On the whole, the result up to this stage of the operations is fairly satisfactory. The figures thus far deduced apply with equal force to any case of electric traction, except where storage batteries are employed—a method which will be treated by itself in a following chapter.

Where distribution is by means of underground conduits instead of aerial lines, the conditions are generally the same that we have been discussing. The losses by leakage are likely to be considerably greater, but this can in part be compensated by a slight increase in the cross-section of the conductors. The difficulty thus far with conduit systems has been not leakage properly so called, but a condition of things that has frequently approximated a short circuit. With improved conduit construction this may be remedied. Experience has shown thus far that the danger here is not so much low general insulation as complete breaking down of the insulation at one or many points. When it becomes possible to save conductors from actual short circuit, it will probably also be possible to avert any disastrous amount of leakage. Third-rail and similar systems of supply are often, though to a less extent, open to the same objections that apply to the conduit; but if the insulation can be kept up at all, it can probably be brought to a point where the losses in the line will not greatly exceed those now found in overhead systems.

The next subject for consideration is the efficiency of the electric motor as a converter of electrical into mechanical energy.

EFFICIENCY OF MOTORS.

So far as losses in the machine proper are concerned, the railway motor in most of its forms gives a very satisfactory efficiency. Other things being equal, a given dynamo electric machine is somewhat more efficient as a motor than when used as a generator, provided the conditions are nearly the same. One reason for this is that in the motor whatever magnetizing power the armature coils possess is added to the effective magnetization of the machine, while in the generator it is subtracted from it. For this reason motors are frequently given an armature relatively more powerful than would be adopted in a generator, particularly if it is desirable to construct a motor of high output compared with the weight. As an example of the effect of armature reaction in helping the efficiency of a motor, we may give the following cases derived from the experiments of Hopkinson. Two large dynamos precisely similar in construction and dimensions were used together, one as a generator driving the other as a motor. The efficiency of the first machine was 87.1 per cent. as a generator and 92 per cent. as a motor. The second machine had an efficiency of 84 per cent. as a generator and 89 per cent. as a motor.

Street-railway motors as now generally built are series-wound machines; but for the necessity of using a high electromotive force and reducing the weight of the machine to the smallest practicable figures, they could be made exceedingly efficient; as it is, there is little to complain of in this respect, although the losses from the resistance of the field coils and from hysteresis, in rather large armatures, are generally more considerable than they would be in dynamos of a similar output.

The principal sources of loss in our present street-railway motors are the regulating devices and the gearing. In starting a 500-volt motor under heavy load, it is necessary in some way to interpose resistance to prevent a flow of current so great as to endanger the armature. It would evidently be bad practice to have a normal field resistance high enough to accomplish this end, for after the motor has gathered headway this resistance becomes quite unnecessary and is distinctly harmful. So, in the motor systems most generally employed, special regulating coils are thrown into series with the motor at the moment of starting, and then more or less gradually cut out, so that by the time the armature has reached its full speed the resistance of the field coils is little more than that necessary to secure the requisite magnetizing force. In the old Sprague motors each limb of the field magnet was wound with three separate coils, the aggregate resistance of which was sufficient to serve the purpose mentioned. After starting, the coils at the first in series were thrown into various combinations, and were all finally parallel with each other; so that the final resistance was thus reduced to a reasonable amount.

In the old type 7½-horse-power motors the initial resistance, with the field coils all in series, was 20 ohms for each motor, the final resistance 3⅓ ohms. The later type, of 15 horse-power, was wound with much heavier wire, with greater precautions against heating; and the resistance varied from 8.6 at the start to about 1.6 ohms when all the coils were in parallel.

Later, since the Sprague system has passed into the control of the Edison Company, a starting coil of about 6 ohms resistance has been placed in series with the motors, but is immediately cut out after starting, and the combinations then proceed as before, although the relative resistances have been slightly changed. The Thomson-Houston motor system, pursuing substantially the same course, employs a rheostat, the resistance of which can be varied within quite wide limits, and also two coils on each limb of the machine, which are first used in series, but finally one of them is cut out, leaving a single coil of low resistance in circuit.

The Westinghouse device for regulation is like the Thomson-Houston, except that the rheostat is made in a few sections instead of a large number.

The uncertainty regarding the real efficiency of street-railway motors as they are now used in practice comes from the varying resistance of the regulating devices, and only a rough estimate can therefore be formed of the average losses that occur from heating. In most forms of motor the hysteresis losses are quite heavy, being, at the average speeds, not far from 200 watts for each motor. Some careful dynamometer tests made by one of the authors* upon Sprague motors of the later type give an average commercial efficiency for a single motor of 83.6 per cent.; with two motors the hysteresis losses are doubled, the power is generally not increased in the same ratio, and the final result for the average efficiency of the pair was 77.6 per cent.; this does not include, however, the losses in gearing. The figures for all the various systems are not easily attainable, but it is safe to say that 75 per cent. would be a fair average, taking into account the fact that some of the systems use one motor and some two, and that the electrical efficiencies and various sources of loss vary considerably, according to the type of motor used.

The figures thus given bring us at once face to face with the question of employing a single large motor in the place of the two generally used. That the efficiency would be raised by such a substitution is well-nigh self-evident, and is amply borne out by experience. The causes of this are, first, the advantage in efficiency of a single large machine over two small ones; the electrical efficiency may be made somewhat higher, owing to the advantage of substituting one magnetic circuit for two; and the losses from hysteresis and friction are perceptibly lessened.

As the practical aspect of this problem is inextricably involved with the efficiency of the gearing, the latter will be discussed before proceeding to further consideration of the question in hand. With regard to the possible average efficiency of street car motors, it is quite within bounds to say that with a properly designed single motor 80 per cent. is quite attainable; and even this efficiency may be somewhat raised, though it is hardly likely to reach 90 per cent., on account of the present necessity for regulating devices of some kind or other and the need for a comparatively light

* *The Electrical World*, May 31, 1890.

weight machine. In most street car systems, even to-day, the armature speed is somewhere nearly 1.000 revolutions per minute, at usual car speeds; and consequently there must be a reduction, by gearing, to one-tenth or one-twelfth the armature speed, between the armature and the car axle. This has been until recently generally accomplished by a double set of spur gears. The efficiency of such gearing is usually given in books on engineering at over 90 per cent. for each set of gears, and with reasonable precautions against dirt this figure may be reached readily enough.

Boxed-in gearing is obviously more advantageous in this respect than open gears exposed to the weather. In the test of Sprague cars just mentioned, the efficiency of the gearing belonging to a single motor was found to be a trifle over 85 per cent. Even the best spur gearing, protected from dirt, is not likely to give 90 per cent. for the double reduction required. The joint efficiency of the gearing of both motors, in the same series of experiments was found to be about 72 per cent.; showing both that the two sets of gears do not run in perfect harmony, and the effect of doubling the fixed losses.

It is very doubtful if a pair of motors can ever be made to run perfectly together; for even if the two machines were originally designed to be as nearly alike as practicable, there will probably be slight differences between their respective speeds, sufficient to throw them out of harmony with each other and increase both the losses in the motor and those in the gearing; and this is more emphatically true if, as generally happens, the motors are now and then taken apart for repairs and the magnetic circuit thereby impaired. Lack of general cleanliness about the motors, and penetration of oil into the joints of the magnetic circuit, have a distinctly deleterious effect, and of themselves are sufficient to produce injurious differences between the two motors. It is not probable that a pair of motors can ever be brought to work in harmony for any great length of time; and there is, therefore, a constant strain upon the gearing that makes itself evident experimentally, as has just been indicated.

During the last six months all the prominent manufacturers of electric street railway motors have brought out low-speed machines with but a single speed reduction between the armature and the axle. This result is reached either by unusual arrangements of the magnetic circuit, or, more often, by multipolar motors. The armatures of these machines have, generally speaking, a speed of between 350 and 450 revolutions per minute, at ordinary car speeds. The motors themselves are of about the same weight as the older double-reduction motors, have a gear efficiency much higher—reaching probably fully 90 per cent. in most cases—and the only uncertain factor is the electrical efficiency. The best information now obtainable indicates that it is hardly as high as in the higher speed machines, although experience in designing has taught makers to obtain a better average efficiency at varying loads. These single-reduction gear motors unquestionably *may* gain enough in the abolition of one set of gearing to more than compensate for the loss in electrical efficiency resulting from low armature speed. This gain in total commercial efficiency is probably trifling, seldom over 5 per cent., but the reduction of noise and of wear and tear on the gearing makes this type of motor a favorite.*

Methods of reducing the axle speed other than spur gearing are not now generally employed. In this connection should also be mentioned the use of connecting-rods driving directly both sets of wheels from a single very low-speed motor, a case which has already been discussed in Chapter III. Belts and sprocket chains are now and then used, but have been for the most part abandoned by reason of their uncertainty. Worm gearing is employed in a few cases—principally storage systems—and has probably about the same efficiency as the double-spur gearing which has already been discussed.

---

* Some recent careful tests indicate that there has been in some cases too great a sacrifice of electrical efficiency in obtaining low-armature speed. In a trial of three types of single reduction gear motors, the speed and power required for a round trip over a long and nearly level track being accurately determined in each case, the co-efficient of traction being assumed at 25 pounds per ton, an unusually high figure, the commercial efficiencies were as follows: Motor I., 37 per cent.; motor II., 58 per cent.; motor III., 75 per cent. Motor I. was undoubtedly underloaded, but the figures disclose no material gain, to say the least, over the older types of motor.

GENERAL EFFICIENCY WITH ONE AND WITH TWO MOTORS.

The experiments of Hale* indicate an even greater discrepancy between the efficiency of a single motor and of a pair than the experiments already mentioned. Fig. 99 shows the losses from the various parts of an electric street railway system as given by Hale. The diagram explains itself in part; it is only necessary, in addition, to state that the differ-

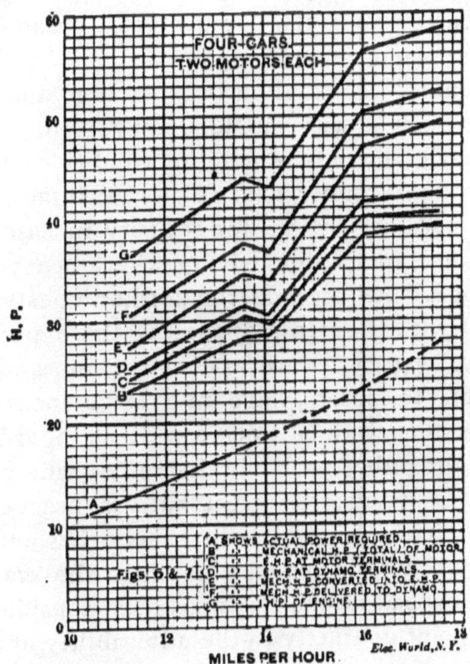

FIG. 99.—PROPERTIES OF MOTORS AND GEARING.

ence between the curves G and E represents the loss in engine and dynamo friction; that between E and D, the electrical loss in the dynamo; that between D and C, the loss in the line; that between C and B, the electrical loss in the motors; and, finally, that between B and A, the loss in friction of motors and gearing. The figures must be taken

* *The Electrical World*, May 17, 1890.

as merely approximate; but they are certainly instructive, and are sufficiently near the truth to emphasize the facts already mentioned.

From what has already been said, however, it is clear that quite different results might be expected according to the particular form of motors investigated. The total efficiency of the combined motor and gearing, as found by one of the authors, was for one motor 71.2, and for the pair 55.9 per cent. This is about a fair average. It is very certain, at all events, that a single motor with its gearing has a commercial efficiency more than 10 per cent. greater than in the case where two motors are employed.

This amount is certainly worth saving, and there is no reason why such a loss should be incurred. In the early days of electric traction, when the conditions of work were not so thoroughly understood as now, two motors were supposed to be necessary; first, for security in case of accident to one; second, for better driving where heavy grades were to be mounted. As regards the former question we are of opinion that with the latest apparatus these precautions are unnecessary, as greater care in construction, and precautions against injury to motors, have in a large measure obviated the dangers that were at first feared; as regards the second count, it is altogether probable that any grade which should be attempted by electric street cars is practicable with a single motor connected to both axles. It should be stated, however, that there is one case in which the employment of two motors instead of one possesses such considerable advantages as to point distinctly to the advisability of its introduction into practice. It frequently happens that for suburban service a decidedly high speed of the car is necessary—for example, between fifteen and twenty miles per hour. Such speed is undesirable and would not be permitted in the crowded part of the city through which the same cars may have to pass. These conditions are beautifully met by employing two motors, and operating them in parallel for the high-speed work and in series during the necessarily slower progress through the city. The average efficiency of the apparatus can be very perceptibly raised by this device, which is already in use at several points. The general features of this case have been discussed in Chapter III.

As regards gearing, it is doubtful if any device for transmitting the power from the armature to the axle with a reduction of speed can be made more efficient and more thoroughly reliable than a single set of spur gearing running in oil. Mention should here be made of the practice of placing the armature directly upon the axle, so as to avoid the use of gearing altogether. This obviates at once the present losses in gearing, and thus saves a very perceptible amount of power. Were the speeds employed on street railway lines considerably higher than they are at present, it would be a perfectly simple matter to put the armature upon the axle; and for railway work and railway speeds this is undoubtedly the best practice, and has been followed with success in the City and South London Railway inaugurated about a year since.

Three types of gearless motor for street car service have been brought out during the past year, and have been employed on a rather extensive experimental scale. They are described elsewhere, in Chapter III., and illustrate three radically different solutions of the low-speed problem.

The Westinghouse gearless motor has its armature core keyed directly to the axle, without any attempt at cushioning of any kind.

The Short gearless motor is provided with a cushioned support quite independent of the car axle, with which, however, the armature is concentric, but mounted on a hollow shaft with about an inch clearance all around the axle. The power is transmitted to the wheels by driving spiders heavily cushioned with rubber.

The Eickemeyer motor is flexibly supported midway of the truck, and drives both sets of wheels by outside connecting-rods.

No tests on any of these machines have been published, but the best data attainable indicate that the total commercial efficiency obtained is no greater than with ordinary double and single reduction gear machines—in other words, everything that is gained in abolishing the gearing is lost in the purely electrical efficiency. The reduction of the armature-speed to as low as from 100 to 150 revolutions per minute means very serious difficulties in getting up a counter electromotive force to insure good efficiency and torque enough to handle the cars readily upon grades.

In none of the forms of gearless motor mentioned is it entirely evident on examination how these two difficulties are met. All three run well in practice, and possess the common advantage of very smooth and noiseless operation. Eickemeyer employs one large motor instead of two small ones, probably gaining thereby electrically, and losing mechanically through the connecting-rods. It is doubtful if any of the forms mentioned can show as high an average efficiency as the best single-reduction gear motors that are now rapidly coming into use. On the contrary, it appears that the efficiency of the gearless machines is less under the present conditions than that of their competitors.*

They do, however, enjoy the advantage of very quiet running and freedom from repairs to any gearing and its connections, and are capable of doing very good service, provided they run under conditions that insure a rather high average speed, as for instance in suburban work. Where two gearless motors are employed on a single car, their operation in series is decidedly advisable, at ordinary car speeds; with this modification, they have even now a considerable field for usefulness. After the design of such low-speed machines becomes more familiar through experience, it is probable that their commercial efficiency can be raised nearly to the value now possessed by the single-reduction gear motors, but up to the time of writing the gearless motors have not come into sufficiently extensive use to show properly by experience their vices and virtues.

Serious difficulties to be met in this case are those involved in regulating devices for the purpose of starting and varying the speed. In the City and South London Railway a rheostat is employed, and this would also be fairly successful on street cars as, indeed, experiment has shown.

* A rather careful approximate test of one of the gearless motors indicated a commercial efficiency of about 40 per cent. at a car speed of between 8 and 9 miles per hour.

Another solution of the same difficulty may be found in the various plans that have been devised for connecting the armature spindle flexibly with the axle, so that the motor may run almost free at the moment of starting and the load be assumed gradually. Several varieties of clutch, epicyclic and hydraulic gearing, have been used for this purpose. Any of them would dispense with the steady use of the rheostat, and with the consequent loss of efficiency; but at the expense, perhaps, of serious mechanical difficulties. These cannot be told *a priori*, and the use of such mechanisms has been up to the present hardly more than experimental. Their advantages cannot be doubted; and were a thoroughly practical apparatus devised, it would probably find its way into very considerable use.

Summing up, then, the subject of motor and gearing efficiency, we find that from 80 to 90 per cent. electrical efficiency is attainable, though not generally attained in street railway motors; that a little over 90 per cent. is quite within reach in gearing efficiency, and that all losses due to gearing may be dispensed with by the use of a motor placed directly upon the axle. In the actual practice of to-day these efficiencies together are generally reduced to less than 75 per cent., and often to between 60 and 70 per cent.

The gains that may well be made are, first, a gain in motor efficiency, especially in the low-speed machines; second, a considerable gain from the use of a single motor; and third, another important gain in the use of only one set of gearing, or its complete abolition. If a single motor geared to both axles be employed, about 80 per cent. is the highest probable average commercial efficiency, but if high-speed work over a long line is to be attempted the maximum efficiency mentioned is within reach. The difficulties inherent in slow speeds on street railway work, however, are so great that only under extraordinary conditions can high efficiency be obtained. Better tracks, allowing the free use of a single motor, and better all-around construction in the driving machinery, are the means by which improvements are to be secured.

We are now in a position to sum up the total commercial

efficiency of electric traction, that is, the ratio between the indicated horse-power at the station and the power actually delivered at the car axle. The two complete tests referred to frequently in this chapter, at Lafayette, Ind., and Syracuse, N. Y., give respectively 25 and 37 per cent. as this final efficiency. We may, however, unite the several efficiencies of station, line, and car machinery to obtain an approximate figure. Sixty-five per cent. is probably a high average for station efficiency in the roads now built. The line efficiency may be taken at about 92 per cent., giving a total efficiency up to the motor of approximately 60 per cent.

With the motors and the gearing generally employed, the average commercial efficiency of the combination is probably not often in excess of 65 per cent., giving a total commercial efficiency for the system, from engine to car wheel, of 39 per cent.; this, of course, is but an estimate. But taking all the factors into consideration it is probable that the average of the roads now in operation would fall quite nearly to the point indicated. In very few cases would it fall below 30 per cent.; in still fewer would it rise much above 40 per cent.

With regard to the efficiency that may be reached by more careful station designing and better motors, we can give but an approximation. From the figures that precede we may take the practicable station efficiency at 75 per cent., the line at 95, the motors at possibly as high as 90, and the gearing, if used, as averaging about 90; giving a possible average efficiency of between 55 and 60 per cent. for the entire system. Anything over 50 per cent. can be attained only by the utmost care in design and construction, and it is very doubtful if this point is passed by any line now operated.

It will be thus seen that there is plenty of room for improvement in commercial efficiency, and that while to-day most electric railroads are decidedly inefficient, it is practicable to improve them to a very considerable extent. Fifty per cent. in total efficiency places the electric roads at least on a par with most cable roads, as regards the question of efficiency alone, with the additional great advantages of independent motor units, variable speed, and the ability to back.

Much improvement can readily be made, but it is in the mechanical rather than the electrical parts of the system; the chief losses, as we have already seen, are in the station and in the car machinery. The former may be obviated by careful designing of the station equipment, the latter by improvement in the mechanical means for transmitting the power from the motor armature to the car axle.

No stronger argument can be adduced in support of electricity as a motive power than the fact that, taking the electric railway as it is to-day, even though nearly two-thirds of the power generated at the station is often lost on the way, though the expenses for repair are sometimes heavy, and the first cost of equipment great, it still stands quite unapproached as a method of cheap and effective rapid transit. Every improvement that is made in the apparatus means an additional advantage for electric traction. It should be noted, too, that the very improvements in motors and gearing which will raise the general efficiency also tend to diminish the repair bills.

In closing this chapter it should properly be stated that although increase of general efficiency is highly to be desired and will result in a considerable saving, it must not be supposed that the power station expenses, with which efficiency has chiefly to do, are a very prominent factor in the total operating expenses of an electric road. On the contrary, the cost of fuel is in nearly every case less than 10 per cent. of the total expenses of the road, so that the whole amount that may be *directly* saved by improvements of various sorts tending to increase the efficiency probably does not exceed 5 per cent. of the total operating expenses. This is, however, enough to make the difference between losing money and paying a fair return on the investment.

CHAPTER VIII.

STORAGE-BATTERY TRACTION.

A CHAPTER on the applicaton of the storage battery to electrical traction must of sad necessity resemble that famous treatise written by a returned sailor on the manners and customs of the Fiji islanders: "Manners, none; customs, beastly."

When in 1880 the voltaic accumulator invented by Planté about twenty years previously was modified into a more practical form and brought to public notice, it was received with every symptom of unwonted joy; for its advent seemed to open a magnificent vista of electrical energy in portable form, ready always and anywhere to supply power for all sorts of industrial purposes. It was puffed by scientific authorities, lauded by the press, and vigorously exploited by the trade.

More than ten years have now passed, and the promised land overflowing with volts and ampères is still just beyond our reach; not that the accumulator has been a failure; on the contrary, it has been a partial success; for, under favorable conditions and for certain purposes, it has done, and will continue to do, admirable work.

To define the storage battery is not altogether easy; and the distinctions between secondary and storage cells—purely fanciful, and introduced into the art principally for the purpose of befogging the mind with legal quibbles—have been a fruitful source of misunderstanding. Speaking in a broad, general way, *we may define the storage battery to be a voltaic couple of such materials that the current-producing chemical reactions may be more or less completely reversed electrolytically.* It differs from the primary battery only in that the materials consumed in the latter, and of necessity replaced to continue the electrical action, may in the former be brought back nearly to their initial state by electrolysis.

As a matter of fact, certain primary batteries are almost identical in their chemical properties with certain storage batteries; although the usual form of accumulator is composed of materials that cannot be economically employed in forming a primary cell. The accumulator, therefore, is subject to many of the same general laws as the ordinary voltaic cell; it is coupled up in precisely the same way, and gives current and electromotive force not widely different from those of a primary cell of similar area of plates.

It has, however, the immense advantage that its component materials after discharge can be brought back to nearly their original state on passing a current through the cell; and it is this feature that has given the apparatus the name of storage battery. The most familiar form of accumulator is that in which the voltaic element is composed of two lead plates coated with oxides of lead and plunged into dilute sulphuric acid. Originally the necessary materials were formed on the plates by electrolytic action; Faure, however, introduced a so-called pasted cell, in which two plates of lead, perforated or roughened so as to permit of more readily retaining the active material, were coated artificially with oxides of lead—one plate with peroxide, the other with protoxide. These oxides mixed with dilute sulphuric acid are plastered upon the plates, and, on drying, the paste sets firmly; and the plates are then immersed in the electrolyte ready for charging, or, rather, the process of forming, which name is given to the preliminary charging and discharging that is necessary to reduce the active material to the proper chemical condition before it is put to real service. The chemical actions that take place in the accumulator are very complex and vary somewhat with circumstances. Speaking in a general way, during the discharge the active material is largely reduced to sulphate of lead; and during a charge this is decomposed, forming on the one plate peroxide of lead and on the other reduced lead. This charging and discharging can be repeated over and over; and experience has shown that of the electrical energy spent in charging the cell, between 80 and 90 per cent. can, under favorable conditions, be regained as electrical energy on the discharge. The electromotive force of lead accumulators such as have just been mentioned is in most cases when the cell is actively beginning its voltaic action about 2¼ volts, and during the action of the battery falls—slowly at first, very rapidly after a time. On charging, the potential difference rises rather rapidly until it

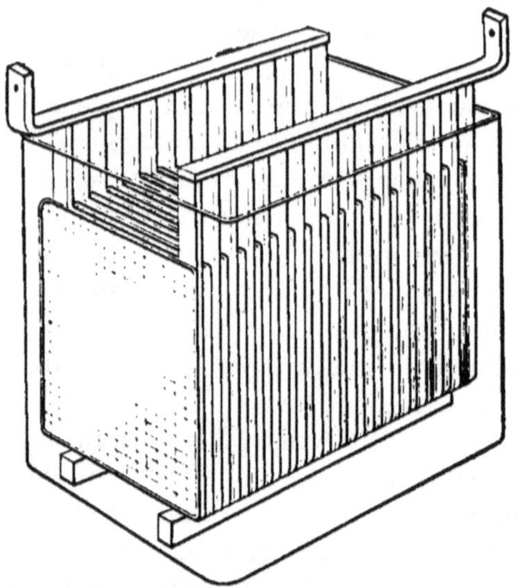

FIG. 100.—TYPICAL STORAGE CELL.

reaches about two volts, and then more slowly until it reaches 2.3 or 2.4 volts; the change in voltage being similar to that which appears during the discharge.

The ordinary storage battery shown in Fig. 100 consists of a number of lead plates formed as just described, connected in parallel into two sets, one of positive and the other of negative plates, and plunged in alternating order in a vessel of dilute acid. The plates that contain the active material, that is, oxides of lead, are usually reticulated as shown in Fig. 101, so that they are full of small square or polygonal cells, in which the paste that forms the active material may conveniently stick. When the cell is in working order, the process of charging converts the salts of lead on the positive plate into oxide of lead and reduces those on the negative plate, for the most part, to a somewhat spongy mass of metallic lead. On discharging, most of the material on both plates goes back into lead sulphate. If it were possible to prepare cheaply and in available form plates of peroxide of lead and spongy lead, we should be possessed of a primary battery in every respect similar to the secondary battery during the process of discharging; as a matter of fact, such plates would be inconvenient to prepare mechanically; whereas it is very handy to form them electrolytically, by passing a current through the cell after it is discharged.

This, in the rough, is the principle of the secondary battery most frequently in use to-day. A large number of other combinations of materials possess in a greater or less degree the same power of forming a voltaic cell the actions of which during the process of charging may be reversed electrolytically; and such combinations form accumulators of greater or

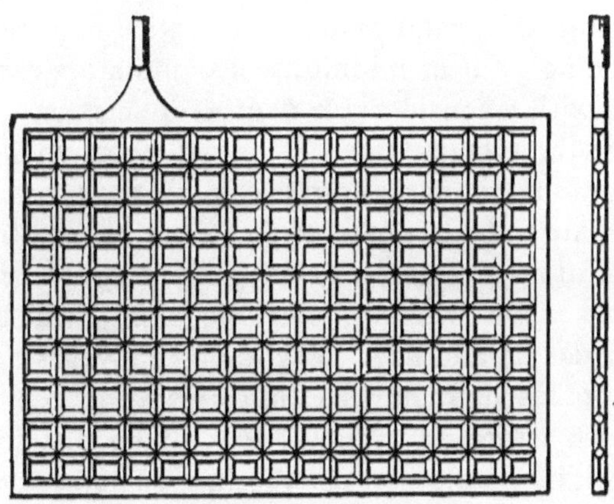

FIG. 101.—TYPICAL "GRID."

less usefulness, of which several will be mentioned further along in this chapter.

Having thus obtained some idea of the general character of the accumulator, we may appropriately pass to a special consideration of its application to practical apparatus, particulary in electrical traction. Experience has unfortunately shown that while these applications are in the abstract very simple and beautiful, in practice there are very serious inherent defects in the accumulators of to-day—so formidable, indeed, that the ingenuity of inventors has been sorely taxed to obtain an accumulator that shall be reasonably efficient, moderately reliable, and tolerably durable.

The accumulator gives an opportunity of utilizing a dynamo for storing up chemical energy at any convenient time or place, and permits it to be transformed back again into electrical energy whenever and wherever desired. The possibility is a most tempting one, and immediately the Faure cell was brought out in 1880 it was put to use for all sorts of industrial purposes. When, as frequently happens, it is desirable to obtain light at times when no power is available, the accumulator can be charged whenever an engine and dynamo chance to be running, and then employed to supply electric lamps as they are needed. In installations for the lighting of private houses, as well as in some central stations, it is very inconvenient to keep the dynamos in action all night; and for the small amount of service required during the late evening or after midnight, accumulators can be, and are, worked with a considerable degree of success. Into this field of work the accumulator was put at once, and has continued in gradually increasing use until the present. It is also to a certain extent employed for running small motors, electro-medical apparatus, and similar very light work, under circumstances where the cell must be charged at one place and discharged at another. Eight or ten years ago it was frequently proposed to have a central station occupied exclusively with charging accumulators to be distributed for furnishing light and power to regular customers. It was soon found, however, that the great weight of the batteries precluded this otherwise very beautiful possibility.

Hardly had electrical traction been proposed, than it was seen that with a durable and efficient accumulator most ideal conditions could be realized; it would be possible to charge the batteries at a point on the line where a power station could be very economically maintained, put them into the cars, and keep the road in steady operation, replacing the batteries with freshly charged ones at every trip, or whenever the original set should become nearly exhausted.

Experiments tending to this end were carried out as early as 1880, and in 1883 a car was put into service at Kew Bridge,

London; this car was equipped with a Siemens dynamo set to run as a motor, and under the seats were stored fifty large storage cells weighing in the aggregate about 4,000 pounds. The car could carry over forty persons, and ran over its level track very smoothly at the rate of about six miles an hour.

Encouraged by this success, a considerable number of experimenters went actively to work; and during the next two or three years cars were put into operation in London, Brussels, Paris, and elsewhere. The motors were as a rule of from 4 to 8 horse-power, and the weight of storage battery varied, according to the power supplied, from 2,000 to 5,000 pounds. The energy stored was capable of running the car from thirty-five to fifty miles without changing the batteries; at the end of this period fresh cells had to be substituted, and as these were of considerable weight the operation was not easy. In Brussels a number of accumulator cars were put into regular service about five years ago. In this country the work has been taken up by several different companies; and during the last three years experiments on a commercial scale have been carried on in New York under the auspices of the Julien Company, which for a considerable time operated ten or twelve cars on the Fourth Avenue road; in Philadelphia, in New Orleans, in Washington, and elsewhere.

At the present time the accumulator cars in New York, Philadelphia, New Orleans, and Dubuque have been abandoned, and not a single road is to-day commercially operated in the United States by storage batteries. The nearest approach to it is an honest experiment in Washington, D. C., the results of which are as yet uncertain.* There are besides a number of experimental cars of all sorts and descriptions not in regular service. It will thus be seen that in spite of the work that has been done, the results have not been altogether satisfactory. It is worth while looking in detail into the matter for the purpose of seeing what the trouble has

* We make this statement advisedly, since not even one road, so far as reported, is operated independently of the storage-battery companies. No safe commercial conclusions can be drawn from cars that are continually under the supervision of skilled electricians employed by the *installing* and not by the *operating* company.

been, and if possible forming some idea as to the outlook for better results in the future.

Much ingenuity has been expended in working out the details of storage-battery traction; almost every conceivable form has been given to the lead plates intended to hold the active material; several of these are shown in Figs. 102 and 103. The intent has been to secure a light supporting structure capable of holding the plugs of active material very firmly.* The cells used on cars are almost universally placed in trays under the seats, and either slid out endwise or taken out sidewise by lifting up doors at the sides of the car. One

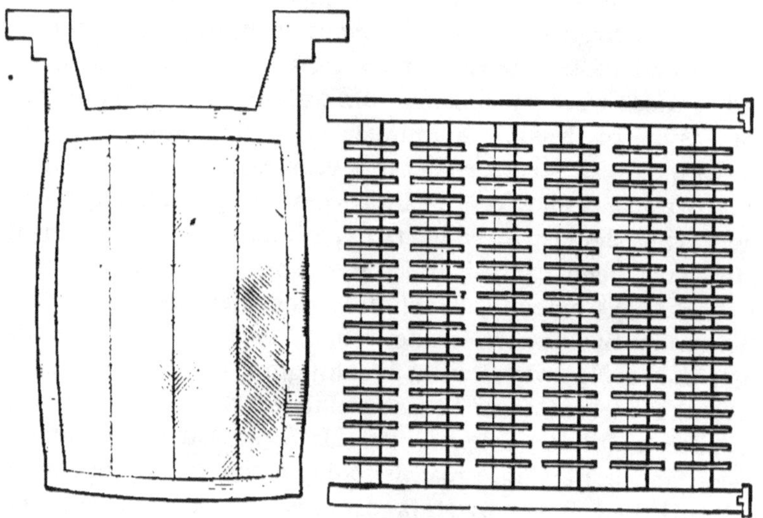

FIG. 102.—SPECIMEN GRIDS.

very neat and successful arrangement for accomplishing this has been in use in connection with the Julien cars on the Fourth Avenue road in New York, and is shown in Fig. 104.

It consists, as will be seen, of something very like a dumb waiter with several shelves placed on either side of a space just wide enough to permit the car to run in. The car is brought between the two series of shelves, the exhausted bat-

* It is unfortunate that lead, the only cheap metal capable of resisting the attacks of sulphuric acid, is so undesirable from a mechanical point of view, and so heavy. If the plugs of active material are in intimate contact with the grid they are almost certain by change in volume to warp the weak lead support, while it is difficult to "take up" expansion without encountering electrical difficulties.

teries shoved out on the empty shelves, and the dumb waiters lowered so that fresh trays of charged accumulators can be

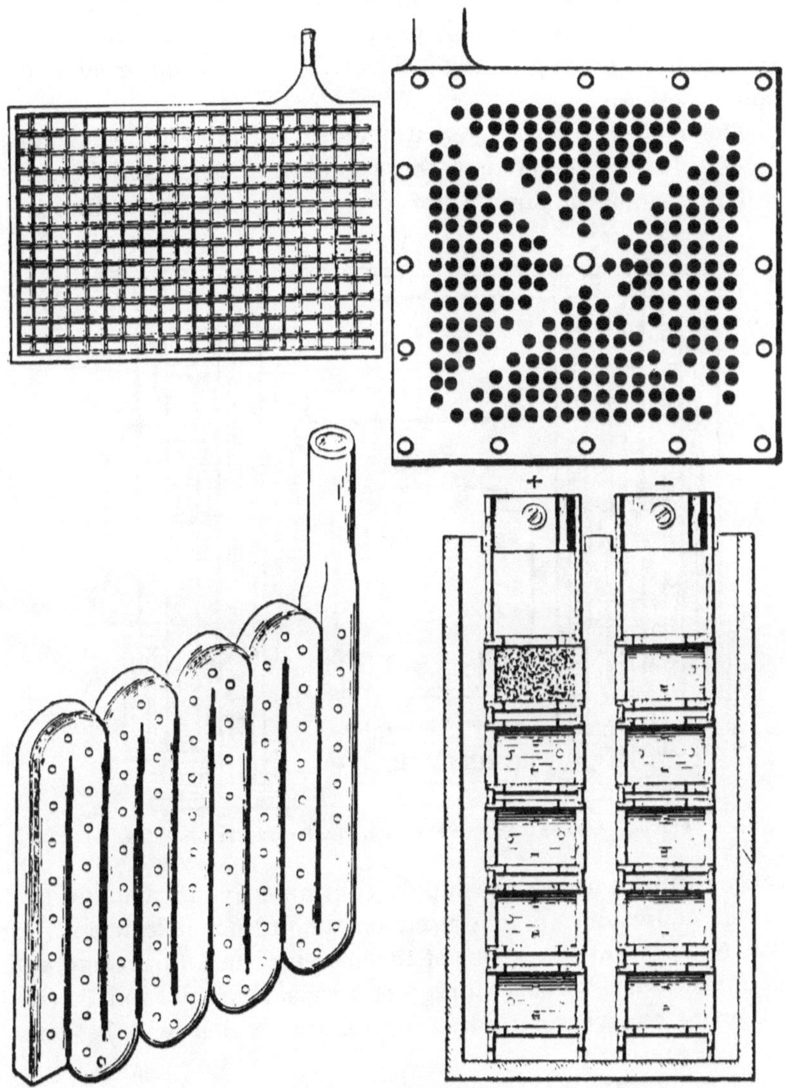

FIG. 103.—SPECIMEN GRIDS.

thrust into place. The operation can be very expeditiously performed by this means. Some very ingenious special regulating devices are employed, both in the station and on the cars; and in the latter case the power possessed in storage-battery traction, of using more or less cells on the motor, as required, enables the latter to be used in a quite economical way—rather more efficiently, doubtless, than the ordinary machines where regulation is effected by a rheostat or by field commutation.

The motors for storage-battery use are almost always decidedly less powerful than those generally employed in the direct systems of supply, for the very good reason that to

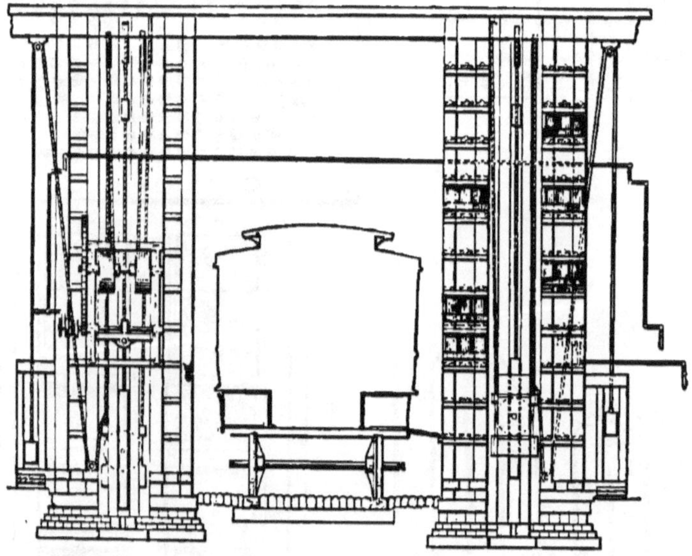

FIG. 104.—APPARATUS FOR REPLACING THE BATTERIES ON A CAR.

carry batteries enough to supply continuously an amount of power equivalent to that used on the ordinary electric railway is at present almost out of the question; and, furthermore, is unnecessary, because storage-battery cars are not well fitted for work on severe grades, for reasons which will be explained presently.

There is little that is unusual in the appearance of electric cars fitted with storage batteries; the only thing to be remarked is the absence of the trolley or any visible collecting device, and sometimes a slight widening of the lower body of the car for the sake of receiving the batteries. Altogether the experiments with storage-battery traction have met with only partial success; for reasons not for the most part dependent on any lack of care or skill on the part of those who have engineered the various attempts, but on account of defects inherent in the batteries—not confined, as a rule, to any particular type, but common to the entire class. As long as an accumulator car is favored with expert care and handled with great judgment and discretion it is likely to operate well, and, so far as successful running is concerned, to compare favorably with any other form of automobile car.

When these conditions are not fulfilled, the time comes when the storage batteries exhibit deterioration, they are not properly replaced, and failure is the uniform result. To investigate the difficulties that have interfered with the development of storage-battery traction we must begin with the root of all the evils that befall them, that is, with the accumulator itself.

Following the classification adopted by M. Reynier we may divide accumulators into four genera, first, the lead and sulphuric acid genus, including the original Planté cell—which, by the way, is not to be despised, even to-day—and the Faure cell with all its modifications, including the great bulk of accumulators that are in use. Second, the lead-copper genus, consisting practically of plates coated with lead oxide immersed together with copper negative electrodes in a solution of sulphate of copper; these are not used in anything more than an experimental way, and are simply interesting theoretically. Third, a very similar cell composed of lead positive plates, zinc, and sulphate of zinc, which possesses an electromotive force slightly higher than the ordinary lead cell and has a high capacity in proportion to its weight, but is open to the serious fault of losing part of its charge on open circuit, and shares many of the inherent defects of the ordinary Planté type. Fourth, the alkaline zincate genus, the analogue of the Lalande-Chaperon battery, better known in this country in its modified form of the so-called Edison-Lalande cell.

In this case the positive plate is of porous copper, the negative plate iron—often gauze—and the liquid sodium or potassium zincate. This type of cell has some admirable properties and has been applied to traction with tolerable results, although not yet in use long enough to justify anything like a final judgment as to its merits. The first genus contains nearly all the species in practical use, their differences being mainly constructional. The reactions are about the same in all, and the defects the same in kind, differing only in degree. If it were possible to make the chemical action in an accumulator perfectly reversible, we should be possessed of an invaluable adjunct in all electrical operations, but unfortunately this is not the case. The fundamental difficulty is one not confined to the accumulator, but one inherent in nearly all chemical reactions that are at all complicated. It is generally true that all except the simplest reactions are not quantitative in character; that is, more or less material is converted into a form that cannot be utilized, or by-products are produced in greater or less quantity.

The amount and character of the by-products is very largely regulated by the speed or the temperature at which the reaction is carried on; it often happens that a slow reaction and a rapid one, albeit of the same general character, lead to different results. Theoretically, the chemistry of the storage battery is that which has been already stated—a reduction during discharge of oxides of lead on the positive plate, with the formation of lead sulphate on both plates. During the charge, the lead sulphate is decomposed electrolytically, forming lead oxide on the positive plate and spongy lead on the negative plate; unfortunately this is not what is known to chemists as a "clean" reaction. On the discharge, free oxygen and hydrogen are formed in the solution, together with ozone and hydrogen peroxide, that attack the plates without materially assisting in the production of the current.

There is local action of a useless and injurious character between the support of the active material and the different parts of the active material itself. Other and more complicated substances—basic sulphates of lead, and the like—are also produced in the cell; besides these there is the inevitable heating owing to ohmic resistance, and losses due to overcharging; and among all these from 10 to 15, or even as high as 30 to 40, per cent. of the energy may be lost in the process of charging and discharging, the amount depending

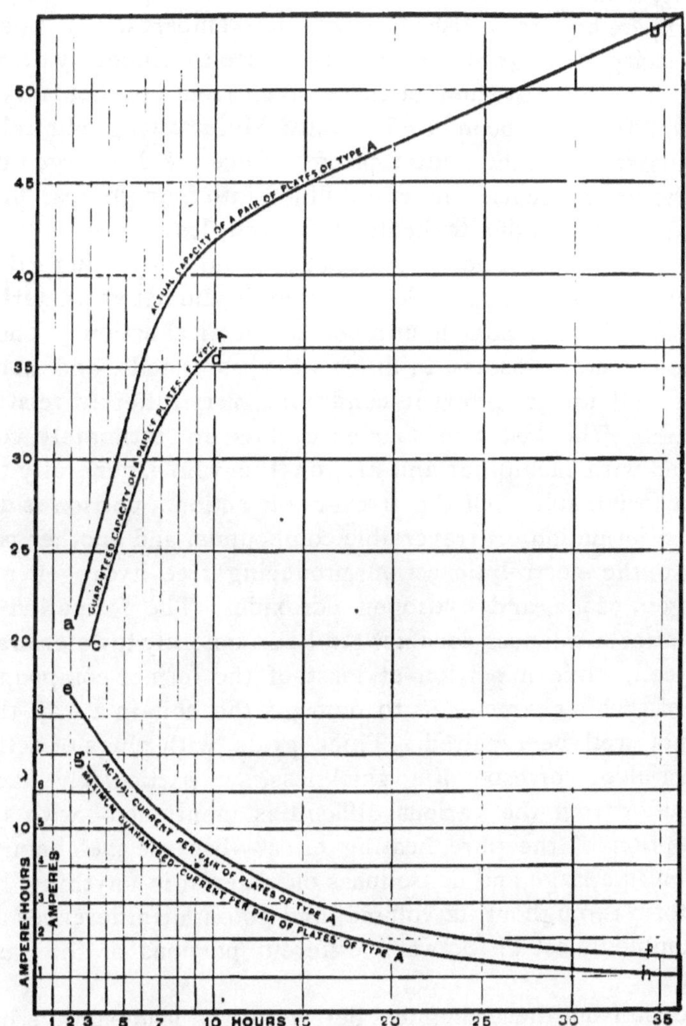

FIG. 105.—CURVES SHOWING VARIATION OF CAPACITY WITH DISCHARGE RATE.

largely on the rate at which the reactions go on—the faster the charge and discharge, the more by-products and the lower efficiency.

In fact, the capacity of the battery, that is, the amount of electrical energy it is capable of giving out, is a function of the rate of discharge, as well, of course, as of the rate of charge. In some forms of storage cell this relation is almost linear, so that if a given accumulator has a storage capacity of 60 ampère hours at a discharge rate of 10 ampères, it will have a capacity of but 30 when 20 ampères are continuously drawn from it. Fig. 105 shows a curve taken from a particular type of cell that has been used extensively, showing admirably the disastrous effect of attempting to force the discharge rate.

The losses incurred may be divided into four groups: first, the direct losses due to heating: second, losses due to local action between the active material and the supporting grids; third, the losses due to local action in the active material; and, fourth, losses due to unreversed chemical action. These various factors possess, in different types of cells, and in the same cell under different conditions, very different relative values. The last two sources of loss are intimately connected with each other and are, on the whole, generally the most formidable. Of the irreversible action a portion is due to the formation of irreversible compounds, and another portion to the electrolytic action producing free hydrogen and oxygen, ozone, and hydrogen peroxide. The formation of the latter substances does not tend permanently to deteriorate the cell, while a portion at least of the former compounds are of such a character as to damage the cell in a way that cannot well be repaired. Thick grids, with plugs of active material of corresponding thickness, are particularly likely to suffer from the various difficulties mentioned—with the exception of the pure heating effect—because the chemical action in a large and dense mass of material is anything but uniform throughout its volume, and potential differences are undoubtedly set up between different portions of the same plug.

Aside from these chemical defects of the lead storage battery there are others of a mechanical nature. Chief among these may be mentioned the well-known and very serious buckling, to which the positive plates in particular are very subject; the cause is the expansion and contraction of the plugs during charge and discharge. This is to a considerable extent unavoidable, although efforts have been made with considerable success to eliminate its ill effects by so forming the plug as to permit of expansion in its own plane, thereby partially, at least, avoiding any tendency toward lateral distortion. The final result of buckling is a short circuit and rapid destruction of the battery. Aside from this, some of the lead sulphate formed during discharge is very frequently lost, falling down into the bottom of the jar and thereby being put out of useful action. At very rapid discharge rates, too, there is a strong tendency toward disintegration of the plugs of active material. Sometimes this action is very violent, producing radiating cracks from the center of the plug, almost as if explosive force were at work from a point within.

Aside from all this, the lead accumulator is heavy; the total weight per horse-power hour of energy stored is in most types from 100 to 125 pounds. By employing thin plates of active material upon unusually light supporting grids considerably lighter cells have been formed, giving 1 horse-power hour of output on approximately 50 to 75 pounds total weight, but these light cells have by experience been found to be too fragile to stand the wear and tear of continued use, and have consequently been generally abandoned except for experimental purposes. From what has already been said it will be easily understood that a very high discharge rate means decreased capacity, and hence a greater necessary weight of accumulator per horse-power hour stored. The figures given are for the so-called normal discharge rates, running usually, in the sizes of cells ordinarily employed, from 10 to 30 ampères. Of course the limiting factor in the weight of battery necessary is the practicable discharge rate. So much for the general properties of the lead accumulator.

The second and third genera of accumulators mentioned previously have not come into commercial use, and there are no signs of their development. The alkaline zincate cell, however, has very recently come to the front, and has accomplished some excellent work. The starting point of this cell, as previously mentioned, was the Lalande-Chaperon primary battery, which was composed of an agglomerated plate of oxide of copper, a zinc plate, and a solution of caustic potash. The action in this cell is the gradual consumption of the zinc, and the reduction of the copper oxide to metallic copper through the agency of the oxygen set free. The oxide of copper serves, then, as the depolarizing agent. Lalande and Chaperon discovered the fact that this form of cell was reversible, although they did very little to elaborate it.

In the hands of MM. Commelin, Desmazures and Bailhache the Lalande-Chaperon battery was developed into a quite efficient accumulator. The positive electrodes of this battery are composed of porous copper plates formed by the compression of finely divided electrolytic copper upon a nucleus of copper gauze; the negative electrodes are made of amalgamated, tinned, iron wire gauze; the positive electrodes are surrounded by parchment-paper cells, the office of which is supposed to be to prevent any cupric oxide—which is slightly soluble in caustic alkalies, but does not dialyze readily—from becoming mixed with the potassium zincate; the resulting accumulator is converted by charging into what is virtually a Lalande-Chaperon primary cell. It possesses under discharge the low electromotive force of only about .8 of a volt per couple, but nevertheless has a high weight efficiency, 50 to 75 pounds being the total weight required for 1 horse-power hour of output.

These batteries were employed with success in some experiments on submarine torpedo boats carried out under the auspices of the French Government, noticeably in the exceedingly interesting trials of the torpedo boat *La Gymnote*, in the harbor of Toulon. The type, however, has not come into general use as yet.

The efficiency appears to be just about the same as that of lead accumulators; but practical trials like the one just mentioned showed that on forcing the output, there was, as there is in lead accumulators, considerable loss due to irreversible action, and a certain tendency to disintegration. The chemistry of the cell, however, has not been sufficiently studied to enable the exact character of these losses to be definitely ascertained. The weight is about the same as that of the lightest types of lead battery. The alkaline zincate cell, however, is of somewhat greater volume for equal output, though doubtless of greater strength and somewhat greater duration of life than lead batteries of the same weight efficiency. It has been desirable to go somewhat into detail in regard to this matter, inasmuch as this type of cell has, in the hands of three American inventors, Messrs. Waddell, Entz, and Phillips, undergone some modifications, and has been applied by them with fairly good results to electric traction. The form of electrodes they have employed is shown in Fig. 106. The positive plates are composed of a folded porous copper wire, having a solid nucleus and surrounded with a woven envelope to serve a purpose similar to the function of the parchment-paper cell in the French form of battery. The negative plate is of iron, and the weight efficiency and general efficiency of the battery seem to be about the same as those of its prototype; but how far the difficulties of unsatisfactory life of the plates, irreversible action at high discharge rates, running down on open circuit, and short circuits from a growth of "zinc tree," have been avoided, only time can show.

FIG. 106.—POSITIVE PLATE OF ALKALINE ACCUMULATOR.

Taking up, now, the application of the storage battery to the specific problem of electric traction, one is at once confronted by the very considerable weight of accumulators absolutely necessary in the present state of the art to secure continued service. The experience of a very large number of electric roads extending over several years has shown that the work required per car mile on ordinary 16 or 18 foot cars is a little below 1 horse-power hour; at the car perhaps three-fourths of a horse-power hour is generally sufficient; this for ordinary grades and the usual rates of speed of from eight to ten miles per hour. We can therefore form a very good idea of the amount of stored energy that must be carried for ordinary service on an accumulator road. Assuming the average car mileage per day at 100, which is very nearly the mean of the roads now running, one sees immediately that the weight of batteries carried for an all-day run would have to be in the vicinity of 7,500 pounds. Ordinarily it has been the practice to change the batteries two or three times a day, but this smaller battery power carried means, as has already been seen, on attempting to force the output for certain emergencies, a considerably diminished storage capacity, so that it has been found practically necessary to provide from 3,500 to 4,500 pounds of battery for the regular service of a car.

This dead weight is carried about continuously. The single car that has been experimentally operated with alkaline zincate batteries has considerably reduced this figure, carrying only about 2,500 pounds of battery, as might be anticipated from the greater weight efficiency; but it has not at the time of writing been in service long enough to permit a fair judgment of the results. As a matter of fact, the effect of forced output on batteries in such service is at times so great that even the weights of battery mentioned are not designed to furnish the power ordinarily used on overhead electric roads; in other words, most of the storage-battery cars that have been operated in this country and elsewhere have secured reduced weight of battery by skimping on power.

A very brief computation founded on a catalogue of storage batteries will show that to produce the 10 or 15 horse-power necessary on even quite moderate grades, either a weight of battery considerably in excess of that mentioned must be carried, or the discharge rate must reach a point very considerably above that for which the rated capacity of the cells is given. There is really no occasion for carrying battery power enough for long runs, for batteries can be changed just as horses are; the difficulty is, however, that a battery, even if charged every trip, must be capable of a certain maximum rate of output; and this at present entails the use of heavy batteries, however often they are changed. The weight of accumulators to be carried can only be materially decreased by securing a higher efficient discharge rate.

It is readily understood that, in electric traction, in starting a car, and on curves and grades, there is always violent demand for current, lasting anywhere from a few seconds to several minutes, and demanding from the battery, if one is used, an especially heavy output. Every such call for a large current tends to lessen the efficiency and life of the cell, so that while in the laboratory the storage battery can be worked to nearly 90 per cent. efficiency—a figure which can be approximated in electric-light service—for traction purposes this percentage must be much reduced. Naturally, those most immediately interested are very chary about giving details as to commercial efficiency; but from the best sources of information obtainable, the results of experiments on a considerable scale with storage-battery cars, one is justified in concluding that in every-day traction work the mean efficiency of the cell is probably between 60 and 70 per cent. —in some cases even lower. Seventy per cent. is certainly a favorable estimate for the battery; and this lowered efficiency means also a lowered weight efficiency, hence more frequent changes of battery or a greater amount carried.

If it were not for the great weight of battery the low efficiency could be tolerated, but the result is a considerable increase in the dead weight that must be carried about; and dead weight in an automobile street car of any pattern is an unmitigated objection. The two tons of battery placed in a car require a very much stronger structure than would otherwise be necessary; and although the motors employed in storage-battery work are usually lighter than those used in the systems of direct supply, for the reason that it is intended to supply less power, the total weight of the car equipment is brought up to a figure unpleasantly high. A standard 16-foot car, with its two 15-horse-power motors, weighs about five tons; while the storage-battery cars of the ordinary pattern such as have been used on the Fourth Avenue road in New York weighed about seven tons, of which 3,600 to 3,800 pounds was battery.

The carrying capacities of the two, however, are practically equal; so that, per unit of useful work, the necessary supply of power required in the direct and the storage systems of supply is in the ratio of five to seven. With a similar motor equipment, that is, the same capacity for doing work, the ratio would be about four to seven. That the two tons of extra weight carried about by storage-battery cars means a considerable disadvantage in the matter of power, is a question that admits of but one answer. With the ordinary co-efficient of traction of about twenty pounds per ton and a car speed of eight miles per hour, the additional power continuously needed to drag around the storage battery amounts to nearly 1 horse-power: this figure is for a level track: if grades are to be attempted in addition, the additional power required would be a little less than 1 horse-power for each per cent. of grade.

As regards the actual commercial efficiency of traction by storage batteries, it is not unfair to say that the systems of direct supply have at least the advantage consequent on less power needed for unit useful work. In the station, the storage-battery road has somewhat the advantage; because the dynamos are of the same efficiency in either case, while in charging batteries both dynamos and engines can be worked at an approximately regular load and near the point of maximum efficiency. Taking the friction of the engine at about 10 per cent. of its rated full load, which is an estimate in accordance with the facts, the efficiency from steam to electrical energy at the charging station may be reasonably taken as about 75 per cent.

This is something like 10 per cent. better than is likely to be obtained on any but the very best overhead systems. The efficiency of the motors will not vary widely in the two cases, although the regulation, by changing the grouping of the batteries—which is readily accomplished in the storage system—gives it a slight advantage. Another small gain may be secured by using the motors as dynamos to restore a certain amount of energy to the batteries during the operation of stopping the car. To offset this we have on the one hand loss in the line, on the other the inefficiency of the battery; the former is somewhere in the vicinity of 10 per cent.; the efficiency of the battery, as just mentioned, is hardly likely to rise above 70 per cent., and is more probably below this figure. So that, taken altogether, we find that the storage battery in station and motors gains perhaps 15 per cent. in efficiency, and loses between 20 and 25 per cent. in the comparison between losses in battery and losses in line. As a net result, the total commercial efficiency from indicated horse-power at the engine to brake horse-power at the wheel is probably nearly, or quite, 10 per cent. less than in the system of direct supply. A similar figure may be reached by considering in detail the efficiency of the series of transformations that occur in a storage-battery plant.

Taking the efficiency of the station at 75 per cent. and the commercial efficiency of the motors at the same high figure, and combining these with the efficiency of the batteries—estimated at 70 per cent.—the total commercial efficiency of the system appears to be about 40 per cent.; while the various tests made on the *very best* direct systems indicate a total commercial efficiency of a trifle under 50 per cent.

Aside from this, we still have the greater power required on account of carrying around the batteries, which is of the magnitude previously stated; this latter disadvantage could be for the most part obviated if it were possible to obtain a light storage battery. In the storage car using the alkaline zincate batteries the total weight of car and batteries is just about the same as that of the standard motor car of the ordinary sort, although the power supplied is not equivalent in the two cases.

To find the relative commercial availability of the storage system, various other factors in the expense must be considered.

Accumulator traction undoubtedly has the advantage of requiring a smaller expenditure, and hence smaller interest charges, at the central station. From the regular way in which the dynamos and engines can be operated the capacity of the charging station may be reduced to one-half, or at all events to two-thirds, that required for operating the same number of cars by direct supply. On the other hand, we have to consider the relative cost of the equipment and maintenance of the battery and of the overhead system necessary for distribution. In the first cost the ratio between them will depend on the number of cars per mile operated by the overhead wires, for the battery charges in the accumulator service vary almost directly with the number of cars. With direct supply the cost of the distributing system increases a little more rapidly than the mileage when the number of cars is uniform. In a long run with few cars the advantage would unquestionably be on the side of the storage battery; on a line with a large number of cars, quite in the other direction.

As regards cost of maintenance the subject is open to no debate whatever. If an overhead system is properly put up it depreciates but slightly; 5 per cent. per annum would be a large estimate, while 3 per cent. would cover it in many cases. The depreciation of the batteries is an indeterminate quantity, but is undoubtedly large. Figures on this point are hard to find, for there is a very wide discrepancy between the estimates of the makers on the depreciation of a given battery and the results found when it is put in service for traction purposes. There have been so few storage-battery roads operated, and of these so few have given to the public any definite report, that the question is a puzzling one.

When the storage system in Brussels was abandoned as more costly than traction by horses, after two years' experience, it was stated that the life of the positive plates was about two hundred days, that of the negative plates considerably longer. An extensive experience in this country with batteries used in lighting railway trains—a far less severe service than traction—showed that the positive plates were, as a rule, destroyed within a year. In Brussels the cost of maintenance of the batteries proved to be 2¾ cents per car mile.

A very recent report from Birmingham, England, where the storage battery has been given a pretty careful trial, showed the cost of maintenance for a total car equipment to be 4 cents per car mile. The result of our present experience with systems of direct supply seems to show that the maintenance of motor and car apparatus amounts to between 1 and 2 cents per car mile—under favorable conditions, near the former figure. Assuming this amount as a fair estimate for the Birmingham line, we reach about 3 cents per car mile as the cost of the maintenance of batteries in that case. One of the American storage-battery companies recently estimated the maintenance of the storage-battery equipment at $700 per year per car. This is probably low, as in Brussels the contracting company's estimate for a similar type of battery was 1⅛ cents per car mile instead of the 2¾ cents actually found. Reports from the late Dubuque, Ia., road gave the life of the batteries at much less than a year. All this would indicate a maintenance account of dangerously near 100 per cent. per annum—certainly over 50 per cent.—an amount far in excess of the depreciation in overhead lines and motors on the trolley system.

We are therefore driven to the conclusion that, considering everything, the storage-battery method, at present, is decidedly more expensive than the usual system of electric traction, and that, unless the batteries are substantially improved in duration of life, it is likely to remain so. Nevertheless, this conclusion does not prove that the storage battery cannot now and then be applied under favorable conditions. Two years ago the experience in Brussels showed that its cost was greater than that of traction by horses, while the recent report from Birmingham shows a small balance on the other side of the account.

There is a very strong prejudice against the use of overhead wires in large cities, and if electric traction is to be introduced under such circumstances the choice lies between conduits and storage batteries. The former will be taken up in a later chapter; the latter certainly has considerable advantages in its favor. Each car is an independent unit, and therefore no accident is likely to cripple the entire system, as might happen with overhead lines. The service, with proper care, can be made reasonably reliable and efficient; for heavy grade work it is at present out of the question, but over good and level track and under favorable conditions as regards traffic, storage-battery traction even to-day has a field for usefulness. So far as we may be permitted to learn from experience, its cost does not differ very widely from that of horse-car service; with everything favorable the advantage may sometimes lie on the side of the storage battery; under other conditions it has often been, and may still continue to be, in favor of horses. In every case, with our present batteries, its cost is greater than electric traction by the overhead system, doing the same amount of work.

In closing, it should be pointed out that while in any electric system a carefully laid and substantial track is necessary, where storage batteries are employed it is doubly so; for not only do the irregularities of rough track and the consequent jolting injure the accumulators, but the heavy cars pound the track in a way to cause serious deterioration. The storage batteries of to-day are unfitted for heavy grade work; their place is on a comparatively level track. The experiments that are now going on, noticeably at Washington, will undoubtedly teach a very useful lesson as to the present state of the subject.*

Until definite reports from them are received one is not justified in counting too much on the hopeful results of the preliminary experiments. An approximate statement from the Dubuque road gave 1.5 horse-power hour per car mile as the output required at the station. This figure affords a basis for a rough comparison with the trolley system. The average station output in the latter case is quite nearly 1 horse-power hour per car mile. Remembering that on account of the greater weight of an accumulator car about a third more power is needed to propel it, the relative efficiencies of the two systems are shown by the above estimates to be not far from the figure deduced elsewhere.

Finally, it is only fair to say that if the accumulator can be made durable or of considerably higher weight efficiency than is now usual, it will have a wide field of operations wherever overhead wires are objectionable. But, as remarked by one of the authors more than two years ago in discussing this question, the time is not yet, and it is by no means sure that any of the present types of accumulator will share in the ultimate success of the system.

* A report from the late Dubuque road, based on eight months' operation, gave the following results: Cost of power at station, including labor, coal, oil, waste, repairs, etc. 1.80 cents per car mile; batteries, renewal expenses of every kind, 3.51 cents per car mile; labor connected directly with the batteries, shifting, cleaning, etc., 1.78 cents per car mile; repairs to electrical machinery, 0.13 cent per car mile. Total motive power expenses, 7.22 cents per car mile. These figures agree quite closely with those already mentioned. A thorough investigation of the two experimental roads in Washington, D. C., by a competent engineer, has shown the motive power expenses to be not less than ten cents per car mile.

CHAPTER IX.

MISCELLANEOUS METHODS OF ELECTRIC TRACTION.

TAKING into consideration—as is purposed in the present chapter—only electrical traction as it exists to-day, we must recognize the fact that about 95 per cent. of it is accomplished on the overhead system of supply generally known as the "trolley" system, and that the parallel method of distribution is well-nigh universally employed. Of the remaining examples of electric roads a portion are operated by the third rail, or ordinary track, system of distribution; a portion by storage batteries; a smaller portion by conduits of various sorts; and the remainder is composed of miscellaneous telpherage systems in a somewhat undeveloped state. The so-called series method of distribution with constant currents has been tried at various times and places, and has been, with the exception of one or two small roads, abandoned. Perhaps the most elaborate trial granted this plan was that by the Short Electric Railway Company in Denver, Colorado, and to some extent elsewhere. The result of this experience has been a complete change to the parallel distribution system.

There are various good reasons for this. The system of switches and cross-connections necessary in operating several motors in series on an extended line has proved of so forbidding a character as to deter further pursuit of the method. Another reason, and one which ought to have been recognized from the very start, is that economy in conductors and efficiency in the motor system compels the systematic use of the highest voltage that can be employed with entire safety to life.

This is true of whatever system of distribution is employed; and if a given amount of energy must be supplied for a fixed line the series system fails in point of economy, if it be limited to a reasonable voltage. Our present electric-railway lines are operated at a safe pressure, but one that probably cannot be considerably increased without reaching the danger limit. Hence, wherever bare wires are to be employed, the series system, being limited to the same total voltage, must transmit the same current if an equal amount of power is to be employed, and is likely to lose both in efficiency of the motors and regulating apparatus, and in the complications before mentioned, without gaining anything in the line.

As regards the three-rail method of distribution referred to, it was employed with a fair degree of success in some of the earlier roads, but proved inconvenient, owing to the position of the bare conductor with reference to the earth. If the third rail be supported above the level of the ground on insulators it gets in the way of general traffic; if placed on a level with the tracks it still menaces, under ordinary circumstances, vehicles and foot passengers, and becomes the seat of insufferable leakage at any of the voltages now generally used; hence it is generally inapplicable. The same objections hold in case of using the ordinary rails as the conducting system.

An exception must be made, however, in favor of lines occupying a road-bed peculiarly their own, as is the case with underground and elevated structures, together with such surface roads as may in rare cases be entirely isolated from ordinary traffic.

Under these circumstances the third rail becomes a very available method of supply, and where the current to be distributed is very considerable, working conductors of the requisite size are perhaps most easily maintained in this way. A noteworthy example is to be found in the City and South London Railway—that pioneer of high-power and high-speed electric traction. One or two elevated and underground roads now under contemplation in this country will very probably be operated in a similar manner.

The miscellaneous systems of supply we may conveniently divide as follows: (*a*) open-conduit electric roads; (*b*) closed-conduit electric roads; (*c*) telpherage lines automatically controlled from one or more points, and either underground or on trestles or aërial lines.

(*a*) The slotted conduit, somewhat similar to that used on cable roads, has been over and over proposed and tried for sheltering the conductors—for the most part with very indifferent success. The principal exponents of the system have been Bentley & Knight in this country and Siemens & Halske in Germany. Two or three small conduit roads have in addition been operated in England, notably at Blackpool and Gravesend, the latter employing, strange to say, the series system of distribution.

The Bentley-Knight roads in this country have now been totally abandoned, after long and costly experimenting; and there is to-day no conduit road in operation in America, unless we except an experimental one in Chicago.

Abroad, Siemens & Halske have met with considerable success, particularly at Buda-Pesth; but how far this result is attributable to the system employed and how far to the less exacting conditions of the European climate, it is impossible at present to say; although the comparatively slight differences between foreign roads and those abandoned here would at least suggest that climate and accidents of location have much to do with the matter.

The fundamental difficulty with all slotted-conduit electric roads is the enormous difficulty of proper insulation. This arises from the very nature of the case, for the conductors are placed in a tube of limited diameter in free communication with the open air through the slot. Water, dirt and mud inevitably find their way in, and sooner or later the result has been either a positive short circuit at a single point or general leakage along the line in sufficient quantity to paralyze its operation.

A large number of more or less meritorious and ingenious forms of conduit have been proposed, varying in the material and arrangement of the various parts that compose the substructure, in the position of the conductors therein, and the methods of insulation employed. The features possessed by all in common are the slot and a pair of bare working conductors, placed in various positions with reference to the slot, but comparatively close to each other. Conduits with a single working conductor and rail return have been proposed, but inasmuch as any connection between the working conductor and the conduit produces a severe short circuit, they have not passed beyond the experimental stage. Perhaps the most elaborate trial given the conduit in this country was in the city of Boston, under the auspices of the Bentley-Knight Company. The construction adopted is shown in Fig. 107. The road was nearly five miles in length, and was in operation less than one year, during which time the continual and irrepressible series of electrical misfortunes showed that climatic conditions forbade commercial success.

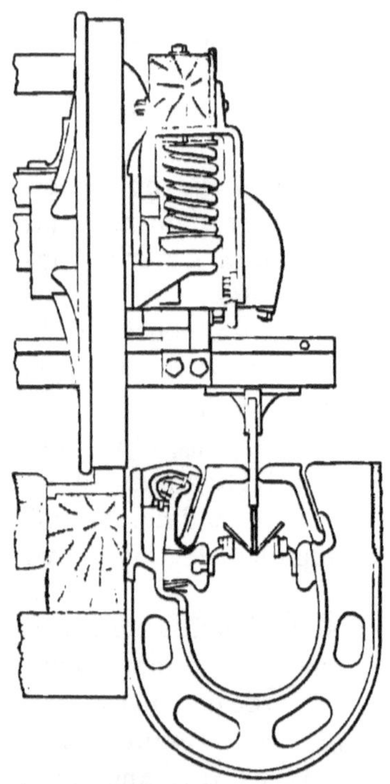

FIG. 107 — BENTLEY-KNIGHT CONDUIT

The Siemens & Halske road in Buda-Pesth, Austro-Hungary, is worthy of a little more extended description, as it is almost unique in its success.

The total length of the road is now 12.4 miles, and 50 cars are in operation. The first section—a trifle over a mile and a half long—was thrown open for use in July, 1889, and the extension has therefore been very rapid. The line was built at the expense of the firm of Siemens & Halske, and may be regarded as the practical demonstration of their system on a large scale.

The conduit is shown in Fig. 108; it is placed directly under one of the rails, and in the main is formed of concrete; cast-iron yokes with flanges 7 inches wide are placed about every 4 feet, supporting the insulators and solidifying the entire construction. The oval conduit is 11 inches wide at the widest point and 13 inches deep; while the total depth of the conduit foundations below the top of the rail is 27½ inches. The slot consists of a pair of beam rails, without any inside lower flange, secured to the conduit yokes by wrought-iron angle pieces. The slot itself in the finished construction is an inch and five-sixteenths wide, nearly twice as wide as would be tolerated in this country. In each yoke there is a socket on each side of the conduit at its widest point, and in these sockets are carried the insulators to which are fastened the working conductors. These latter are made of angle irons, and are supported by the Y-shaped projections from the insulators.

As will be seen by a glance at Fig. 108, these are quite high above the floor of the conduit—nearly a foot—and, furthermore, are sheltered by the upper portion of the oval so that they cannot be readily touched from the outside, and are protected from the access of rain. At convenient intervals there are settling boxes, connected with the city sewers; and whatever water enters the slot is collected at the lowest points, and is thus carried away. The distribution of power is at 300 volts, and the mains, instead of being carried on

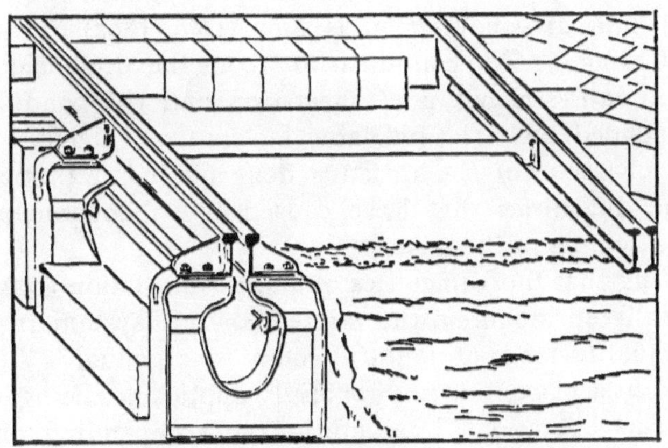

FIG. 108.—SIEMENS & HALSKE CONDUIT—BUDA-PESTH.

poles, are of lead-covered cables, protected with iron bands and laid in the earth along the line. At convenient points there are junction boxes, from which feeders follow the course of the road and are connected to the working conductors in the conduits.

The road has grades up to about 1.5 and 1.6 per cent., and a considerable number of curves—none of them, however, shorter than 84 feet radius. One curve of 148 feet radius is on the steepest grade.

The speeds permitted are from 9, or even 12, miles per hour in the less densely populated portions of the city to barely 4 miles an hour over street crossings. The daily car mileage per car during the 16 hours of service is from 75 to 90. The road has certainly proved a commercial success, and one of the engineers of the company is authority for the statement that the cost of running, exclusive of taxes, is only 37 per cent. of the income. This Buda-Pesth tramway is almost the only conduit road in existence to-day of which it can be said that its regular operation has been a success.

As will be seen from the description just given, the conduit itself possesses no very extraordinary features; it is simply well made, substantial, and carefully drained. This latter fact is probably the most important factor in the very good results obtained. The concrete conduit is of itself a fair insulator, and serves to afford additional security against grounds.

A half-mile of conduit road is now (Dec. 1892) in operation in Chicago. The conduit is of about the dimensions of that at Buda-Pesth, but is of sheet-iron and the conductors are suspended by rigid insulator blocks from above. The general character of the structure does not insure immunity from the difficulties that have caused the abandonment of similar attempts in the past.

Granting that the difficulties attending insulation could be successfully surmounted, the slotted-conduit system in any of the usual forms is of doubtful commercial value.

In the first place it is not generally applicable, being confined in its operations, by conditions not depending on its special excellence, to large cities; and, second, even in these, to locations where drainage is at least reasonably good. The former objection is true also of the cable roads, which have proved brilliantly successful commercially. But difficulty of drainage, while disastrous to electric roads, is only a slight practical inconvenience to the cable.

The most serious objection to the underground electric conduit of the open type is its cost, the necessities of drainage compelling the employment of a tube of considerable dimensions. The cost of the substructure is high, $20,000 a mile being a moderate estimate. Doubtless many inventors of such systems may claim an expense vastly less than this, but if they guarantee that limit, they or their backers must be supplied with very well-filled pocket-books.

This high initial cost puts the conduit road at a very serious disadvantage, compared with overhead roads, and perhaps even with storage-battery systems; for the latter, as has been already shown, can be installed at an initial cost little exceeding, in many places, that of overhead systems; and were the batteries to be improved in even a moderate degree, the interest charges and maintenance of the conduit and its conductors would be relatively so large as to give the storage battery the advantage.

In this connection, however, should be mentioned incidentally a type of slotted-conduit road that has recently been proposed, which obviates many of the difficulties of insulation, but unfortunately substitutes for them others, which will be discussed a little later on in connection with closed conduits. The type of road alluded to is one in which the working conductors are insulated, except for a series of projecting studs furnishing current to an arrow carried under the car, by the passage of which they are thrown into connection with the working conductor through a series of switches. The type is an interesting one, although at present it is purely experimental, and on a very small scale at that.

(*b*) Of closed-conduit systems there are many, elaborated by Pollak, Lineff, Gordon and others, abroad, and Wheless and others in America. The violent opposition to the erection of overhead wires in foreign countries has compelled greater activity in this direction there than in America. These systems have very much in common, the fundamental idea of all of them being to put conductors permanently underground, and to connect them automatically—by the passage of the car—to short sections of working conductors on, or immediately under, the surface and underneath the car itself. By this means the amount of bare conductor steadily subject to leakage or short circuits is very much reduced. The switching-in of the several sections is accomplished generally by electro-magnetic means, through current supplied from the car.

These systems, in first cost, are intermediate between overhead lines and the regular slotted-conduit type. Like the latter, they may be very seriously interfered with by climatic conditions, and are therefore somewhat limited in their application; there is also an additional difficulty, which will be appreciated by every working electrician, due to the multiplicity of switches intended to be automatic in their action.

The real seriousness of this difficulty cannot be definitely told until some such road has been in operation for a considerable time. It will be readily understood, however, that it is by no means negligible, and would probably be the source of a considerable amount of trouble.

In addition to these closed conduits or modified track methods of supply there exists a class of conduits either

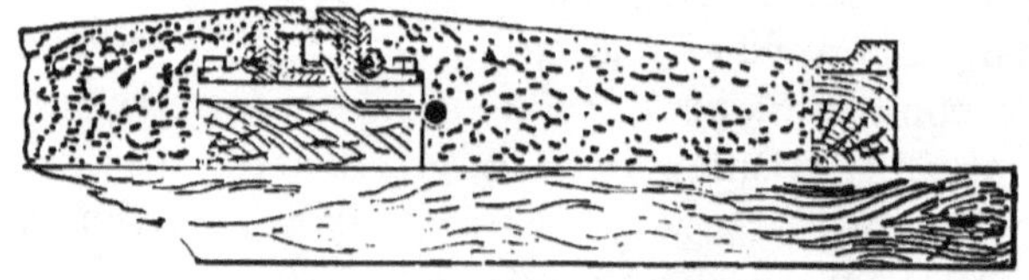

FIG. 109.—VAN DEPOELE CONDUIT WITH FLEXIBLE LIPS.

closed, but with flexible walls, or partially closed, having only flexible lips to shut the slot, except when the arrow carried by the car opens it.

A good example of the latter construction, due to Mr. C. J. Van Depoele, is illustrated here (Fig. 109). The former arrangement is one wherein the leads are inclosed in a casing having flexible walls. The upper portion of this casing is taken up by the bare working conductor, and as the car passes along, carrying a trolley wheel on either side, the pressure exerted on the working conductor is sufficient to force it into contact with the main.

These variations on the conduit theme would, if mechanical difficulties could be overcome, be very valuable adjuncts to the methods of distributing current now available. They are, however, of somewhat dubious practicability; for the conditions to be fulfilled by a flexible insulating wall or insulating lips are so severe as to be almost forbidding. Of course, any of these arrangements can employ either a continuous conductor or one divided into sections automatically switched into circuit, as in the Lineff and other systems previously mentioned.

The whole group of devices that have just been described are interesting both mechanically and electrically, but have none of them passed the experimental stage; and while it is perhaps not too much to expect that something practical and valuable will be evolved from the amount of ingenuity that has been spent upon them, the time is not yet. This line of investigation, however, is of more than usual promise.

The number of such modified conduits is too great for anything like elaborate description. Figs. 109–113, however, give a fair idea of the Van Depoele, Wheless, Harding, Lineff, and Pollak-Binswanger plans, which are fairly typical of the directions in which a successful solution of the great distribution difficulty has been sought. We can only suspend judgment on their merits until they have been proved by experience; for while the difficulties are evident, the practicability of the means taken to avoid them can only be told by protracted trials.

The use of alternating currents has now and then been proposed for electrical traction, and it is worth while saying something, at least, about their employment for such purposes.

The alternating current has in general two very marked advantages over the continuous current for all sorts of electrical distribution and service. The first and most important is the great ease of transformation from higher to lower voltage and *vice versa*. This is an immense convenience, and can be made to effect a very great saving in the cost of distribution; since it is possible to distribute the electrical energy in the form of current of enormous voltage and then to employ it in the various translating devices at as low a voltage as may be desirable for the particular purpose in hand.

Inasmuch as it is difficult to insulate dynamos and motors for very high voltages, while transformers can be easily insulated, it becomes possible by their use to accomplish the

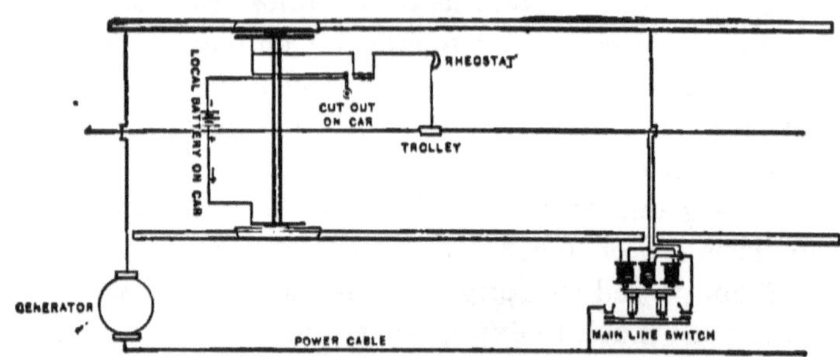

FIG. 110.—WHELESS CLOSED-CONDUIT SYSTEM.

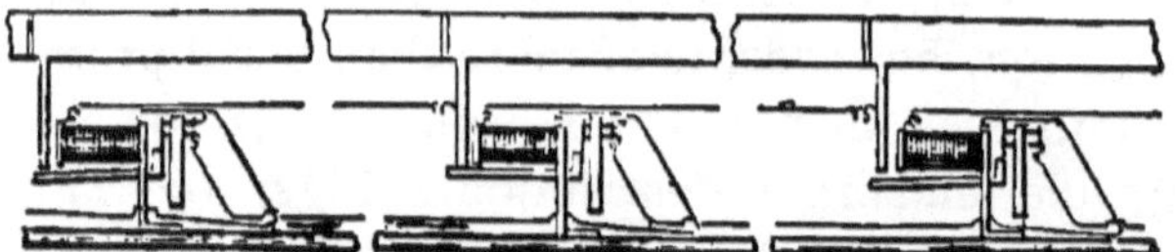

FIG. 111.—HARDING CLOSED-CONDUIT SYSTEM.

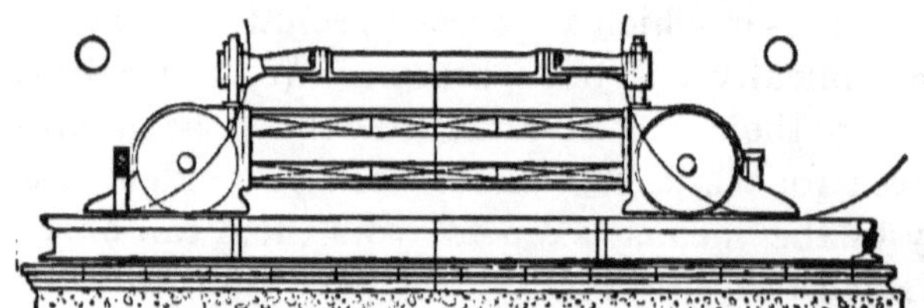

FIG. 112.—LINEFF CLOSED-CONDUIT SYSTEM.

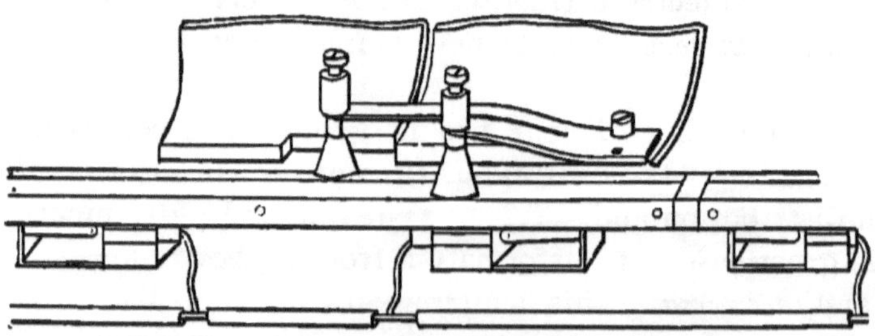

FIG. 113.—POLLAK-BINSWANGER CLOSED-CONDUIT SYSTEM.

distribution of the current at any desired voltage, and to utilize it at a voltage practicable for the ordinary translating devices.

The second great advantage of the alternating current depends on its regulation in connection with inductive resistances wasting only a trifling amount of energy.

If there were at present a thoroughly efficient and durable alternating-current motor which could be run on the ordinary two-wire system of distribution and would start readily under heavy load, it would be possible to employ alternating currents for electric-railway work with the greatest advantage, at least upon long lines; for a very small amount of copper would be required for the distributing conductors, which would feed the working conductor at intervals by means of transformers. More than this, the motors on the cars might be conveniently and economically regulated by a coil of variable self-induction, thereby doing away with one of the difficulties which is at present considerable.

It must not be forgotten, however, that the cost of transformers is not an inconsiderable item; so that we should not be able to reap the full advantage of the alternating current, except in systems of distribution where the cost of the mains aside from the working conductor is very considerable; the longer the line, the greater the advantage thus to be obtained.

Just at present, however, from the lack of a suitable motor, little headway is being made toward alternating-current railway systems. The "drehstrom" motor has been suggested for this purpose; but the complications introduced by the multiple leads required are considerable, probably so great, when applied to general traction purposes, as to overbalance the other good features of the machine.

It is not impossible, however, that some form of alternating-current motor may soon appear that will lend itself readily to traction work; and in this case it will be likely to have a considerable field in long-distance electric railroading.

Various other suggestions for the use of alternating currents have been made, some of them even going so far as to propose a primary inducing circuit in a conduit and a secondary coil carried on the car. The great difficulty of insulation and the lack of an efficient inductive relation between the two coils under such circumstances is enough to dismiss the scheme without further comment.

It will thus be seen that there exist at present, aside from the regularly employed systems of electrical traction, a considerable number of arrangements for the distribution of current, which have not passed the experimental stage and which are possessed of more or less merits and demerits, but no one of them is yet sufficiently important to entitle it to anything more than casual mention here. Some day one or more of these may assume a position of commercial importance, but until that time comes we can do nothing more than bring in the old Scottish verdict of "not proven" against their claims of great practical usefulness.

Passing, then, from these we reach (*c*) a group of devices for electrical traction that involve comparatively few new electrical principles, but are noticeable on account of certain useful properties of their own. We refer to the various systems of automatically controlled electric roads intended in the main for freight and express service, and designed to simplify the conditions of traffic by permitting automatic control of an electrically driven train.

There are two classes of such devices that may well be separated from each other. First, we may consider the very ingenious systems of telpherage intended for the carrying of freight at a low cost and in a very simple manner. Later we shall have to give some brief mention to the plans for high-speed service of a similar kind, intended to supplant the pneumatic tube and similar apparatus rather than merely to serve as a substitute for carting.

Telpherage owes its origin to the late Prof. Fleeming Jénkin, who both devised the system and coined its name. It consists, in brief, of a wire-rope tramway carrying suspended cars or trains of cars driven by electric motors, to which current is supplied from the supporting rods or cables. Similar apparatus, driven by wire rope, is now frequently used—principally for the purpose of transporting earth in making excavations—but Professor Jenkin conceived the idea of extending such a service, and enabling goods and material to be transferred across country cheaply and rapidly.

The appearance of a telpher line resembles that of the wire-rope devices just mentioned, consisting of a double line of cable or rods suspended from stout posts and carrying trains of small cars. The general appearance of such a line is shown in Fig. 114, which represents the system at Glynde, England, which has been in operation for several years, carrying crude cement from the pits out of which it is taken to a railway siding, whence it can be carried to the cement works.

The convenience of such an arrangement is obvious, although its application may be somewhat limited. Its greatest merit is the cheapness of the construction and the ease with

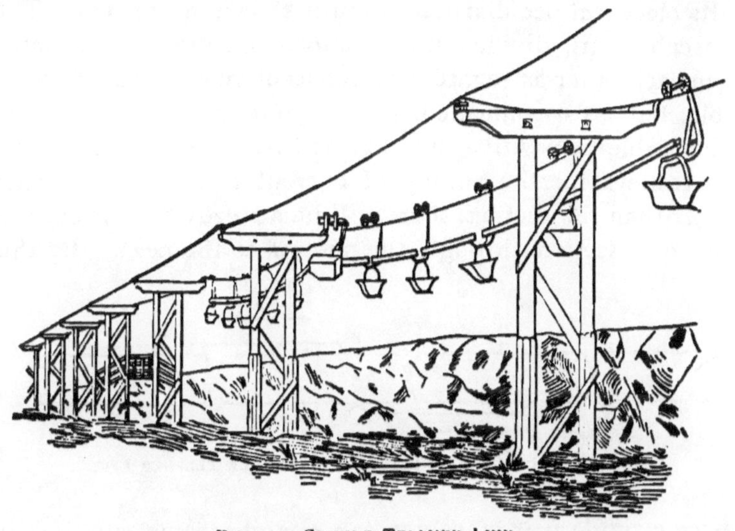

FIG. 114.—GLYNDE TELPHER LINE

which it can be operated where the laying down of track would be inconvenient.

The special feature in telpherage systems of this class is the method of supplying current to the car. This is, of course, done through the supporting rods, but to get contact with both positive and negative conductors requires no little ingenuity.

Of course, with a pair of cables supporting each car, or with a cable and a trolley wire, the problem would be very straightforward; but in most cases—for cheapness and mechanical simplicity—it has been thought advisable to use only a single supporting cable and no subsidiary conductor; hence a little electrical complication is necessary to enable two cables to act both as positive and negative conductors to trains running in both directions, one on each cable. Two systems are employed for this purpose—the one, series distribution at nearly constant current, such as has been proposed for ordinary electric tram-car lines; the other, a very ingenious cross-over parallel system, such as is used in the Glynde telpher line.

Its electrical peculiarities are fully shown in Fig. 115. The two cables are divided into sections, electrically insulated from each other as regards the consecutive sections of either cable, but cross-connected, so that if any given section of either cable be positive the next will be negative. By using two trolleys for the supply of current to a train somewhat longer than the section, there will be a steady flow of current from one section through the motors to the next. By this

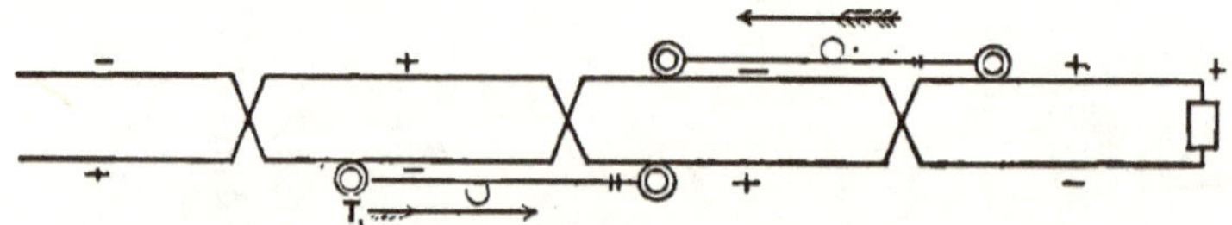

FIG. 115.—ELECTRICAL CONNECTIONS OF GLYNDE TELPHER LINE.

device a single pair of cables can operate trains running in both directions. A suspended electric locomotive drags several cars, and is very readily controlled as it approaches the ends of the lines. In the Glynde line ten small trough-shaped cars are used in each train, and the speed reached is four or five miles an hour.

The telpher system has never come into any use in this country, but there are at present three lines operated in England, the longest being a mile and a half. There are, of course, a considerable number of ingenious minor devices for the securing of constant speed and the proper control of trains, and besides, an automatic block system is provided, so that there will be no danger of rear-end collisions.

Satisfactory work can be done in this line of automatic transportation of goods, and it would not be surprising to see it come into occasional very efficient use in localities where ordinary tramways are, from considerations of cost or difficult country, impracticable. The supporting structure can be made comparatively light; the cables or rods need be less than an inch in diameter, and a large amount of goods can be carried at a comparatively small cost. Such lines are perhaps at their best in serving as feeders to railway lines, connecting them with manufactories, mines, quarries, and the like.

In America, however, the system has not yet been introduced practically, although small experimental roads have been operated in years past by Mr. Daft and Mr. Van Depoele. Nothing commercial has come from these experiments, and just at present there is no sign of further development.

The motors and electrical arrangements generally of telpher lines present no specially striking features, the only apparatus in any way peculiar being safety and controlling devices, which may be arranged in a large number of ways, according to the fancy of the individual inventor who has the problem in hand.

Owing to the beautiful manner in which electrically operated vehicles can be controlled from any given point, the idea of an automatic electric railway, with great capabilities for speed and of a simple and cheap construction, has proved peculiarly attractive to the minds of inventors.

The telpher system of which we have just spoken, is the mere rudiment of automatic electric traction, and development in the direction of higher speed and greater permanence of service was very natural. It would seem at first sight to be quite within the bounds of possibility to lay a narrow-gauge track upon a comparatively cheap elevated structure, and to run upon it trains suitable for the transportation of mail and express matter; substitute rails and permanent structures for the cables or rods of the telpher line, increase the speed, and the problem would appear to be solved.

The possibility of running an electric train in some such fashion at the rate of 100 or 150 miles an hour has been recognized by many of those connected with the development of electric traction. There are, however, a few electrical difficulties, more mechanical ones, and commercial considerations of a most forbidding character.

The telpher line carrying its burden of ore or luggage of various sorts over a comparatively short distance is relatively practical; but where a train is to be run at a speed of 100 miles an hour over a distance of 100 miles, more or less, peculiar conditions are encountered. In particular, such high-speed service is only desirable for the transportation of things comparatively small in bulk and comparatively great in value, therefore the class of goods which the owners would least care to trust to an automatic service, where accidents would be especially serious in their results.

It is very doubtful, indeed, whether the United States Government would ever consent to intrust its mails between New York and Boston, for example, to any automatic system whatever, on account of the great ease with which it could be interfered with and the mails robbed, and the serious delays that even a trifling accident could produce.

As soon as such a service ceases to be automatic it loses much of its distinctive character, and might as well be extended into a general high-speed electric road intended for transporting passengers as well as express matter. Such a development is highly desirable and quite practicable, but an electrical high-speed package express running unguarded and unwatched over long distances is well subject to the ancient criticism of being "neither fish, flesh, nor good red herring."

Nevertheless, just such schemes have been brought forward over and over again; and one of them—the Weems project—is described in Chapter X. in considerable detail, and has proved of great value to the cause of electric traction by demonstrating on a small scale, but none the less effectively, the possibility of enormous speeds.

The logical line of development from the experiments is the electric lightning express, as sketched in the same chapter. Very high speeds require remarkably well-constructed track and road-bed, and this is really one of the most serious difficulties that has to be encountered. A road-bed and line equipment of suitable character is expensive in a prohibitive degree, unless the traffic is to be heavy and will pay unusually well.

Furthermore, the line must be of considerable length in order to render high speed worth attaining. At the rate of 100 miles per hour, getting up speed and stopping requires so much space and power that the advantages of so great a speed are almost thrown away unless the road extends over a considerable distance.

All these considerations point to the true electric railway rather than to any automatic system. Nevertheless, several of the latter have been proposed. For the most part they embody no striking features, save a number of ingenious devices for the proper governing of speed, automatic slowing down and stopping, and the regulation of the train from a distant point. These only possess interest in so far as they have been carried out, and they have not been carried out except in the Weems road before mentioned and in one other, of which we will now speak.

The only automatic high-speed road which is in even experimental operation to-day is that strange freak of an inventor's ingenuity—the Portelectric system. Abandoning all idea of driving wheels with their motor and motor armature, the inventor of this scheme has proposed to utilize the direct attraction of solenoids for a hollow steel car running on a suitable track. There are required a continuous series of solenoids over the entire length of the line, each of them powerful enough to exert a strong attractive pull on the car and furnished with current at the proper time by automatic switches opened and closed by the passage of the flying armature, a plan suggested long ago abroad.*

Working models were constructed, and, of course, worked well, as model systems always do; and, encouraged by this, still further experiments were made, and there has been until recently near Boston an oval experimental track about half a mile long. It consisted of an upper and lower rail, the car being provided with flanged wheels above and below. The car itself was a wrought-iron elongated shell with ogival ends, 12 feet long, and weighing 500 pounds. The solenoids were placed six feet apart, and each was composed of 20

* See English patent No. 58, of 1862.

pounds of No. 14 copper wire, having a resistance of about 5 ohms; the successive solenoids were thrown into circuit in front of the car by the action of a contact wheel mounted upon the latter and running upon the upper rail, which was divided into sections and utilized as a conductor.

It was expected that enormous speeds would be reached; this belief, however, has not been justified, as nothing even roughly approximating the speed obtained by one of the authors in an experimental run on the Weems system has ever been reached. The efficiency of such apparatus is obviously, from the character of the electro-magnetic action involved, low; and the sparking from the inductive currents set up in the helices as the contact is broken has proved to be most severe, although this might be partially obviated by short-circuiting the helices instead of cutting them out.

An idea of the actual efficiency of the apparatus may be obtained by considering the fact that between 9 and 10 electrical horse-power were required to drive the car at a speed of 33½ miles per hour—50 feet per second. Portions of the track were level, and in other portions there were grades of small length, but rising as high as 4 per cent. The total weight of the car being 500 pounds, assuming any reasonably plausible traction co-efficient, it is evident that the efficiency obtained was quite low. This much space is devoted to the discussion of this very unusual system simply by reason of its striking peculiarities, and not from any probability that such a device can ever supplant the electric motor proper; unless, possibly, on a small scale, as a substitute for the pneumatic tube.

As it is, the present chapter has been necessarily somewhat disjointed in character, having dealt with a wide range of subjects—treating schemes practical and impractical, possible and abandoned as hopeless.

To the minds of the authors the most hopeful of the methods mentioned are some of the modified conduits. There are no signs as yet that any of these deserves more extended mention than we have accorded to it here. Nevertheless, their development should be very carefully watched; for the direction is one from which something important may be looked for, and from which something practical may be reasonably hoped.

The slotted conduit, although it has been shown from experience that it may be successful, is both commercially and electrically somewhat discouraging in our American climate; possibly its very best chance for giving good results is in the conversion of some of the existing cable roads into electric roads. With conduits so large as those employed for cables the difficulties of drainage would be reduced to a minimum, insulation would probably be practicable, and, at least on roads of considerable length, as good efficiency could be obtained as with cables. With respect to any and all of the other devices, we can only advise our readers to suspend judgment awaiting further evidence.

CHAPTER X.

HIGH-SPEED SERVICE.

UNDER this heading it is intended to consider such service as involves greater speeds, distances, and, usually, greater loads than occur in street-railway work, properly so called.

To-day there is little to record of actual accomplishment in this direction. The City and South London Railway stands as a unique example of rapid transit for large cities, in which several cars are run in a train, on a special roadway, permitting higher than any allowable street speeds.

In 1886 Mr. F. J. Sprague conducted experiments on the New York Elevated Railway, having in view the replacement of the steam locomotives in a service requiring five loaded cars, each about 35 feet long, to be hauled at a maximum speed of 30 miles per hour. Mr. Leo Daft, during the years 1888 and 1889, made similar experiments. The failure to attain immediate success may be traced in both cases to several causes —immaturity of motor design at that time, limitation of money available for making needed changes, difficulty of experimenting without disturbance of regular traffic, and occupation of the experimenters in other absorbing work.

In the light of the experience of to-day no one of the competent engineers who has had to do with the electric-railway work of the last few years would hesitate, if supplied with the proper amount of money, to undertake the successful installation of an electric rapid transit system meeting all the requirements of the New York elevated railway or London underground service. The conditions are in many respects much simpler than those met with on the streets. Especially is this true in regard to the placing of the conductor system in the common street, where the convenience of the electrical engineer must be subordinated to every immemorial right of man, beast, or tree.

And the motors may be placed where dust and mud do not corrupt and stones do not break in the fields.

A brief description of the City and South London plant will illustrate the comparative simplicity of this class of work. The propelling power for each locomotive comes from two motors, the armatures of which are concentric with the axles and keyed directly thereto. Each armature is rated at about 50 horse-power, at a normal speed of 25 miles per hour. The motor frames are inclined upward as shown in Fig. 116. For the outgoing side of the circuit the working conductor, divided into convenient sections, is of channel steel. Current is taken from this through a sliding contact shoe attached

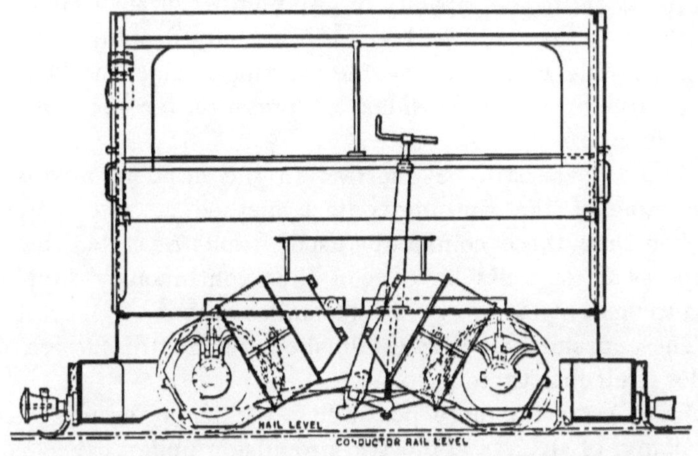

FIG. 116.—ELECTRIC LOCOMOTIVE OF CITY AND SOUTH LONDON RAILWAY.

to the locomotive. For the other side, the rails are used as in ordinary practice. The channel steel is insulated from the road-bed by glass insulators, the general line pressure being 500 volts. The motors are series wound and regulation is effected through a rheostat placed in the main circuit.

We have been able to obtain but little information concerning the actual operating expenses of this line. Reports indicate that they have at least been satisfactory to the bondholders.

The subject of the relative economy of steam and electric train service was discussed by one of the authors in a paper read before the American Institute of Electrical Engineers in May, 1890. As the matter was then taken up quite in detail, and as no later developments have been found to alter the conclusions then reached, we draw largely upon that paper for this chapter.

As will be seen, the question in regard to electricity is not "Can it be done?" but "Can it be done more economically than by steam?"

In the light of present achievements we may state without argument the following propositions:

First—It is possible to construct motors capable of doing the maximum work required to-day in transportation.

Second—It is possible continuously to generate electrical energy equal to the capacity of any number of such motors.

Third—At any desired loss and over any desired distance it is possible to supply by the running-contact method the necessary current, at considerable pressure, for the working of such motors.

Should there still be question in the mind of any as to the value of the running-contact method at speeds much higher than those commonly used, it may be stated that 75 ampères at 500 volts have been thus continuously supplied to a car moving at more than 110 miles an hour.*

These premises being established, further discussion divides itself into three parts:

First, As to the mere possibility, without reference to the economy, of steam and electric propulsion under given conditions.

Second, As to the relative cost of exerting in a locomotive any unit of power by electric, as compared with direct steam, motors.

Third, As to the relative amount of power required by the two agents to transport a given paying load under given conditions.

In using the word steam as above we have in mind only the direct application of steam power on the tracks. The case of cable propulsion is not here compared, as that has

---

* See "Report of High-Speed Electric Railway Work," read Feb. 24, 1891, before the American Institute of Electrical Engineers by O. T. Crosby.

within its restricted field of application already often been compared with horse, steam, and electric power.

The limiting possibilities of locomotion may be understood by considering a prolongation of the lines of present practice in the direction, first, of loads handled; second, grades climbed; third, speeds attained; and, fourth, length of continuous runs.

Since the effect of grade as compared with load is simply to increase the tractive effort required for a given load and speed, it need not be separately treated, except that there has been some question of increase of adhesion in the ground-return method, which in extreme cases might appear as an advantage for electric propulsion. The matter is not of great importance; and we will refer to it only so far as to say that general experience and some special tests show that the adhesion co-efficient is not increased in any practical degree by the mere passage of the current from wheel to rail.

The capacity of an electric engine, like that of a steam engine, to haul any given load is measured by the tractive effort possible to be produced and the relation between weight and adhesion for given track conditions. The ready multiplication of cylinders in the one case and armatures in the other, while maintaining mechanical unity and the ready coupling of distinct locomotive units, renders the whole question of capacity to exert a given horizontal effort, without regard to the time element, unimportant. It goes without saying that, if desired, a single armature may be constructed capable of exerting as great a drawbar strain as any locomotive now in use.

As to limiting speeds, it is not easy to-day to make even "an educated guess," either for steam or electric propulsion. The high figures for steam that have been recently presented both from England and America are higher than the limiting figures as they would have been given by many competent authorities only a few years ago. Eighty-six miles per hour in England, on the Northeastern Railway, eighty-seven miles per hour, and later a rate of 90.45 miles per hour, in this country, on the Reading Railway, have been reported since

Jan. 1, 1890. These runs are noteworthy not only for the fact of unusual speed, but because, as shown by indicator cards in the English case, and as may be deduced from the consideration of the maximum cylinder power in the American case, the train resistances are far below the values that would have been predicted by even the most liberal of the received formulæ on the subject. The total resistance per ton, as per indicator card, in the 86-mile run was only 13.4 pounds. According to Searles' formula, adopted by Wellington, it should be 69 pounds, engine and tender being taken at 50 tons. The load of 347 tons was carried at 86 miles per hour by an expenditure of 1,068 horse-power—this on a level. The engine was compound.

Would it be possible to attain a speed twice as great, or say 150 miles per hour?

A driver 24 feet in circumference must revolve 550 times per minute in order to travel 13,200 feet per minute, or 150 miles per hour—this without slip. Since in the case considered the revolutions per minute reached 309, and since in the Reading case a much higher rate must have been reached—the drivers being smaller—and since on stationary engines a speed much above 550 revolutions per minute has been attained, it seems beyond question that from this point of view the supposed case is quite possible.

Considering the matter of steam supply, we are again brought to consider the whole question of train resistances at all speeds.

Total resistance to motion should be sharply divided into two classes: the resistance due to motion through air, and that due to friction and blows between vehicle and track and to friction and blows between parts of the vehicle.

For the most part, those who constructed the formulæ now found in the text-books worked on road-beds far inferior to the best work of to-day, at speeds much less than those now attained, and with wrong values for at least one of the species of resistance—the atmospheric. On this point there has recently been presented, as the result of experiments at high velocities, a formula showing the pressure to be approximately a function of the first, instead of the second, power of the velocity as ordinarily assumed.* A convenient datum point may be given, stating that at 100 miles per hour the pressure on one square foot, normal to the direction of motion, is 13 pounds; while proper shaping of the front may reduce this to 6.5 pounds. The absolute values given, while corresponding quite closely with those of received formulæ in the neighborhood of the velocities heretofore experimentally attained, depart widely from those assumed for velocities higher than 30 miles per hour, and calculated by the quadratic relation between velocity and pressure. Using the more trustworthy values, we are able to separate more nearly than heretofore has been possible the atmospheric from all other resistances met at high velocities. Some inaccuracy still remains, by reason of the difficulty of obtaining exact measure of the resisting areas in a train; but by study of careful tests made by others on the New York Central and on English roads we find that over the range from about 40 up to 80 miles per hour the tonnage co-efficient seems practically constant at 8 pounds. This, of course, applies only to first-class road-bed and rolling stock. Whether this co-efficient remains constant at higher speeds we do not know. There is no reason to assume, as has often been done, that it increases with the square of the velocity; and, on the other hand, it will not be safe to assume constancy. From experiments made with a single 2.5-tons car at about 100 miles per hour, the tonnage resistance at that speed seems to be about 20 pounds per ton.† Though this value seems quite high as compared with the eight pounds at 80 miles per hour, the difference is in large part to be explained by the poor condition of the track used for the experiment and a constant curvature which would call for about four pounds per ton. Until better evidence can be had it will be safe, at least, to assume a value of 20 pounds per ton, on a first-class track, with good rolling stock, at 125–150 miles per hour.

Having made this necessary digression, we may return to the matter of steam supply, and state that by reducing weight

* "An Experimental Study of Atmospheric Resistance," by O. T. Crosby, London *Engineering*, May 30, June 6-13, 1890; New York *Engineering News*, May 31, June 7-14, 1890, and *The Electrical World*, May 24, 1890.

† "Report of High-Speed Railway Work" above mentioned.

TABLE I.

Speed.	Tonnage.	Areas exposed per ton in sq. ft.						Rate of work in H. P., 10 per cent. loss.						H. P. at 20 per cent. loss.			H. P. at 40 per cent. loss.		
		1.0	0.75	0.5	0.25	0.2	0.1												
		Horizontal effort.						2 and 3	2 and 4	2 and 5	2 and 6	2 and 7	2 and 8	2 and 3	2 and 5	2 and 8	2 and 3	2 and 5	2 and 8
20	8	2.6	1.95	1.3	0.65	0.52	0.26	0.61	0.58	0.534	0.501	0.49	0.48	0.70	0.07	0.55	0.93	0.8	0.73
40	8	5.2	3.90	2.6	1.3	1.04	9.52	1.56	1.40	1.23	1.068	1.05	0.99	1.77	1.40	1.12	2.37	1.86	1.5
60	8	7.8	5.85	3.9	1.95	1.56	0.78	2.77	2.43	2.09	1.75	1.67	1.54	3.15	2.17	1.77	4.20	3.2	2.37
80	8	10.4	7.80	5.2	2.6	2.8	1.04	4.32	3.70	3.00	2.46	2.35	2.00	4.90	3.50	2.37	6.53	4.6	3.2
100	12	11.0	9.75	6.5	3.25	2.6	1.3	7.32	6.35	5.40	4.45	4.27	3.90	8.32	6.15	4.42	11.1	8.16	5.9
120	15	15.6	11.70	7.8	3.9	3.12	1.56	10.79	9.39	8.03	6.65	6.37	5.84	12.21	9.12	6.66	16.3	12.2	8.85
140	20	18.2	13.65	9.1	4.55	3.64	1.82	15.67	13.7	11.88	10.06	9.75	8.95	17.80	13.5	10.17	23.6	18.0	13.5
1	2	3	4	5	6	7	8	9	10	11	12	13	14	15	16	17	18	19	20

Speed.	Tonnage.	Coal and water for one hour. 5 lbs. coal. 15 lbs. water.						Locomotive and tender at 100 lbs. per H. P.			Weight: Machinery and fuel, and load. Steam. 10 per cent.			Weight: Load, and weight of machinery. Electric. 10 per cent.			Weight. Load, and weight of machinery. Electric. 20 per cent.			Weight: Load, and weight of machinery. Electric. 40 per cent.		
		2 & 3	2 & 4	2 & 5	2 & 6	2 & 7	2 & 8	2 & 3	2 & 5	2 & 8	2 & 3	2 & 5	2 & 8	2 & 3	2 & 5	2 & 8	2 & 3	2 & 5	2 & 8	2 & 3	2 & 5	2 & 8
20	8										1927	1931	1943	1964	1968	1971	1958	1964	1967	1944	1952	1956
		12.3	11.66	10.65	10.12	9.9	9.6	61.6	58.3	48.4	73	69	57	36.6	32.	28.8	42.	36.	33.	55.8	48	438
40	8										1815	1836	1881	1906	1926	1941	1894	1916	1933	1858	1888	1910
		30.8	27.9	24.6	21.30	21.07	19.8	154	138.7	99.0	185	164	119	93.6	73.8	59.4	106.2	84.	67.2	142.2	111.6	90
60	8										1668	1715	1815	1834	1875	1908	1811	1858	1892	1738	1808	1858
		55.4	48.6	41.8	35.0	33.4	30.8	277	243	154	332	285	185	166.2	125.4	92.4	189.0	142.2	106.2	262	192.	142.2

80	8	86.2	73.9	61.8	49.3	47.1	42.1	431	370	210	1483 517	1568 432	1748 252	1741 259.2	1820 180	1880 120.	1706 294	1790 210.0	1858 142.2	1608 391.8	1724 276	1808 192
100	12	146.5	131.2	108.2	89.0	84.3	77.8	732	636	389	1122 878	1256 744	1523 477	1561 439.2	1676 324	1766 234	1500 499.2	1631 369	1735 265.2	1334 666	1510 489.6	1646 354
120	15	215.4	187.88	160.	133.0	127.4	116.8	1078	939	584	707 1293	901 1099	1299 701	1354 646.2	1518 481.8	1650 350.4	1268 132.6	1453 547.2	1600 399.6	1022 978	1268 732	1469 531
140	20	313.5	276.0	237.	201.0	195.1	179.1	1567	1380	895	120 1880	383 1617	926 1074	1060 940.2	1287 712.8	1517 483.	932 1068	1190 810	390 610.2	584 1416	920 1080	1190 810
		21	22	23	24	25	26	27	28	29	30	31	32	33	34	35	36	37	38	39	40	41

Speed.	Tonnage.	Power for one ton freight and car. Steam. 10 per cent.			Power for one ton freight and car. Electric. 10 per cent.			Power for one ton electric at 20 per cent.			Power for one ton electric. 40 per cent. loss.			H. P. Steam. / H. P. Elec. Loss on both 10 per cent.			Ditto. Loss steam 10 per cent. Loss elec. 20 per cent.			Ditto. Loss steam 10 per cent. Loss elec. 40 per cent.		
		2 & 3	2 & 5	2 & 8	2 & 3	2 & 5	2 & 8	2 & 3	2 & 5	2 & 8	2 & 3	2 & 5	2 & 8	2 & 3	2 & 5	2 & 8	2 & 3	2 & 5	2 & 8	2 & 3	2 & 5	2 & 8
20	8	0.633	0.55	0.49	0.62	0.54	0.48	0.715	0.68	0.56	0.95	0.82	0.75	1.01	1.02	1.02	0.88	0.81	0.87	0.66	0.67	0.65
40	8	1.72	1.34	1.05	1.63	1.27	1.02	1.88	1.46	1.15	2.55	2.0	1.57	1.05	1.05	1.03	0.91	0.92	0.91	0.67	0.67	0.66
60	8	3.32	2.44	1.70	3.02	2.23	1.61	3.48	2.55	1.87	4.8	3.54	2.55	1.1	1.09	1.05	0.95	0.95	0.91	0.69	0.68	0.66
80	8	5.83	3.82	2.28	4.96	3.3	2 13	5.74	3.91	2.55	3.12	5.34	3.54	1.17	1.16	1.07	1.02	0.98	0.90	0.71	0.71	0.65
100	12	13.05	8.60	5.12	9.38	6.44	4.42	11.08	7.54	5.09	16.6	10.8	7.17	1.39	1.33	1.16	1.17	1.14	1.00	0.78	0.8	0.71
120	15	30.4	17.8	9.0	15.9	10.5	7.08	19.25	12.5	8.35	31.8	19.25	12.0	1.91	1.7	1.27	1.58	1.42	1.08	0.95	0.62	0.75
140	20	260.8	62.0	19.3	295.6	18.46	11.8	38.2	22.7	14.63	80.8	39.1	22.7	8.8	3.35	1.64	7.00	2.7	1.39	3.2	1.6	0.85
		42	43	44	45	46	47	48	49	50	51	52	53	54	55	56	57	58	59	60	61	62

The figures "2 and 3," "2 and 5," etc., indicate combination of the tonnage co-efficients in Col. 2 with the value for atmospheric resistance shown in Col. 3 or Col. 5, etc., as the case may be. Similar reference is made in Table V.

and area both to something less than one-half the existing values in the 86-mile run, the same effort would produce the speed 150 instead of 86. The area cannot be thus reduced, but by assuming a greater reduction in weight—say to 100 tons, or to little more than engine and tender—maintenance of the higher speed might become possible, with nearly the same steam expenditure as in the recorded case.

. To attain that speed, from rest, might require such original weight of fuel and such length of favorable track as to make the feat practically impossible with steam. This leads us to inquire into the dead weight necessary for hauling, say, one ton at different speeds. Table I. gives the horse-power required for exerting the tractive effort for one ton at various speeds, at various efficiencies, with various values of cross-section per ton, and with the two agents—steam and electricity.

Column 1 shows speed in miles per hour from 20 to 140; column 2, corresponding tonnage co-efficient, or resistance, in pounds per ton, exclusive of atmospheric resistance. Columns 3 to 8, inclusive, show horizontal effort needed for overcoming atmospheric resistance under various assumptions as to area exposed per ton, from 1 square foot to 0.1 square foot per ton. The former figure corresponds nearly to the case of a heavy locomotive propelling itself alone. As load is put on behind it other ratios are formed. Oblique surfaces are supposed to be reduced to equivalent normal surfaces. Columns 9 to 14, inclusive, show rate of work in horse-power per ton for the various cases of area exposed to atmospheric resistance, efficiency of locomotive being taken at 90 per cent. Columns 15 to 17, inclusive, show horse-power per ton for the extreme and middle cases of exposed area, and for locomotive efficiency of 80 per cent. Columns 18 to 20, inclusive, show same for efficiency of 60 per cent. Columns 21 to 26, inclusive, show weight of coal and water per ton carried for one hour, assuming 5 pounds of coal and 25 pounds of water per horse-power hour on a steam locomotive. The coal figure is very close to actual practice. The water figure is less, but makes allowance for scooping water at convenient intervals. Continuous scooping is not considered practical or economical. Columns 27 to 29, inclusive, show weight of steam locomotive and tender required to generate the required horse-power per ton, under the assumption of 100 pounds per horse-power and 90 per cent. efficiency. Only three cases of exposed area are taken; that is, one foot, one-half a foot, and one-tenth of a foot per ton. The weight of steam locomotives is not calculated for any other efficiency figure than 90 per cent., as this seems to be quite constantly attained or surpassed. The assumption of 100 pounds per horse-power is closely true for many good types of locomotive when working at speeds from 60 to 80 miles. At lower speeds this figure is too low, but it is assumed that for any ruling speed engines may be built of minimum weight for that speed. In passing through lower than ruling speeds both electric and steam motors work at low output per pound of weight; hence the assumption of constant weight per horse-power will not introduce error materially affecting comparative results. At higher speeds than 80 miles existing engines would show less than 100 pounds per horse-power; but as their boiler capacity is reached at that speed the necessary increase for any regular work would carry the weight figure to very nearly the figure given for the 60 to 80 mile running. Columns 30 to 32 show weight of steam locomotive and tender, plus weight of fuel and water, per ton hauled; also weight of load—freight and freight car—that may be hauled by such weight of motive power; the load figures being obtained by subtracting the motive power weights from 2,000 pounds. Columns 33 to 41 show corresponding figures for electric locomotives, under the assumption of 60 pounds per horse-power, and at the three efficiencies—90, 80, and 60 per cent. In 30 to 41, inclusive, weight of load is written first, weight of locomotive (plus tender, fuel, and water for steam) being written below.

The 60 pounds per horse-power covers weight of the containing car for motors. We cannot here go into detailed figures on this point, but believe that any investigation will find the figure safe, supposing always that the unit be, say, 25 horse-power or more. Columns 42 to 53, inclusive, show the horse-power required to be exerted for hauling one ton of load, *i.e.*, freight and freight car, the relation between these two being taken as the same for either steam or electric propulsion, hence not necessary to enter here. These columns apply to steam at 90 per cent. and to electricity at 90, 80, and 60 per cent., and for the three cases of exposed area. They are readily obtained from the previous columns by making allowance for the horse-power necessary to haul that part of every ton, total weight, which must go into motive power, machinery, and fuel. Columns 54 to 62, inclusive, show the ratios between horse-power required by the two agents for hauling a ton of load (freight and freight car) at the different speeds, efficiencies and area relations.

It is plain that if we can now obtain the ratio of cost per horse-power, as given by the two agents, in the corresponding cases, we can easily determine the speeds at which the one or the other agent becomes the more economical.

Let us first obtain the cost of electric propulsion.

TABLE II.

ELEMENTS OF THE COST OF ONE H. P. HOUR, ELECTRIC, IN CENTS.

Capacity	100	300	500	800	1,000	1,500	2,000	3,000	4,000	5,000	6,000
Engineer	0.4	0.13	0.08	0.05	0.04	0.04	0.04	0.04	0.04	0.04	0.04
Fireman	0.3	0.10	0.06	0.037	0.03	0.03	0.03	0.03	0.03	0.03	0.03
Dynamo man	0.4	0.13	0.08	0.05	0.04	0.04	0.04	0.04	0.04	0.04	0.04
Helper	0.25	0.08	0.05	0.031	0.025	0.025	0.025	0.025	0.025	0.025	0.025
Superintendence	0.30	0.10	0.06	0.037	0.03	0.02	0.015	0.01	0.01	0.01	0.01
Coal	0.475	.475	0.475	0.475	0.475	0.475	0.475	0.475	0.475	0.475	0.475
Oil, waste, and water	0.15	0.15	.15	0.15	0.15	0.15	0.15	0.15	0.15	0.15	0.15
Interest and depreciation steam plant	0.057	0.051	0.044	0.033	0.028	9.022	0.022	0.022	0.022	0.022	0.022
Ditto electric plant	0.057	0.051	0.044	0.033	.028	0.022	0.022	0.022	0.022	0.022	0.022
Ditto building	0.028	0.026	0.022	0.017	0.014	0.011	0.011	0.011	0.011	0.011	0.011

For this, form Table II., showing the elements in the *cost* of one horse-power in stations of various capacity—from 100 to 6,000 horse-power. Engineers and dynamo men are assumed to receive 40 cents per hour, and to superintend a maximum of 1,000 horse-power. This would produce in some cases fractional engineers, as for a 1,500-horse-power plant; but such complication has been avoided by assuming a constant value per unit of power in the pay-roll element in plants exceeding 1,000 horse-power. Firemen and helpers are taken at 30 and 25 cents per hour, respectively. Superintendence at 30 cents per hour is apparently low, but is equivalent to 60 cents for daylight hours, since the plant is able to run 24 hours with the same general superintendence as for 12 hours. It is further supposed that the total of this item will not require increase until the capacity reaches 3,000 horse-power, beyond which it remains constant per unit of power. As no other element of cost is supposed to vary beyond this point, the table shows here a minimum total cost per unit and a constant cost beyond it.

Coal is assumed to cost $3 per ton, and to be consumed at the rate of 3.2 pounds per electric horse-power hour in the dynamo. A slight error is made in taking the rate of consumption as constant while changing the capacity of the engines. The cost of the steam plant is taken to vary from $50 per horse-power, in a small plant, to $20 in a plant of 1,500 horse-power. The dynamo plant is taken to vary from $50 to $20 per horse-power, in going from 100 to 1,500 horse-power.

TABLE III.

TOTAL COST OF ONE H. P. HOUR.

Output in per cent. of capacity while working.	Hours of work per day.	Capacity of Station.							
		100	300	500	800	1,000	1,500	2,000	3,000
00	24	2.42	1.29	1.06	0.9148	0.860	0.835	0.830	0.825
	18	2.52	1.30	1.115	0.938	0.888	0.855	0.849	0.829
	12	2.85	1.52	1.228	1.028	0.955	0.95	0.895	0.867
90	24	2.62	1.36	1.12	0.947	0.88	0.86	0.85	0.83
	18	2.77	1.45	1.17	0.98	0.01	0.88	0.87	0.853
	12	3.10	1.61	1.29	1.06	0.99	0.94	0.925	0.89
80	24	2.87	1.45	1.18	0.987	0.92	0.91	0.88	0.86
	18	3.06	1.55	1.23	1.03	0.98	0.936	0.926	0.90
	12	3.42	1.73	1.37	1.13	1.06	1.000	0.98	0.95
70	24	8.19	1.57	1.26	1.04	0.96	0.925	0.91	0.89
	18	3.42	1.68	1.32	1.09	1.01	0.952	0.94	0.91
	12	3.82	1.90	1.48	1.19	1.09	1.02	1.01	0.97
60	24	3.62	1.73	1.36	1.11	1.01	0.975	0.96	0.94
	18	3.85	1.85	1.44	1.16	1.06	1.015	1.00	0.965
	12	4.31	2.11	1.63	1.30	1.17	1.085	1.07	1.015
50	24	4.22	1.95	1.52	1.20	1.10	1.045	1.03	1.00
	18	4.50	2.10	1.61	1.27	1.15	1.09	1.076	1.039
	12	5.09	2.40	1.83	1.43	1.28	1.19	1.17	1.114
40	24	5.11	2.28	1.73	1.35	1.21	1.15	1.13	1.10
	18	5.47	2.48	1.88	1.44	1.33	1.25	1.23	1.17
	12	5.20	2.82	2.17	1.64	1.50	1.38	1.35	1.28
30	24	6.63	2.84	2.10	1.59	1.40	1.325	1.30	1.26
	18	7.08	3.02	2.27	1.71	1.51	1.40	1.38	1.32
	12	8.06	3.52	2.65	1.98	1.73	1.57	1.54	1.44

The cost of buildings is taken to vary from $25 to $10 per horse-power. This is the most indefinite item. Interest, maintenance and taxes are roundly assumed at 10 per cent. per annum on the whole plant.

With Table II. as a basis, Table III. has been calculated, giving the total cost per horse-power hour in stations of various capacities, working at various percentages of their full capacity and for 24, 18 and 12 hours per day, respectively.

A glance at the table shows that in a 100-horse-power plant the cost varies from 2.42 cents for a 24-hour run at full capacity, to 8.06 cents for a 30-per-cent. output continued only 12 hours per day.

This extreme case would doubtless be ameliorated by dropping the superintendence and combining engineer and fireman—though to this the unions might object. This capacity is smaller, however, than need be considered.

The minimum cost given by the table is 0.816 cent; this is for a 3,000-horse-power plant, working at full capacity 24 hours per day.

The next element of cost—that of the conductors for the current—is obtained from Table IV., showing the investment in dollars for the copper required to transmit one horse-power a distance of one mile, at varying initial and final pressures. The constants for this table were thus obtained: Taking a well-known line wire, it is found that from No. 4 to No. 000 B. W. G. the average weight ratio of insulated to bare wire, per unit of length, equals 0.844. When bare copper sells at 15 cents per pound this insulated wire sells at 20 cents to 22.5 cents on the copper alone, when insulated. If, therefore, we take copper at 25 cents per pound we provide for a very good insulation. The cost of one mil-mile is thus found to be $0.004. Combining this with the familiar formula

$$\text{C M} = \frac{16{,}600 \times \text{h. p. transmitted} \times \text{distance in feet.}}{\text{E. M. F. at motor} \times \text{volts lost} \times \text{motor efficiency}},$$

the tabulated values have been determined from the resulting formula:

$$\text{Cost} = \frac{760{,}320}{(E - V)\,V}, \qquad (1)$$

in which E = E. M. F. at station, V = volts lost on line; motor efficiency is taken at 90 per cent. and distance at 5,280 feet. The tabular figures give the investment. To reduce to actual cost per horse-power hour, the rates of interest and depreciation and the ratio between power transmitted and power possible to be transmitted must be had. For one year the possible horse-power hours = 365 × 24. Take annual interest and depreciation at 1-16 the investment and ratio of actual to possible power transmitted per annum at 1.0, 0.8, 0.6, 0.4, 0.2, 0.1, 0.05, then the divisor of the tabular number becomes 140,160, 112,128, 84,096, 52,064, 28,032, 14,026 and 7,008, respectively.

It remains to obtain values for the distance of transmission.

Let n = number of miles of line supplied from one station.

h = horse-power required for unit locomotive.

K = maximum number locomotives per mile at any time.

n h K = total power to be transmitted at time of K.

A = percentage of dynamo power lost on line.

$$\frac{n\,h\,K}{100 - A} = \text{maximum power to be generated.}$$

$$\frac{n\,h\,K}{100 - A} \times \frac{A}{100} = \text{'' \quad '' lost on line.}$$

b = cost in cents of generating one horse-power hour in station.

r = ratio of average power required to maximum power required.

L = interest and depreciation on supporting structure.

This last may be omitted from the calculation determining division of line into sections, since it remains the same whatever that division may be. Omitting this, the expression for cost of transmission becomes:

$$\text{Cost} = \frac{760{,}320\, n^2\, h\, K\, r}{(E - V)\, V \times 140{,}160} + \frac{b\, n\, h\, K\, A}{r\,(100 - A)\, 100}$$

$$= \frac{5.42\, n^2\, h\, K\, r}{(E - V)\, V} \times \frac{b\, n\, h\, K\, A}{r\,(100 - A)\, 100} \qquad (2)$$

Supposing that the time schedule of trains, horse-power required per train, and efficiency of motors be known, and that the initial E. M. F. be in all cases taken as high as the state of the art permits, the only variables remaining in this expression are n, b, and A. This latter, the value for the drop on the line, will generally be determined by conditions other than those of strictest economy as shown by getting a minimum value for cost of transmission. We must have a reasonably uniform E. M. F. all along the line, in order that the motors may work satisfactorily. It will not be wide of the mark to assume 10 per cent. as a limiting variation of line potential. This is the figure generally assumed in calculating wire for street railways.

The cost of one horse-power hour, b, must vary with the capacity and conditions of working of the station; hence it is a function of n—*i.e.*, length of unit section, of A, and of that inexpressible variable—the conditions of the service. This stands in the way of obtaining any perfectly general definite expression for n, which, but for this, would result from placing the first differential co-efficient of C, with respect to n, equal to zero, and solving to find the value of n giving a minimum value for C. If trains be run at very short intervals, increase of n would be followed by proportional increase in capacity of station; but, as above shown, this need not go

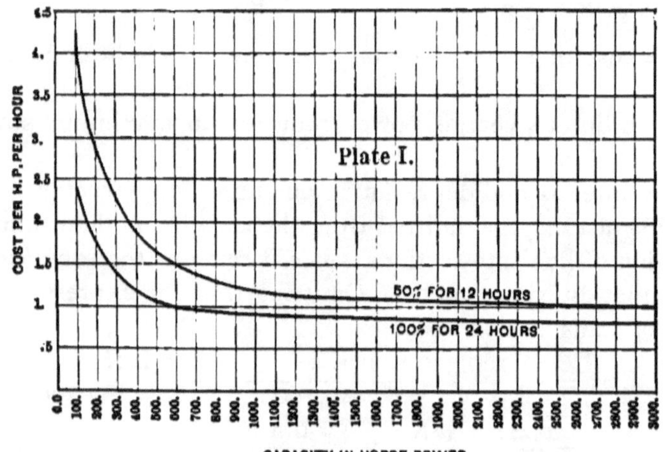

FIG. 117.—VARIATION OF COST OF CURRENT WITH CAPACITY OF STATION.

beyond 3,000 horse-power unless the service be so heavy as to require practically contiguous stations of that capacity. If we suppose a case of this short interval service, so short that a change in n will not be followed by any change at the station in the relation between maximum capacity and average output, or in the number of working hours, but only in the normal capacity, the relation between *b* and *n* for two cases of average output and working hours is shown by Fig. 117. The equation of the 50 per cent. curve from 200 to 1,000 horse-power seems to be very nearly

$$b = \frac{n^2 - 98n + 1{,}000}{12\,(n - 100)}.$$

If the trains be run at very long intervals we may require no greater station capacity for 20 than for 10 mile sections;

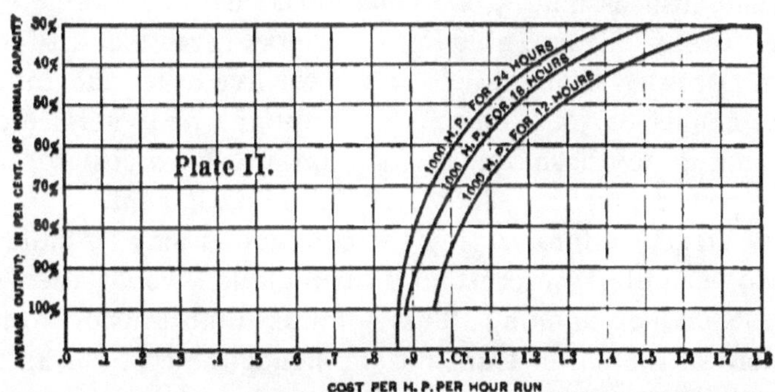

FIG. 118.—VARIATION OF COST OF CURRENT WITH CAPACITY OF STATION.

but the relation of output to normal capacity will vary, and possibly also the number of hours during which the working force would require to be kept on pay.

Taking the case of a change in average output only for a 1,000-horse-power station, we have the curves in Fig. 118 for 24, 18, and 12 hours' work. The equations are nearly of the same form, and are approximately the equations of arcs of circles referred to an origin outside the circle. But each would contain different constants, and would vary more or less from the exact formula for the circle.

If in the application to a particular case the algebraic expression for b above given, or any other resulting from any of the possible progressions through Table III., be substituted in equation (2), we may, by differentiation, solve that particular case for the most economical value of n. If the relation between b and n cannot be algebraically expressed, then the proper value for n may be determined by a few trial values, the corresponding values of b being taken from the table.

For the present purposes of comparison we will assume a case not more favorable than might often be met on busy steam lines—*i.e.*, a station of 2,000 horse-power normal capacity, working 18 hours per day at 40 per cent. of its normal output, the cost per horse-power being 1.25 cent.

To obtain the cost of the line we will assume that the average distance of transmission is five miles. This would correspond to one station for every twenty miles of road. We will also assume 5,000 volts initial E. M. F. and 10 per cent. "drop." From Table IV. the copper investment is found to be 34 cents for one mile. Then for five miles the investment equals $8.50. Making assumptions as to service corresponding to those for the station, we have cost for one horse-power equaling $8.50 ÷ 20,000 = 0.042 cent.

The structure for carrying the conductors may be built for $2,000 per mile. Interest and depreciation would then become $200 per annum. This total is almost wholly independent of the power transmitted; hence the cost per unit of power will vary inversely with the number of units transmitted. Assuming a constant distribution of one 500-horse-power train for every 20 miles of line, we have the cost of this item for one horse-power hour:

$$20{,}000 \div 365 \times 24 \times 25 = 0.09 \text{ cent.}$$

Reaching the locomotive, we must add, supposing an average output of 500 horse-power, 0.08 and 0.06 cent, respectively, for driver and his assistant. The latter is necessary only as a substitute for his principal in case of emergency, but as such he would doubtless always be placed on trains of considerable value.

Repair on electric locomotives is not as yet well defined. That the repair bill must be far less than in the case of steam locomotives follows almost necessarily from the great reduction in the number of parts, especially of moving parts.

From Mr. Arthur Wellington's very valuable work on railways, we take the figures showing percentage distribution of locomotive repairs by parts.

Boiler, 20 per cent.; running gear, 20 per cent.; machinery, 30 per cent.; lagging and painting, 12 per cent.; smoke-box, etc., 5 per cent.; tender (running gear, 10 per cent.;

TABLE IV.—FOR DOUBLE METALLIC CIRCUITS.

COPPER INVESTMENT, IN DOLLARS, FOR TRANSMITTING ONE H. P. ONE MILE.

Motor efficiency = 90 per cent.

Drop on line, volts.	Initial E. M. F.															
	500	1,000	1,500	2,000	2,500	3,000	3,500	4,000	4,500	5,000	5,500	6,000	7,000	8,000	9,000	10,000
100.....	19.00	8.44	5.42	4.00	3.20	2.60	2.20	1.80	1.60	1.44	1.40	1.28	1.10	0.98	0.84	0.76
200.....	13.66	4.80	3.42	2.12	1.64	1.34	1.14	1.00	0.88	0.78	0.70	0.64	0.56	0.48	0.42	0.38
300.....	19.00	3.60	2.12	1.50	1.20	0.94	0.80	0.68	0.60	0.54	0.48	0.44	0.36	0.32	0.28	0.26
400.....		3.20	1.72	1.20	0.92	0.72	0.60	0.52	0.44	0.40	0.34	0.32	0.28	0.24	0.22	0.20
500.....		3.00	1.50	1.00	0.76	0.60	0.50	0.40	0.36	0.34	0.30	0.28	0.22	0.20	0.18	0.16
600.....		3.20	1.40	0.92	0.66	0.54	0.42	0.34	0.32	0.28	0.26	0.22	0.20	0.16	0.15	0.14
700.....		3.60	1.34	0.82	0.60	0.46	0.38	0.32	0.28	0.24	0.22	0.20	0.16	0.14	0.12	0.10
800.....		4.80	1.34	0.80	0.56	0.42	0.34	0.30	0.26	0.22	0.20	0.18	0.14	0.14	0.10	0.10
900.....		8.44	1.40	0.76	0.54	0.40	0.32	0.28	0.22	0.20	0.18	0.16	0.12	0.12	0.10	0.08
1,000.....			1.50	0.76	0.50	0.36	0.30	0.24	0.20	0.18	0.16	0.15	0.12	0.10	0.08	0.08
1,200.....			2.12	0.80	0.48	0.34	0.28	0.22	0.18	0.16	0.14	0.12	0.10	0.08	0.08	0.07
1,400.....			5.42	0.90	0.48	0.33	0.26	0.20	0.18	0.14	0.12	0.10	0.08	0.08	0.06	0.06
1,600.....				1.12	0.52	0.34	0.24	0.20	0.16	0.14	0.12	0.10	0.08	0.06	0.06	0.056
1,800.....				2.12	0.60	0.34	0.24	0.18	0.16	0.12	0.10	0.09	0.08	0.06	0.06	0.05
2,000.....					0.76	0.36	0.24	0.18	0.15	0.12	0.10	0.08	0.06	0.06	0.054	0.046
2,400.....					3.20	0.52	0.28	0.20	0.14	0.12	0.10	0.08	0.06	0.05	0.048	0.04
2,800.....						1.34	0.38	0.22	0.16	0.12	0.09	0.08	0.06	0.05	0.04	0.038
3,200.....							0.78	0.30	0.18	0.12	0.10	0.08	0.06	0.05	0.04	0.034
3,600.....								0.54	0.22	0.14	0.10	0.08	0.06	0.05	0.04	0.032
4,000.....									0.36	0.18	0.12	0.08	0.06	0.046	0.038	0.032
4,500.....										0.60	0.16	0.10	0.06	0.04	0.036	0.030
5,000.....											0.30	0.15	0.06	0.04	0.038	0.030
6,000.....													0.12	0.06	0.04	0.032
7,000.....														0.10	0.054	0.036
8,000.....															0.08	0.046
9,000.....																0.008

In using the table it will be convenient to draw curves of equal cost. They will be, roughly, arcs of circles convex toward the left.

body and tank, 3 per cent.), 13 per cent.; total, 100 per cent.

Of these items we may at once and with certainty strike out boiler, smoke-box, etc., and tender—thus dropping 38 per cent. of the total. Having no boiler to carry, the running gear will be less in quantity.

The wear will be less, due to the use of rotary instead of reciprocating effort. It will then be fair to reduce this item by half, making another saving of 10 per cent. So in the machinery item there can be no question that with the rapid advance toward slow-speed motors, reducing gear, and sound insulation methods, the great advantage of having only *one moving part* in the motor itself must operate to reduce very largely the repair figure, probably to half its value in a steam locomotive, leaving it at 15 per cent. In lagging and painting, the omission of boiler and other parts must again effect a reduction, say to 6 per cent. The total reduction thus plainly indicated must be then very nearly 70 per cent.

The actual cost of repairs to-day on steam locomotives is, on the Pennsylvania Railroad, nearly 0.75 cent per horse-power hour. Reducing as above, this figure becomes 0.22 cent for electric traction. This refers to engines of considerable power, say from 400 to 1,000 horse-power capacity. The figure for both steam and electric motors would go up for smaller powers.

The interest charge results from considering the cost of an electric locomotive as $50 per horse-power, and the duty as six hours per day, full capacity. The average duty of steam locomotives is only about three hours. The higher figure results from a smaller number of repairs necessary, due to greater simplicity of parts. Then

$$\text{Interest for h. p. hour} = \frac{50.00 \times 0.05}{365 \times 6} = 0.11 \text{ cent.}$$

Before summarizing we must know something of the efficiency of the system. If the locomotive be of 90 per cent., or 80 per cent., or 60 per cent. efficiency, we must generate, respectively, 1.25, 1.4, and 1.85 horse-power hour in the station for every horse-power hour actually delivered to drivers (line loss being supposed constant at 10 per cent.). We should then have in the station 1.56, 1.75, and 2.30 cents, respectively. Then, for total:

Station	1.56	1.75	2.30
Conductor system	0.042	0.04	0.04
Structure for same	0.09	0.09	0.09
Wages	0.14	0.14	0.14
Repairs	0.22	0.22	0.22
Interest	0.11	0.11	0.11
Totals	2.16	2.35	2.90

Of the whole amount it is to be noted that the station item is three-fourths. Taking the most favorable case shown by the table—3,000 horse-power capacity, working for 24 hours at 100 per cent. of capacity—those figures become 1.5, 1.62, and 2.00 cents, nearly, the reductions in the other items being still a little indefinite, without making another series of independent assumptions.

The cost of one horse-power hour exerted by steam locomotives is to be obtained only by some circumlocution, the reports of cost being based on train-miles. As quoted by Wellington, the coal consumed per passenger train-mile on the Pennsylvania Railroad is very closely 50 pounds, and an average performance will show 5 pounds coal per horse-power hour while under way. This shows that one train-mile equals 10 horse-power hours. The coal consumption of a freight train-mile is much higher, but the divisor would also be higher, on account of the greater number of stops, in backing and switching, and greater delays while on a trip, thus increasing waste of coal.

Again, the Pennsylvania Railroad reports show a cost of fuel per train-mile of 5 cents. The cost of coal to that company, as nearly as we can learn, is about $1.50 per ton. Hence it would appear that 66 pounds per train-mile are consumed. As the terminal losses of fuel and delay (getting up steam and drawing fires, etc.) are known to be in the neighborhood of 25 per cent. of the total consumption, we have 6.6 (pounds) as the divisor, and the same ratio again appears. The cost of one train-mile—as to motor power alone—is given by the Pennsylvania Railroad as 22 cents. And this is practically equal to the average for the United States. The itemized statement is very carefully made up, and seems to cover everything except interest on the engine investment.

Knowing the annual mileage per locomotive to be about 20,000 and the cost to be $10,000, interest charge per train-mile (not quite, but nearly, equivalent to engine mile) becomes 2.7 cents. Total cost becomes 24.7 cents, or 2.47 cents per horse-power hour.

We may safely use this figure in the comparison to be made, since any positive error in the calculation of coal per horse-power hour or negative error in horse-power hours per train-mile will be offset by the difference in cost of coal per ton to the Pennsylvania Railroad as compared with the value assumed in Table III.

At $3, instead of $1.50, the fuel item would be 10 cents and the total motive power 29.5 cents. Leaving the statistics of actual cost, we may reach, by the method of Table III., nearly the same figure, considering the locomotive as a 1,000-horse-power steam plant of low first cost, burning 6 pounds of coal per horse-power hour, and working at one-third capacity for about three hours per day. The result thus reached is about 10 per cent. higher, but the 2.47 cents seem more reliable.

These values multiplied into the corresponding values in the last nine columns of Table I. give the following values of:

$$\frac{\text{Power units, steam}}{\text{Power units, electric}} \times \frac{\text{cost per unit, steam.}}{\text{cost per unit, electric.}}$$

This value, Table V., when greater than unity indicates greater economy by electricity than by steam, and *vice versa.*

A glance at the table shows the dominating necessity of increasing the efficiency of the mechanism delivering energy from the electric line to the vehicle. We cannot count upon a higher efficiency than 90 per cent. for the motor. Hence, save in the case of putting the armature directly on the axle, we cannot hope to reduce the total loss to less than 20 per cent.—a case permitting one set (*i.e.*, two gears) of spur gearing between armature and axle. As the ordinary relation beween tonnage and resisting area will lie between the second and third columns of each efficiency table, it appears that with a 20 per cent. loss electricity becomes cheaper than steam at about 70 miles per hour, the frequency of service being such as is assumed above.

TABLE V.

Ratios of cost of motive power = $\frac{\textit{cost by steam}}{\textit{cost by electricity}}$.

Efficiency of electric engine.	10 per cent.			20 per cent.			40 per cent.		
Tonnage and area relation.	2 & 3	2 & 5	2 & 8	2 & 3	2 & 5	2 & 8	2 & 3	2 & 5	2 & 8
Speed.									
20	1.15	1.16	1.16	0.92	0.85	0.85	0.56	0.56	0.55
40	1.19	1.19	1.17	0.95	0.95	0.95	0.57	0.57	0.56
60	1.25	1.24	1.19	1.00	1.00	0.95	0.58	0.58	0.56
80	1.33	1.32	1.22	1.07	1.03	0.94	0.59	0.59	0.57
100	1.58	1.42	1.32	1.23	1.20	1.05	0.66	0.65	0.59
120	2.57	1.94	1.47	1.66	1.49	1.13	0.80	0.78	0.64
140	10.03	3.82	1.87	7.35	2.83	1.46	2.72	1.36	0.72

In the case of 40 per cent. loss our new agent betters the old only at 140 miles per hour. And yet this loss—40 per cent.—is about the best we do with our present systems of electric street car propulsion.

Considering the very short life of the art, these results are excellent. Indeed, excellence is shown in the mere fact of success in competing with horses under conditions very trying for any mechanism not made of india-rubber or whit-leather. How great that success has been we need not here proclaim. But the steam locomotive is a foeman more worthy of our steel, or, rather, of our annealed soft iron and 99 per cent. conductivity copper.

Consideration of all that precedes leads to the following general conclusions:

1. A slow-speed armature placed on the car axle would place the electric motor in the lead at all service speeds.

2. For speeds above 70 miles per hour an electric motor of 90 per cent. efficiency, working through gearing of 90 per cent. efficiency, would prove more economical than the steam locomotive—save in cases of very infrequent service on very long lines.

3. On lines for heavy traffic steam would be more eco-

nomical than electricity if motor and gearing have a combined efficiency as low as 60 per cent., up to 100 miles per hour.

4. At speeds of 100 miles per hour and upward neither steam at 90 per cent. nor electric apparatus at 60 per cent. efficiency is commercially practicable.

5. Inasmuch as the saving of coal in stationary, as compared with locomotive, engines is one of the chief causes of the greater economy of electric propulsion at any speed this advantage will increase with that difference and also with the price of coal.

6. Any cause other than inefficiency of motor which increases the power required to haul a ton of freight increases the advantages of electricity, since it enlarges the value of the coal difference and the dead-weight difference.

Thus, bad roadways and large areas exposed to atmospheric resistance, as in street-railway work, lower the speed at which electric motors of any efficiency become cheaper than steam.

7. In descending to small locomotive units the electric motor loses less, relatively, of its advantage—another reason for success on street lines.

8. Multiplying the number of motors should be as far as possible avoided.

9. In special cases cleanliness and compactness of electric machinery may be of great value; in case of very frequent stops the possibility of returning to the line the energy now wasted in brakes may be of considerable value. This, however, can be obtained only by sacrifice in the matter of dead weight, as normal working is implied to be at comparatively low magnetization. Loss due to low efficiency in starting can scarcely be avoided, either in steam or electric engines.

10. Other minor pros and cons may be enumerated, but, considering the general economic results of the two systems, we reach no more definite conclusions.

While only one condition of station working has been taken for final comparison, that case is an average one, and comparative results would be but slightly affected by ordinary variations. Extreme cases may be readily determined from the tables and formulæ presented.

11. Some difference of opinion as to the proper values for

the various constants is to be expected, but these differences taken all along the line would probably nearly balance between positives and negatives, leaving the general results and the method unchanged.

Whether these conclusions be accepted or not, we believe the tables and formulæ here presented will be found valuable.

Consulting Table V., we find that rapid transit service for large cities can be accomplished electrically with a saving in motive power of about 20 per cent. on the cost by steam. This supposes the electric motor to have an efficiency of about 90 per cent., at 30 miles per hour, and contemplates the construction of the armature concentric with the axle. In respect to this point there would, at this writing, be little, if any, difference of opinion. It should be added that in considering city service a more favorable case of station expense might be assumed than that which has entered into Table V. There we have taken a total capacity of 2,000 horse-power working at 40 per cent. of that capacity during 18 hours, the cost per horse-power hour being 1.25 cent.

But in the case of city service the load will be tolerably uniform and the station capacity may be so determined that the average output may be, say, 60 per cent. of the capacity. We might, indeed, fairly take 1 cent per horse-power hour for the station cost. This assumption would be followed by an increase of about one-eighth in the figures given for "10 per cent. loss" in Table V. We might, then, fairly say that an economy of 30 per cent. may be had over the cost of motive power by steam. Now, in the average rapid transit system the motive power expense may be taken at about one-third of the total operating expense. On this total, then, the saving may be said to be about 10 per cent.

The conclusions here drawn from the tables above given have been checked by careful detailed estimates made for special projects now on foot in Chicago, Ill., and by engineers not familiar with the tables. The results were very nearly identical, tending to show that, at least in all general discussion of such work, very satisfactory guidance is given by the figures here presented.

Concerning speeds considerably beyond those now commonly attained by steam, no experiments have been made and reported save those made by one of the authors at Laurel, Md.* The work was done for the Electro-Automatic Transit Company of Baltimore, Md. The organizer of this company was Mr. David G. Weems. Lack of means has thus far stood in the way of accomplishing on a proper scale the demonstration required in order that the world may become familiar with, and practically interested in, the practicable increase of speed of transportation up to 125 or 150 miles per hour.

In the experiments above referred to a car about 30 inches high by 24 inches wide by 20 feet long, carrying two motors with their armatures keyed directly to the axles and weighing about 2.5 tons, was driven around a circular track (28-inch gauge, 2 miles in circumference) at a final speed of about 115 miles per hour. Control was effected from the power station, placed inside the circle. The voltage of the line was 500. An upper rail carried on a framework built over the track served as one side of the circuit, the lower track rails and earth as the return circuit.

Current was received from the upper rail through an upward bearing brush. The track was of exceedingly light

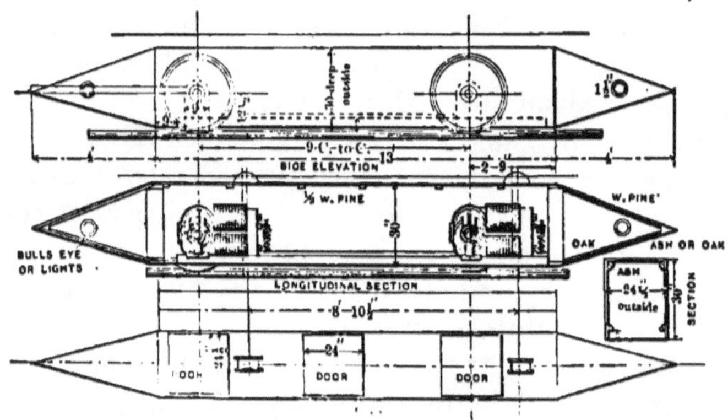

FIG. 119.—PLAN AND ELEVATION OF EXPERIMENTAL CAR.—WEEMS SYSTEM.

construction—quite unfit for the work attempted to be done. Figs. 119 and 120 show the car and station. Valuable infor-

* See "Report of High Speed Electric Railway Work," by O. T. Crosby. Read before the American Institute of Electrical Engineers, February 24, 1891.

mation, however, was obtained concerning atmospheric and tonnage resistance.

In the American Institute paper above referred to these results were reported quite fully, and plans were shown for

FIG. 120.—STATION AND TRACK AT LAUREL, MD.—WEEMS SYSTEM.

a passenger service or passenger and mail service at 125 miles per hour. Since nothing more definite has yet been done in this direction we can only quote some of the conclusions reached, and further refer the reader to the paper itself for details.

A. For purposes of demonstration it was proposed to the company that they should build a track about 4 miles (not less) in circumference, and should run thereon a train of two or three cars drawn by one locomotive. Calculations were based on the following data:

(1) A speed of 150 miles per hour on a level was to be aimed at.

(2) Cross-section of car should be a minimum consistent with the seating of passengers. This was taken at 6 feet by 5 feet—30 square feet (crowning higher along middle of car).

(3) Gauge of track should be standard—4′ 8.5″.

(4) Character of track should be simply the best that art could contrive, rails to weigh from 65 to 90 pounds per yard.

(5) Electromotive force should be as high as the art of insulation would permit.

(6) Whatever might be the number of cars in a train, all were to be so connected as to present a continuous exterior, thus presenting only one cross-section to the atmosphere.

(7) Atmospheric resistance at 150 miles per hour would be taken at 15 pounds per square foot of cross-section, a wedge or parabolic locomotive head being used. This value is 50 per cent. in excess of that indicated by experiments.

(8) Traction co-efficient, exclusive of air resistance, would be taken at 150 miles per hour to be 25 pounds per ton—this for reasons set forth previously.

From the two resistance co-efficients just given, it followed that for every square foot of cross-section, regardless of weight, 6 horse-power would be required, and for every ton of weight, regardless of cross-section, 10 horse-power must be supplied.

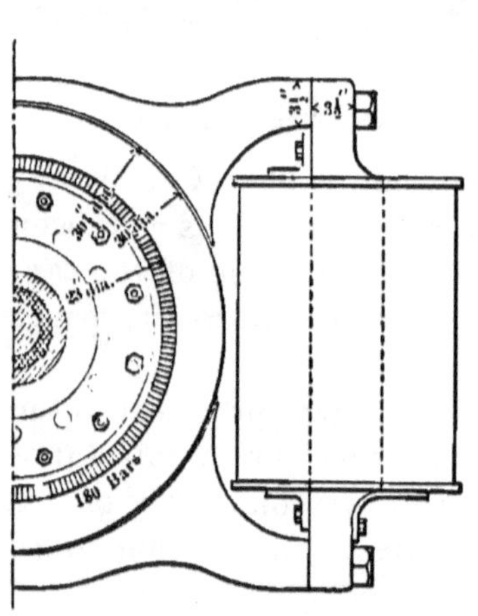

FIG. 121.—HALF SECTION OF MOTOR FOR HIGH-SPEED SERVICE.

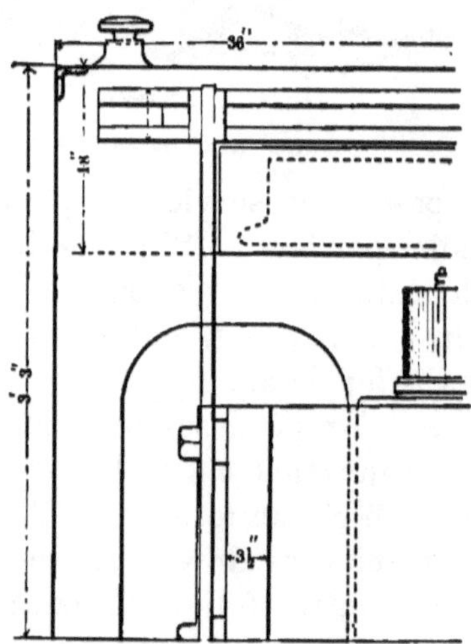

FIG. 122.—HALF PLAN OF MOTOR FOR HIGH-SPEED SERVICE.

For a locomotive of about 600 horse-power, having 30 square feet cross-section, the dead weight was calculated to be about 18 tons. Steel cars of equal cross-section were designed, weighing, empty, about 5 tons, with a carrying capacity of about 5 tons.

The important question as to power required could then be tabulated thus:

	150 miles per hour.	120 miles per hour.
Locomotive alone..........................	360 H. P.	288 H. P.
Locomotive and one car loaded...........	460 "	369 "
Locomotive and two cars loaded..........	560 "	446 "
Locomotive and three cars loaded.........	760 "	528 "

To perform the work here indicated two motors were to be provided, one armature directly on each of the two axles of the locomotive car. These motors, outlined in Figs. 121-124, were of the Manchester type. They were supposed to be at first connected in arc across a 1,500-volt circuit, each taking about 130 to 150 ampères and delivering about 250 to 300 horse-power. Should it be desired to experiment with

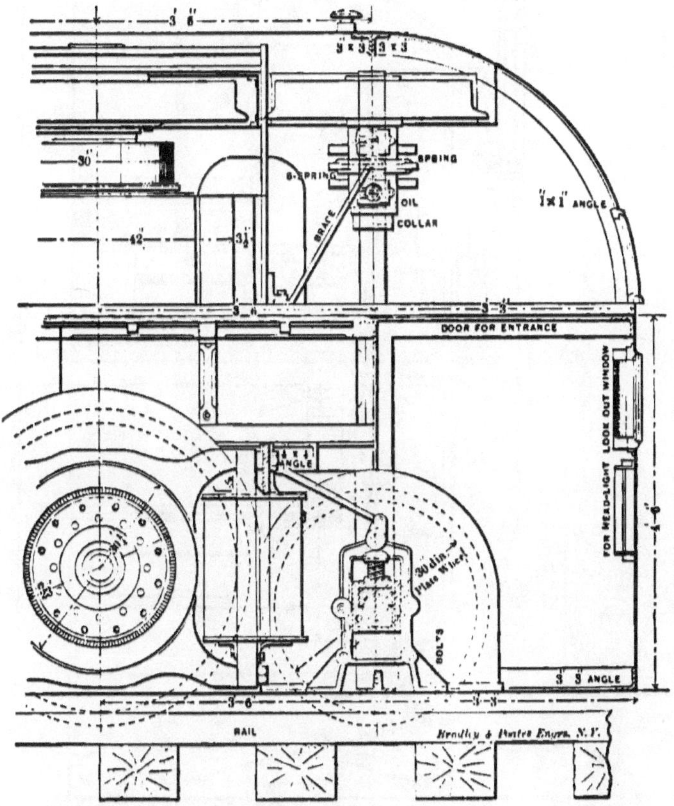

FIG. 123.—HALF LONGITUDINAL SECTION OF ELECTRIC LOCOMOTIVE.

higher line potential, the motors in series would permit 3,000 volts to be easily tried. The armatures of the Gramme type were 30 inches outside, 23 inches inside, diameter. The armature conductors were to be of about 40,000 circular mils cross-section, or from 500 to 600 c. m. per ampère. Many successful machines designed to work indoors go as low as this figure, and since the dissipation of heat must in such case proceed more slowly than in the case here considered, the figure seemed very safe. Moreover, it was intended to introduce a blast of air from the car front, which should continuously flow through the open center of the armature.

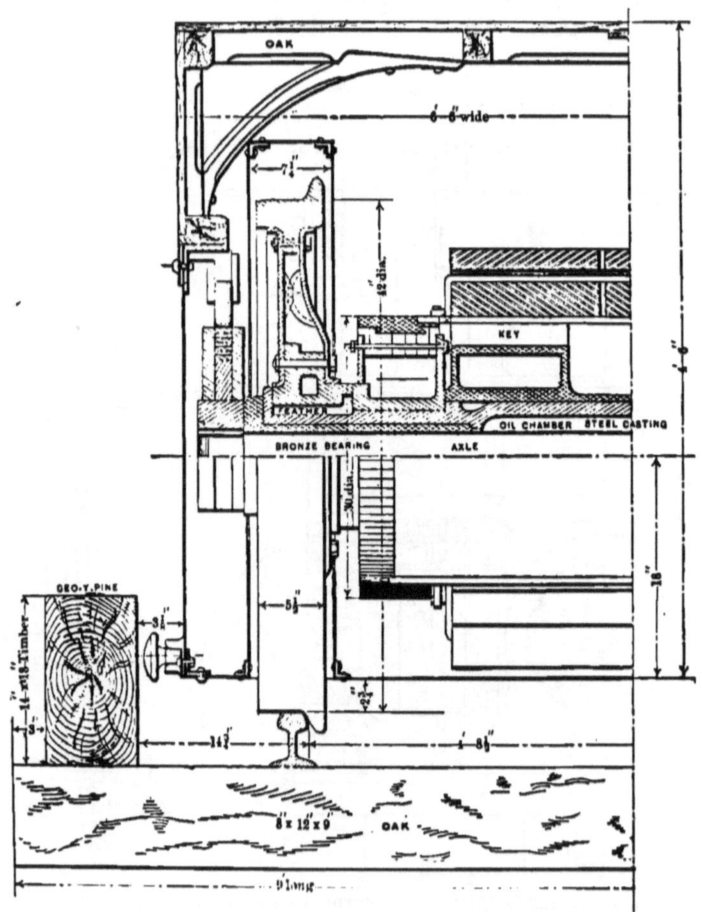

FIG. 124.—HALF TRANSVERSE SECTION OF ELECTRIC LOCOMOTIVE.

How far the rate of dissipation would in practice be carried by such means cannot now be known, but all recent evidence goes to show that the heat capacity would be much increased.

The cross-section of core in the armature was $3.5'' \times 40'' =$ 140 square inches gross, or, deducting 15 per cent. for paper insulation between disks, and 8 per cent. for conductor slots cut in periphery, 110 square inches net. Through this armature it was desired to force 22,000,000 lines (C. G. S. units) or 100,000 per square inch = 15,200 per square centimeter. The magnet coils of the two armatures were calculated to be in series with each other, but in shunt with respect to the two armatures. This variation from present street-car practice was thought to be justified by the following considerations:

First, no system of commutation could obviate the need of a considerable external resistance, which could be easily made sufficient within itself for speed control; second, the maximum torque for a given armature current is always desirable and can best be obtained by permanent saturation; third, this constant and maximum magnetization would tend to diminish sparking—an evil to be specially avoided in long-distance work, involving continuous runs of several hours.

While it was thus intended to use shunt coils at first, a little experience would soon demonstrate whether the advantage lay with this or the series winding.

The commutator, Fig. 124, was designed to be 23 inches in diameter, which would of course give rise to a very high speed relative to the brushes. It was thought better to face the mechanical trouble that might be connected with great circumferential speed rather than to diminish the diameter by decrease of sections, since this in turn might produce sparking.

Without going into the mechanical details of the commutator, we may add that a single bar could be taken out or replaced without removal from the shaft.

The rheostat for the armature circuit was designed to be of iron wire, with the expectation, however, of experimenting also with liquid or carbon resistance. The resistance and heat capacity of this main line rheostat were determined with reference, first, to the current necessary for starting; second, to currents needed for maintaining low speeds.

A small rheostat and a condenser were designed to be placed in the circuit of the magnet coils, and a rheostat also in the lamp circuit, these resistances giving delicate control.

The problem of retardation for a mass of, say, 40 tons running at 150 miles per hour is a serious one. The Galton and Westinghouse tests indicate a very low value for the co-efficient of friction between brake-shoe and wheel at high speeds; they report it in some cases at 60 miles per hour as low as 0.04. As, however, it rapidly increases at lower speeds, an average of 0.1 may be taken. The total retarding effort, *i.e.*, brake resistance and track and atmospheric resistance, must be kept below that which will just produce sliding of the wheels, and the co-efficient of sliding may be taken on an average at 0.08 of the weight on the wheels. If, then, we represent by P the maximum allowable brake pressure on the wheels, by R resistances other than brake friction, and by W the weight on braked wheels, we may write the following inequality: $0.1\ P - R < 0.08\ W$.

The value of P should be determined as that which, with a small margin, will satisfy this condition of inequality.

Following this calculation, it was found that a brake pressure of about 5,000 pounds should be applied to each wheel. This was designed to be produced by magnetic brakes similar to those used by Mr. Daft on the New York Elevated Railway. The form and approximate dimension of those brakes are shown in Fig. 125.

A mass of 80,000 pounds moving at 150 miles per hour represents 64,000,000 foot-pounds of energy. Average resistance due to brakes = 0.1 × 60,000 = 6,000 pounds. Average resistance from track and atmosphere for average speed of 75 miles = 400 pounds.

In addition, the motors may be required to generate a current which shall be wasted over the rheostat. Since the field circuit in the case of shunt motors is independent, magnetization may remain constant, and the average E. M. F. of 1,500-volt motors (*i.e.*, motors generating 1,500 volts E. M. F. at 1,100 revolutions per minute) would be, while the train was coming to rest, 750 volts. Make the external resistance such that, the two armatures being in series, an average current of 200 ampères shall flow. Then the average dynamo resistance, in pounds = 2,000. Total retarding effort = 6,000 + 400 + 2,000 = 8,400 pounds. Dividing 64,000,000 by this last quantity, we have 7,620 feet as the length of run in coming to rest, and the time of stopping about 100 seconds.

For the mechanical construction of the locomotive two plans were contemplated. One had a 12-foot rigid wheel base and no pilot wheels; the other had a 7-foot wheel base for the drivers and a pony axle in front, free to move later-

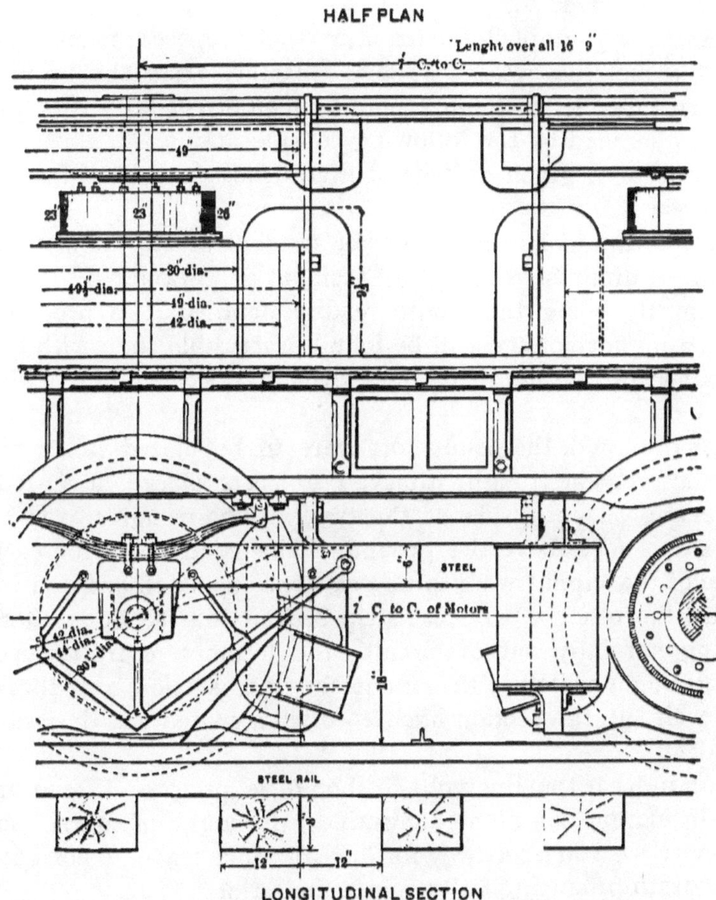

FIG. 125.—ARRANGEMENT OF ELECTRIC BRAKE FOR HIGH-SPEED LOCOMOTIVE.

ally over a certain distance, dragging the drivers in the same direction. This is the general principle of pilot wheels, so largely used on high-speed engines. The second arrangement is preferred, for in it nearly all the weight goes on the drivers.

In the first design of locomotive car the operator was to sit between the two motors, where also would be placed the controlling devices. In the second design the operator would be placed over the pony axle, the devices being chiefly in the cylindrical or parabolic head. Detailed drawings of freight and passenger cars were completed, but need not here be referred to.

As to the supply of current, it was proposed to use the double metallic system. This departure from the ground return, now so commonly employed in street-railway work, seems justified by the following considerations:

1st. For a given voltage it diminishes the cost of insulation, *per se.*

2d. It diminishes the danger of accident within the motors.

3d. It diminishes danger of accident to workmen.

4th. By being thus more readily handled it is probable that a higher voltage will be found practicable than with the ground return; hence, in fact, even first cost might not be greater.

As designed, the conductors were to be placed in an inverted wooden trough, attached to posts placed at 12-foot intervals on each side of the track. The trough was to be about 5.5 feet above the ground. The conductors for each side of the circuit were to be two—one insulated and continuous, the other a bare, flat strip broken into sections, these being normally out of circuit; the locomotive trolley arm operating a switch to throw out the rear section and throw into circuit the section ahead. Such a switch was designed in detail.

As to what the line voltage should be, progress in the art of insulation can alone determine. It would, of course, not be necessary on a roadway for long-distance travel to consider the death-producing voltage as a limitation.

In the important matter of limiting curvatures calculations were made in the ordinary way, using particular values for weight and height of center of gravity. A safe speed was then tabulated, being just one-half the speed that causes the resultant of weight and centrifugal force to pass through the outer rail of a curve. The beneficial effect of super-elevation, neglected in the calculation, would, of course, practically much increase the factor of safety.

The following table resulted from the method described:

Radius in feet.	Safe speed, by rule above, in miles per hour.
1,000	71
2,000	98
3,000	120
4,000	140
5,000	158
6,000	172
7,000	187
8,000	198

Of course, naught save practical trial could determine whether the assumed factor of safety, which is quite large enough at present speeds, would in fact be right for the higher speeds aimed at.

COMMERCIAL ASPECTS.

Copper calculations were made, based on a station potential of 3,000 and also of 6,000 volts, stations being 50 miles apart (thus supplying current 25 miles on each side), and on a service requiring one train on each 25-mile section, the drop on the line being 33 per cent. of station potential. Then, for a line of 1,000 miles, as from New York to Chicago, with copper at 15 cents per pound, and for 3,000 volts initial pressure:

	Per mile.
Cost of conducting system, including frame-work	= $7,000
" station per mile, including steam power	= 3,000
" double track, including depots, etc.	= 55,000
" train equipment, 3 cars each, 50 trains and engines	= 1,000
Total per mile	$66,000

In this, the cost of track construction is based on road-bed figures reported for the Erie Railway Company. To arrive at operating expenses, suppose 20 trains each way each day and a schedule speed of 125 miles per hour, *i.e.*, 8 hours for the trip. Suppose a station development of 800 horse-power per train on the line, or 6,400 horse-power hours per trip. In stations of large capacity working under the supposed conditions the cost of one horse-power hour = 0.90 cent, or, per trip, for power, a cost of \$57.60. For trainmen, let there be two men on each train and one per train in reserve; and suppose \$3 per man per trip of 8 hours, then, cost per train trip $= 3 \times 3 =$ \$9. Interest charge per train trip, \$66,000,000 $\times$ 0.05 $\div$ 365 $\times$ 40 = \$225.42. For maintenance of way take \$1,000 per mile of double track per annum, or per day for the supposed line \$2,700, or per trip, 2,700 $\div$ 40 = \$67.50. For general expense of operating company, suppose \$1,000 per day, or per train trip \$25. For wear and tear, oil, etc., on train itself, take \$10 per trip. Then total $= 56.60 + 9 + 225.42 + 67.50 + 25 + 10 =$ \$394.42.

For receipts we should have about the following:

Suppose the average train be of two cars (such as designed), having each a capacity of about 10,000 pounds freight (considered to be such as express and mail matter), or, say, of 15 passengers. Suppose average train load of freight 15,000 pounds, of passengers 20. Take freight at 33 cents per cwt., passengers at \$25. The income per trip in either case \$500, showing profit over fixed charges of, say, \$100 per trip, or \$4,000 per day. The total daily load of freight, each way, required to justify the service above treated is 300,000 pounds, or if 200 through passengers be carried (*i.e.*, 10 trains for passengers each way each day), then only 150,000 pounds of freight. The present movement between New York and Chicago, in mail and express, is not far from this figure. A 500-mile line connecting Boston, New York, Philadelphia, Baltimore and Washington would be even more profitable. It is to be remembered also that a service of small trains is contemplated, making fixed charges relatively very large. Much larger trains could be, and, we think, will be, run at the speeds here considered; but we believe that the smaller effort would precede the larger. It is further to be noted that the use of 6,000 volts potential and the extension of some existing roadway instead of the building of a new one would diminish first cost. As grade crossings would be out of the question, and as cities would require to be entered above or below the surface (country roads could be carried over on bridges), it is perhaps best to consider only the larger figure already used.

To demonstrate all, or nearly all, that has been outlined here would cost about $300,000, covering the construction and operation for a reasonable time of the 4-mile circular line proposed above. Last year the Baltimore company met unexpected difficulty in efforts to raise the needed money, and their operations were suspended just at the time when plans were completed—even to working drawings.

Unfamiliar as is the project, the reader will perhaps be the more ready to share the confidence of its promoters on reading the words from Prof. Henry A. Rowland and Dr. Louis Duncan, of Johns Hopkins University, to whom were submitted full details of the plans partially described in this paper. After viewing the figures for power required, they say:

"We believe from the data obtained that the values given are not too low, and that the horse-power which Mr. Crosby calculates is *not less than the amount required.* While we have investigated carefully and considered all the data obtainable, yet the existing experiments are not sufficient to accurately fix a limit to the train resistance. *We believe, however, that the value assumed by Mr. Crosby is safe.*

"The motors, the calculations and drawings for which are in the appended statement, will develop horse-power for which they are designed, namely, 500 horse-power. We point out some modifications which will be beneficial. *We believe that they will drive the train* [locomotive and two cars—O. T. C.] *at the required speed* [speed then considered being 120 miles per hour on a level].

"The possibility of a train being derailed by an obstruction on the track increases with the speed. At speeds up to 90 miles, however, there seems no increase in the number of derailments. In the case in question the center of gravity of the cars is very low, and *it would be difficult to derail them on straight parts of the track.* The radius of the curves should of course be great, but not so great as would be required for an ordinary train going at these high speeds. *The question of safety is,* however, *almost wholly a question of track construction.* Considering the form of the proposed train, its comparatively light weight making a less demand on the track, it is certain that with a carefully constructed road it could attain with safety speeds which would be impossible with trains as at present constructed. As these latter have several times made 86 miles, and often made 80 miles, *it would seem that a speed of* 120 *miles or even more, with the electric cars, would not be outside the limits of safety.*

"The plan for supplying current to the motors is feasible.

"The design of the motors is discussed in the detailed report. It is generally good.

"Should it be demonstrated by an actual test that passenger trains can be run safely and economically at a speed over 100 miles per hour from here to Chicago, the financial aspects of the case would certainly be improved. *We are of the opinion that the chances are in favor of this being accomplished by the present scheme.*" (Italics are by the present authors.)

These words, though guarded and accompanied by criticism of the detail, are yet a substantial approval of what was submitted.

A recent paper by Mr. Carl Zipernowski, setting forth plans (not yet well matured) for a single-car, 125-mile-per-hour passenger service between Vienna and Buda-Pesth, shows that the interest in and efforts toward the great step soon to be made are not confined to the American people—usually more adventurous (though perhaps less thorough-going) than their European cousins.*

---

* Since the first edition of this work a very notable contract has been entered into between the Baltimore and Ohio Railway Company and the General Electric Company. This contract calls for the construction of three 80-ton electric locomotives capable of doing service as follows: To pull a 1200-ton train at 15 miles per hour over an 6.8 per cent grade, and to pull a 500-ton train over the same grade at a speed of 30 miles per hour. The locomotives are to operate in a tunnel about three miles long, which the Baltimore & Ohio Railway Company has built through and under the city of Baltimore, Md.

The locomotive will probably be constructed to permit of much higher speed with lighter loads, those mentioned being maximum for the freight and passenger service, respectively, which are contemplated as possible within the next ten years. It is expected that these locomotives, which will be driven by direct-coupled slow-speed generators, will be in operation in the summer of 1893.

The confidence thus shown by two large companies in the future of electric traction on a large scale is most reassuring to those who follow the progress of the art.

CHAPTER XI.

COMMERCIAL CONSIDERATIONS.

In treating of the commercial value of any enterprise we are led to discuss:

1. Original investment.
2. Cost of maintenance and operation.
3. Gross income.

Of the original investment required for a street railway the two most important items are those to which it is most difficult to assign average values that may serve as useful guides. These two items are cost of franchise and cost of track construction. It may be said that the franchise may cost from $0.00 to $500,000 per mile of roadway. Track construction may run from $5,000 to $50,000 per mile of single track.

It should be added, however, that variation in this item is often due to the fact that it is made to bear a part of the charge for franchise, imposed in the shape of restrictions as to the how or where of performing the necessary engineering work. In treating the matter here we shall omit consideration of the franchise cost, and give moderate values for track construction, not applicable where special engineering difficulties are to be met.

The minimum figure above given—$5,000 per mile of single track—cannot often be reached. It supposes a 45-pound T-rail, cross-tie work, with very light grading and no paving.

In the general summary of cost intended here to be presented we shall take $10,000 per mile of single track as a safe average figure. This estimate, however, is not designed to cover the expense required to make of the track a good conductor. Let us suppose for this purpose a No. 0 bare copper wire to be laid and bonded to each rail, as described in

"Instructions to Linemen" (see Appendix B). For this item it is safe to estimate $200, covering labor, and $400 for material.

Going next to the poles, a fair figure for round, trimmed cedar poles, laid down, not set, may be taken at $2.50. Squared poles of Georgia pine may be had for $3.50; and in either case setting, if without concrete or blasting, need not cost more than $2 to $2.50. For a good sawed pole, with one cross-arm, it is safe to say that $600 per mile will cover the cost, the cross-suspension method being supposed. (As the poles are set about 125 feet apart, longitudinally, and one on each side of the street, we count on about 90 poles per mile.)

Iron poles may be had for from $18 to $27 each. The pole of lower cost will stand ordinary strains. At curves or other special points something heavier may be required. The standard round pole, in three sections, of 6, 5, and 4 inch diameter, respectively, may be had for about $22. Setting of these poles in concrete, without blasting, should not cost more than $5 each. Blasting alone may cost from $10 to $20 per pole. For a mile of iron poles (90), set, we take $2,500 as a fair figure.

They will last longer than wood, look somewhat better than the best sawed pole, but are less desirable in one respect, namely, that the liability to "grounds" is somewhat greater than with wooden poles.

The bare trolley wire will cost from $175 to $285 per mile, according to size—the latter figure covering No. 0 B. & S. at present market rates. Cost of span wire, insulators, etc., may be taken at about $100 per mile; labor of erecting wires, superintendence of contingencies, at about $325 per mile. Taking $275 for the cost of the bare trolley wire, we have $700 per mile, covering the trolley wire in place.

The amount of feed wire, of course, is indefinite, as has been explained in Chapter IV.

The following costs per mile of a good insulated wire show how the amount may be made up, having determined the length and cross-section of feeders:

No. 0 B. & S. per mile....................	$500
No. 00 B. & S. per mile.................	650
No. 000 B. & S. per mile................	850
No. 0000 B. & S. per mile...............	975

The cost of erecting this wire runs from $75 to $100 per mile. As a safe general figure we may take $1,000, covering feeder wire in place. Resuming, we have:

Track construction..................	$10,000
Earth circuit..........................	600
Wooden poles..........................	600
Trolley wire, etc......................	700
Feed wires............................	1,000
Total, track and line, per mile......	$12,900

If iron poles be used, add $2,000, or, in round figures, say $13,000 with wooden poles, $15,000 with iron poles.

Car bodies, ready for motors, may be had for from $750 to $1,500, the higher figure referring to those of unusual construction. A very good body may be had for $1,000, for a 16-foot car. The longer bodies cost from $1,250 to $2,000 each. Trucks for these cost about $600. Electrical equipment, consisting of two 15-horse-power motors and accessories, may be taken at from $1,800 to $2,500, according to the make. If one motor be used, from $1,100 to $1,800. Assuming $2,250 for the electrical equipment, the car ready to operate costs $3,500. (All the figures here and elsewhere used suppose goods are purchased for cash. No treatment can here be given to the increase of figures due to various forms of credit.)

Passing now to the station, it is to be noted that the cost per horse-power of the steam and electric plant varies considerably with the size of the installation. For convenience here, let us suppose the steam and the electric plant to be each of 500 horse-power capacity. We may then assume, per horse-power installed, for a complete high-speed steam plant, including boilers and fittings, $50 *—for a complete

* For a low-speed steam plant these figures should be somewhat increased—say 20 per cent.

electric plant, $40, or $90 per horse-power for the combination. It has been shown that there should be provided from 10 to 20 horse-power at the station for each car on the line. Taking 15 horse-power per car, it appears that an investment of $1,350 must be made in station machinery per car to be operated.

The cost of ground space for station, car-barn and office will, of course, vary widely. The cost of buildings will also vary considerably, there being in this item a considerable "personal equation" of the particular management. Since we desire to present some reasonable figure, we find that $50 per horse-power of station capacity may serve in a majority of cases to cover all real estate. Reduced to the investment per car to be operated, this becomes $750. If, now, there were any fixed number of cars per mile of track or mile of street, the whole investment might be expressed at a certain amount per car operated. This ratio, however, varies widely—being principally determined by the size of the city in question.

It may be roughly stated that cities of population from 15,000 to 100,000 show from one to two cars per mile of track, while from 100,000 to 300,000 inhabitants there should be from two to four cars per mile.

So, in still larger cities, a slightly higher ratio.*

Let us take two cases—first, that in which we find one car per mile of track. The investment per 16-foot car then becomes:

One mile track and line, wooden poles.....	$13,000
One car ready to run.....................	3,500
Steam and electric plant per car..........	1,350
Real estate............................	700
Total, per car.........................	$18,550

Then, second, for three cars per mile, iron poles on line, the items are as above, except track and line, which now

* The West End Company of Boston serves about 650,000 people. It has 260 miles of track and 2,745 cars. New York City has 276 miles of track and 2,519 cars.

enter at $5,000 (one-third of $15,000) per car, or the total per car $10,550.

We may now find the interest charge per car-mile, which is a convenient—perhaps the most convenient—unit of operating expense and earning capacity.

Each car may be taken to make 35,000 miles per year—a little less than a steady average of 100 miles per day, for all cars.

At 5 per cent. on $18,550 we have per annum . . $927.50
or, per car-mile, 2.65 cents.
At 5 per cent. on $10,550 we have per annum . . 527.50
or, per car mile, 1.5 cents.

Thus far the reports of operating expenses of electric railways have been widely divergent. Some order is now beginning to appear. It should be recognized, however, that considerable differences as between one company and another must continue to exist until local prices of labor and material shall be everywhere uniform. We have seen many useless contentions as to the proper average figure for operating expenses, one contestant having in mind labor of motor-men and conductors at 22 cents per hour, as in Boston, Mass., the other considering the same item at 11 cents per hour, as in Knoxville, Tenn., and very many other cities of comparatively small population.

So, the one has in mind coal at about $5 per ton, as in Boston, the other coal at about $1.75, as in Knoxville.

Again, one may consider a road of 5 or 6 cars which runs its own independent station at a necessarily high figure per car, while another refers to a similar small road that may have its dynamos in an electric-light station, an equitable arrangement being made for the additional coal and wear and tear—the additional attendance and general expense being almost *nil*.

Let us take up the various parts of operating expense, *seriatim*.

As to the cost of getting upon the line the electric power needed *for one car-mile of work*, it could be, in a station of about 500 horse-power capacity, not far from 1.0 to 1.2 cents, as shown by Table III., Chapter X. (*i.e.*, table of cost of one horse-power hour in various stations). The cost per horse-power hour is a little more than this figure, since one horse-power hour will produce a little more than one car-mile of service, say 1.2 miles. For long cars the figures should be increased by at least 50 per cent.; we would then have a station cost of 1.5 to 1.8 cents.

These figures, too, may hold for a standard 16-foot car drawing a trailer. In the figures here given, taken from the table, coal enters at 0.475 cent per horse-power hour. It is assumed to cost $3 per ton and to be burned at the rate of 3.2 pounds per horse-power hour.

As a matter of fact, the consumption per horse-power hour generally goes far beyond this. Bad design of station, as in wrong relation between capacity and service of engines, poor piping, etc., together with poor firing, may double, even treble, this figure. We must, therefore, be understood as speaking of what *can* be done by *good* practice, not of what *is* done by *bad* practice. The effect upon the cost per car-mile, should coal consumption be trebled, would be to add about 0.8 cent, making nearer to 2.0 than 1.0 as the round figure. We have seen reports of less actual cost than that resulting from the table and mentioned above. But in a majority of cases actual practice is worse than the tabular figure. We have therefore averaged several cases of practice generally called good, and find 1.35 as a fair figure. Many statistics show higher. *

The difference between maximum and minimum values is striking. But it should be noticed that, in the average, roads operating 3 cars are grouped with those operating 140. In many respects, but especially in the cost of power, must there be very wide differences between very small and very large plants. It scarcely seems useful to average results obtained

---

* Mr. J. S. Badger (*The Electrical World*, October 30, 1891) reports as the average of twenty-two trolley roads a cost of 2.32 cents for this item. It is made up as follows.

Maintenance of power plant, repairs on engines, dynamos, etc.

Highest.	Lowest.	Average.
0.86	0.05	0.36

Fuel, wages, oil, waste, etc., at station.

Highest.	Lowest.	Average.
4.95	0.48	1.96

These two items should be summed to get cost of power in the line, making 0.36 + 1.96 = 2.32.

under such widely different conditions. The figures given in the text are not expected to apply to very small plants. They will be useful as applied to companies operating, say, 10 cars or more. See curves (page 286) showing change in cost of one horse-power hour according to change in magnitude of station. The change up to about 500 horse-power is very rapid.

As an example of good station service the East Harrisburg Street Railway Company may be cited. They kindly report to us as follows:

POWER PLANT EXPENSE FOR MONTH SEPTEMBER, 1891.

Coal at $2.10 per ton................	$241.12
Dynamo oil..........................	5.55
Cylinder oil........................	5.80
Waste and coal oil..................	3.81
Engineer and fireman...............	206.63
Sundries...........................	21.24
Total...............................	$484.15

Number of cars, 18; mileage, 63,028. Cost per car-mile, 0.75 cent.

Many contracts heretofore made between lighting companies and railway companies have been on a car-day basis, with some expressed or implied assumption concerning car-miles per day. Unless the parties to such contracts are satisfied to let matters rest on a guesswork basis, it is certainly best to at least come as near to the facts as possible by the car-mile unit. The most accurate guide for proper charge is to be found in watt readings taken at the station, all the peculiarities of service (*pro* and *con* the railway company) being thus taken into account.

When it is remembered that occasional line leaks have been found—in one case reaching a flow of 60 ampères—the wisdom of this course will be better appreciated. True, there are as yet no satisfactory large wattmeters for railway stations: but several small ones may be used in multiple, or frequent reading of the ammeters (checked by frequent reading of the voltmeter) will generally give a closer determination than can be made by estimating the consumption per car-day, or even per car-mile.

In case a contract of this sort is about to be made we would advise the following course: Agree upon a rate per horse-power hour, a minimum of consumption being fixed, or on a sliding scale with respect to amount of power used; have five-minute readings on the ammeter, and fifteen-minute readings on the voltmeter taken during a period of, say, four weeks; keep a close record of the car-miles made during the same period.

In this way a very reliable figure for power per car-mile can be had, and the frequent readings may be discontinued—to be taken up again when the season changes, as from summer to winter, and to be occasionally renewed as a check on any changes in service. As soon as reliable recording wattmeters can be had for all stations there need be no more uncertainty in a contract of this sort than is the case in a contract for the supply of gas to a building.

Some existing contracts call for a certain sum per day for each generator of a given size that may be in operation. This method is of course less accurate than the others. A just relation between service and cost *may* be hit upon and it may not.

For a small number of cars, say six, it is not uneconomical to pay 4 cents per car-mile rather than erect a suitable station. There will not be a great profit in this to the lighting company unless they have already all-day attendance as a part of the lighting service. Four cents per horse-power hour or per car-mile leaves a handsome percentage of profit to the power station, and yet the railway company can scarcely compete with that figure by their independent production on a very small scale.

In the matter of motor repairs the advances of the current year will make a great and welcome change. There has been a wide variation in the experiences of different companies; averages have been less useful guides than will now be the case. Perhaps the widest fluctuation of "luck" has been found in connection with armature repairs; next would follow field windings. It is safe to say that ten roads might be chosen from which, selecting two sets of five each, the average armature and field repair account for the one set would be five times that of the other. Individual instances may be found in which one road has had five times as much intelligence in its management as a neighboring road, and may have had armatures made with five times or with one-fifth of the care given to those that chanced to be received by that neighbor.

The winding of electrical apparatus is largely a matter of conscience as well as skill. Constant supervision cannot be given to every workman, yet a moment of negligence or ill-will on his part may fatally weaken an armature. Shop-testing is important—should never be omitted; yet it cannot bring out all weaknesses as they are brought out by bad handling on the road, or even by the exigencies of legitimate service. In all this matter the most marked improvement recently made is the use of Gramme or hollow-cylinder armatures instead of Siemens or solid-core armatures.

Field windings have been much improved by good detail work and by decreasing the heat developed in them, or increasing the rate of dissipation of that heat. In gears, too, there has been a great change for the better. The number has been diminished, the speed reduced, and those that remain have been carefully inclosed in dust-proof casings. These, besides excluding dust, pebbles, etc., serve also as reservoirs for oil or grease, into which the larger gear (axle gear) dips half an inch or more, thus lubricating itself and the armature pinion. Experience with these improved forms has not been extended enough to show accurately what their repair rate is; but that it will be *greatly less* than that of the earlier high-speed, exposed motors, we have *prima facie* evidence of the strongest character, and also the gratifying results of some months of practical use for some hundreds of such motors as are here described.

The following table, based on the returns of several millions of car-miles of hard work under good management, shows the life of various wearing parts in *double-reduction* motors. These machines were covered on the side, rather loosely, by canvas, and under the bottom a metal pan was hung. Dust was not excluded, but the protection from pebbles, etc., was good.

Article.	Life in car-miles.	First cost.	Cost per car-mile in cents.
Axle gear	29,600	$7.50	0.025
Intermediate gear	28,600	6.25	0.022
Intermediate pinion	10,050	6.75	0.067
Armature pinion	8,260	6.05	0.073
Trolley wheel	5,750	1.40	0.025
Armature bearing	24,200	3.70	0.015
Intermediate bearing	35,100	4.40	0.013
Total			0.240

It may be said that many roads have shown a repair account for motors, gears and trolleys running from 2 to 3 cents per car-mile—some even higher. We can see no reason why, with slow-speed machinery, this item need go beyond 1 cent. It can be shown to be considerably less than this for the period during which the single-reduction motors have been in use.

Maintenance of the line is well covered by 5 per cent. on its cost. Averaging this in iron and wooden pole construction, at $3,000 per car for line investment and at 35,000 car-miles per year per car, this item amounts to 0.4 cent per car-mile. At $1,000 line investment per car it becomes 0.14 cent. It is an interesting fact that on curves, where traffic is very dense, the sides of the trolley wire are worn by the trolley flanges so as to require renewal of the wire in a very few years.

Maintenance of dynamos and engines has already been accounted for in taking the cost of power from the table.

Maintenance of track for electric railways is still a somewhat uncertain factor. Typical, well-constructed electric railway road-beds are not numerous; use of them has not been long; reports have not been full; paving expenses (varying widely according to municipal requirements) have been combined with those of the track proper.

It is found that on steam railways the injury to track is approximately in proportion to the number of locomotive miles, rather than total ton-miles, including train weights. In other words, each passage of a set of driving wheels does a certain amount of injury, and this amount is great in comparison to that done by the car wheels following. In street railway practice car-miles and driving-wheel miles are almost always as 1 to 2 (or 1 to 4), since, except for trailer service (rarely exceeding two trailers), locomotive and carrying car are combined. The West End Street Railway Company of Boston, Mass., operating about 325 electric cars over tracks not yet completely relaid for electric services, reports the track maintenance at 1.08 cents per car-mile. The same figure is reported by the Pleasant Valley Railway Company, of Pittsburg, Pa.* Until more extended figures can be obtained we may adopt these as a fair guide. No distinction is here made between effect of long cars and short cars, though the former are in considerable number on the West End lines.

Maintenance of car bodies and trucks is thoroughly covered by allowing 20 per cent. per annum on their cost, *i.e.*, $250, or 0.72 cent per car-mile.

We have already stated that in the matter of wages of conductors and motor-men wide differences are to be found. The pay per hour is known to vary from 10 cents to 22 cents. Let us take 18 cents as an average. We then have 36 cents for the two men. They will run very nearly 8 miles per hour while under pay, hence the cost per car-mile is 4.5 cents.

On a few electric roads, in rather small cities, no conductors are employed. Even with horse cars, however, it has been found of advantage to have conductors in nearly all towns of considerable size. Of cities in the United States having more than 100,000 inhabitants we now recall only one, New Orleans (population 242,000), in which conductors are not employed. The higher speed made, the greater number of passengers carried, and the occasional attention required for the trolley are all considerations which give greater value to the conductor on an electric than on a horse car.

* Rather curiously, just half of this amount is reported by Mr. J. S. Badger (*The Electrical World*, Oct. 30, 1891) as the average given by twenty-two trolley roads for maintenance of way. The highest figure given in the averages is 1.86 cents, the lowest 0.10 cent. Probably reconstruction of old track will in part explain the very high figure; use of perfectly new track for a short time probably explains the very low figure. Mr. Badger's report appeared after the text above was written, and is a valuable contribution to our knowledge of the actual operating expense. In this particular item, however, we think it safer to adhere to the larger figure, 1.08, reported by companies operating a considerable mileage for a considerable time.

The charge for accidents is another widely varying item On the same road it varies much from month to month, or from season to season. From an examination of a number of cases we find that 0.25 cent per car-mile is not far from a fair average. Insurance companies have recently placed accident risks for a number of electric railway companies, generally for a fixed percentage of gross receipts. This percentage has varied from 0.75 of 1 per cent. to 2.25 per cent., the higher figure in cities having a bad accident record, due to unusually crowded streets. To show the relation existing between the figure just assumed and those held in view by the insurance companies—a certain railway company has gross receipts of about 20 cents per car-mile. Its rate is 1 per cent. (very nearly). This means 0.20 cent per car-mile, a little less than the figure assumed. Three months' average —April, May and June, 1891—for this item on the West End Railway Company of Boston shows .57 cent per car-mile. Mr. Badger, in the report referred to in the last foot-note, takes 0.061 cent. This seems entirely too low.

General expense, covering officers' salaries, office expenses, superintendent, taxes, insurance, etc., runs from 1.0 to 2.25 cents per car-mile. In making general estimates it is best to take at least 2.0 cents.

Resuming, the various items for an average case appears thus:

COST PER CAR-MILE (STANDARD 16-FOOT CAR).

Power delivered on line	1.35	cents.
Repairs on electric machinery of car	1.00	"
Repairs on line	0.43	"
Conductors and motor-men	4.50	"
Repairs on cars and trucks	0.72	"
Maintenance of roadway	1.08	"
General expense	2.00	"
Accidents	0.25	"
Total	11.33	cents.

Throughout the above calculation *good management* is supposed.

An inspection of these items shows the great advantage to be attained from running trail cars whenever their use is justified by the traffic. The consumption of power for the two cars will not exceed 50 per cent. of that required for the motor car alone, since all losses of transmission are met in the former. One active conductor will take care of both cars. Maintenance of line, of track, and of car machinery are scarcely increased at all. Nor does it appear that general expense would be considerably increased. Assuming, however, that this item must always be in proportion to the volume of business done, we have still, on the above basis for a motor car, not more than, say, 5 or 6 cents for the cost of running a trail car-mile.

Comparing the cost of operating a two-car train (motor car and one trailer) with that for one long double-truck car, it would appear that they are very nearly the same. Service based on the long car is, however, much less elastic in meeting traffic demand than is that based on the use of trailers. The additional expense over that for the standard car must be met at all times, whether traffic be heavy or light. More time is required to take on and let off passengers from the long car than from the two shorter ones, or even from a single standard car carrying the same load as the long car.

Let us see approximately what it costs to stop a car, without now considering the loss of schedule speed (or the greater consumption of current, if speed be maintained) due to increase of time for stops. For convenience, take the even figures 12 cents and 16 cents as the cost per car-mile for short and long cars respectively. At 9 miles per hour, schedule speed, we have the cost per hour of running equal to 108 and 144 cents respectively. This, per minute, equals 1.8 cents and 2.4 cents respectively. Suppose the average time lost per stop, from full speed to full speed again, be 10 seconds and 15 seconds, respectively, for short and for long cars. These figures correspond very nearly to the observed facts in several cases. Then for cost of a short-car stop we have $1.8 \div 6 = 0.3$, for a long-car stop $2.4 \div 4 = 0.6$ cent, or just twice as great.

The advantage of the long car over the motor car and

trailer lies in its smoother motion over the track. How far this will enter as an earning factor it is difficult to say. We know that improvement, the refinement of service, generally carries with it increase of travel. The pedestrian, seeing a chance to get a seat, enters the car which, if already crowded, he would not have entered. High speed, carpeted seats, good lights, nice trimmings, general neatness, politeness of employees—all these increase the number of passengers. And all these may be given, together with greater ease in entry and exit, in the motor and trail car. Whether the smooth riding of the double truck will attract additional passengers, only a long experience can determine.* At present there is a reaction toward the trail car.

As to how far it is best to go in furnishing seating capacity, there are, indeed, certain conditions of heavy travel—as when every standard 16-foot car that can be run shows standing crowds—for which it is plain that either the long car or the train must be applied. For traffic less than this, yet quite heavy for the single car, and which might be increased by further comfort in riding, we can only say that each particular case must be determined by trial and good judgment.

Passing now generally to the question of earnings, we can do little more than refer to the facts as disclosed by reports. It is evident from the figures above given how much must be earned in order that the enterprise may be self-supporting. We may say roughly that an electric car must earn 15 cents gross per mile run in order to pay 5 per cent. on the investment—*no account being taken of cost of franchise.* In large cities, where labor is considerably higher, this figure must be increased to 18 or 20 cents.

Since, in the United States, the fare is almost everywhere the same—5 cents—it follows that for every mile run the company should receive from three to four passengers in order to live. Expressed in passengers *per car per day*, assuming 110 miles per day as made by each car, the required numbers are from 330 to 440. The relation between the

* The question of side entrances for these long cars is now considered as a serious one by all the companies who have had any experience in their use. A recent double-decked long car, manufactured by the Pullman Company, of Pullman, Ill., shows both side and end entrances.

population of a city and the total number of people who ride in street cars is an interesting inquiry. It is also full of irregularities. The "lay" of the city, the occupation and habits of its inhabitants, the quality of the service given, are all of great value in determining the number of fares to be taken from a given number of people. This ratio is frequently expressed in this way—that the whole population is carried so many times per year or per day. On pages 324 and 325 we give a table setting forth important commercial

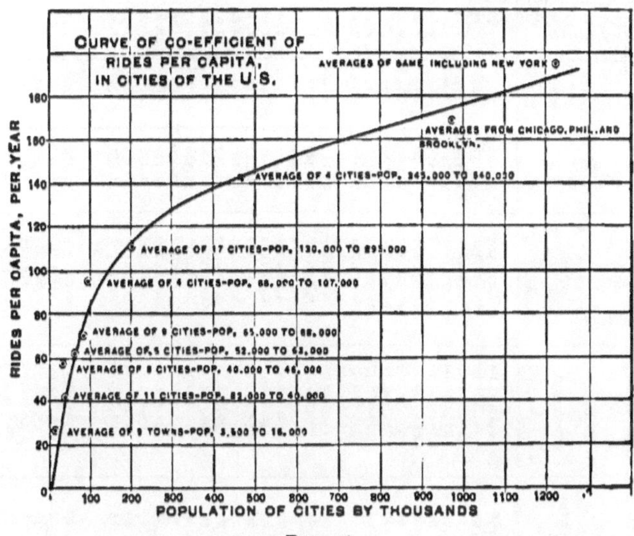

FIG. 126.

facts of the street-railway service in a number of cities of the United States.

The "cars possible to be supported" column shows the number of cars that could make a living, assuming from the above discussion that 400 passengers must be carried per car per day.

The last column shows the actual number of cars reported from these cities, including every car, running or not.

The increase of what may be called the riding co-efficient with increase of total population appears plainly from the table. This relation is displayed more clearly by the above curve, Fig. 126.

STREET RAILWAY STATISTICS FOR FIFTY CITIES.

City.	State.	Population.	Area.	Street Car Trackage.	Gross Receipts from Operation.	Passengers Carried.	Pass. Carried per day.	Rides per Capita.	Amt. Paid per Capita per Annum.	Cars in Constant Operation Possible to be Supported	Cars Reported.	*Box Cars.	Open Cars.
New York City	N. Y.	1,515,301	40.2	276	$10,941,175.89	218,823,517	599,516	144	$7.22	1,499	2,373	1,912	461
New York City	N. Y.	Elevated.		90	9,291,681.60	185,833,632	509,133	123	6.15		921		
Philadelphia	Penn.	1,046,964	129.4	455	7,481,954.27	164,458,842	450,572	157	7.15	1,126	1,691	930	761
Brooklyn	N. Y.	806,343	26.5	275	4,833,945.06	101,705,821	278,646	126	5.99	697	2,486	1,344	1,142
Brooklyn	N. Y.	Elevated.		89	2,504,538.70	49,429,228	135,423	61	3.05		373		
Boston, †	Mass.	640,000	35.3	269	5,685,299.04	114,991,678	315,046	179	8.88	775	2,045	1,002	1,043
Buffalo	N. Y.	255,664	39.	59	761,344.86	16,425,983	45,003	64	2.97	113	207	143	64
Pittsburg, ‡	Penn.	238,617	29.6	146	1,621,501.94	33,241,793	91,073	97	4.71	228	521	287	234
Rochester	N. Y.	133,896	15.6	64	578,658.47	11,370,134	31,151	85	4.32	78	192	127	65
Providence	R. I.	132,146	19.	58	935,630.85	18,473,724	50,613	139	7.08	127	341	227	114
Allegheny City	Penn.	105,287	7.3		See Pittsburg.								
Albany	N. Y.	94,923	11.	31	254,479.90	4,787,847	13,117	50	2.65	33	116	105	11
Syracuse	N. Y.	88,143	20.	61	306,732.40	6,330,158	17,343	72	3.48	43	132	99	33
Worcester	Mass.	84,655	34.	22	232,474.66	4,851,574	13,292	57	2.73	33	73	39	34
Lowell	Mass.	77,696	11.2	27	209,830.45	4,183,392	11,461	54	2.70	29	105	57	48
Scranton	Penn.	75,215	25.	29	146,801.00	2,988,403	8,187	40	1.95	21	52	29	23
Fall River	Mass.	74,398	11.	16	154,747.62	3,143,212	8,611	42	2.08	22	66	30	36
Troy, §	N. Y.	83,388	5.2	27	356,867.68	6,244,276	17,107	102	5.86	43	120	61	59
Reading	Penn.	58,661	7.	13	130,648.63	3,175,979	8,701	54	2.23	22	79	44	35
Lynn, ‖	Mass.	150,420	10.6	59	601,094.32	11,637,089	32,704	77	4.00	82	252	116	136
Lawrence	Mass.	44,654	6.7	11	79,250.50	1,585,506	4,344	36	1.77	11	37	19	18
Springfield	Mass.	44,179	34.5	21	180,990.50	3,628,008	9,940	82	4.09	25	74	42	32
Utica	N. Y.	44,007	6.	32	131,584.20	2,676,790	7,333	61	2.99	19	85	67	18
Manchester	N. H.	44,126	40.	15	46,627.33	888,990	2,436	20	1.05	7	25	13	12
New Bedford	Mass.	40,733	19.	16	143,000.44	3,235,116	8,864	79	3.51	23	72	35	37
Erie	Penn.	40,634	7.	15	74,603.50	1,553,718	4,257	38	1.83	11	38	21	17
Somerville	Mass.	40,152	4.		See Boston.								
Cambridge	Mass.	70,028	5.8		See Boston.								
Portland	Maine.	36,608	3.5	10	136,010.53	2,728,935	7,614	75	3.72	20	54	36	18
Holyoke	Mass.	35,528	12.5	7	33,884.99	693,419	1,900	20	.96	5	18	9	9
Binghamton	N. Y.	35,093	10.	32	75,032.65	1,662,587	4,555	48	2.14	12	46	28	18
Yonkers	N. Y.	31,945	12.5	6	11,706.10	234,122	641	8	.37	2	14	12	2
Salem	Mass.	30,735	7.	41	228,503.73	4,475,950	12,263	145	7.43	31	122	59	63
Long Island City	N. Y.	30,396	10.	38	151,350.28	2,936,847	8,046	97	5.00	21	114	75	39
Elmira	N. Y.	28,070	7.2	14	33,144.06	662,882	1,876	24	1.20	5	31	21	10

CITY.	STATE.	POPULATION.	AREA.	Street Car Trackage.	Gross Receipts from Operation.	Passengers Carried.	Pass. Carried per day.	Rides per Capita.	Amt. Paid per Capita per Annum.	Cars in Constant Operation Possible to be Supported	Cars Reported.	*Box Cars.	Open Cars.
Chelsea	Mass.	27,850	2.5		See Lynn.								
Pawtucket	R. I.	27,502		9	$48,259.49	964,814	2,643	35	1.76	7	23	23	
Haverhill	Mass.	27,322	24	14	56,801.94	936,170	2,565	34	2.08	7	38	21	17
Brockton	Mass.	27,287		18	102,647.25	2,210,706	6,057	81	3.76	15	46	21	25
Auburn	N. Y.	25,887		6	11,833.16	236,663	648	9	.45	2	16	12	4
Taunton	Mass.	25,389		8	37,560.31	767,292	2,102	30	1.48	6	23	13	10
Newton	Mass.	24,357	22	6	16,685.02	348,597	955	14	.68	3	17	9	8
Hempsted	N. Y.	23,517			No Report.								
Newburg	N. Y.	23,263		5	24,794.85	495,897	1,359	21	1.05	4	11	11	
Malden	Mass.	22,984			See Lynn.								
Poughkeepsie	N. Y.	22,836		4	16,954.52	268,622	736	12	.74	2	9	9	
Cohoes	N. Y.	22,432	55		See Troy.								
Fitchburg	Mass.	22,007		6	30,961.92	509,143	1,395	23	1.41	4	13	6	7
Oswego	N. Y.	21,826		3	6,494.13	134,536	369	6	.30	1	5	5	
Lewiston	Maine.	21,668		10	21,309.74	426,195	1,168	20	.98	3	22	15	7
Gloucester	Mass.	21,262		6	34,436.23	650,692	1,783	31	1.61	5	18	8	10
Kingston	N. Y.	21,181		3	29,044.70	446,770	1,224	21	1.37	3	8	8	
Woonsocket	R. I.	20,759		6	17,911.75	358,235	982	18	.86	3	10	10	
		6,978,924	716.6	2,428	$43,810,786.22	997,813,484	2,734,754	2,811	139.79	5,228	13,134	7,160	4,680

* This includes motor cars not designated as box or open.

† Based on the aggregate population of Boston and suburbs through which the West End Street Railway cars run.

‡ Based on the aggregate population of Pittsburgh and Alleghany City; a total of 343,904.

§ Based on the aggregate of Troy and Cohoes.

‖ Based on the aggregate population of Lynn and towns through which the Lynn and Boston Street R. R. cars run.

GENERAL NOTE.—The number of cars reported in each city is much in excess of the number in actual service. This fact is due to the large number of old cars unfit for service, but which are carried on the books of the several companies from year to year or until such time as opportunity presents to dispose of them.

In the city of Utica, N. Y., this fact is clearly demonstrated.

The total number of cars in operation is about 32, whereas the total number reported is 67, showing that the larger portion was surplus cars.

It is personally known to the authors that this surplus was made up of old cars, a number of which were sold for $25.00 each, which fact shows the unfit condition of these cars for service.

This condition of affairs of surplus old cars is in some degree general with all street railway companies.

In endeavoring to treat of any particular case, the wide variations from which the above averages were made should be borne in mind.

It has become a common practice to express the cost of operating as a certain percentage of gross earnings. Such a relation is a useful one to be defined. Yet it is to be remembered that unless the service in a given city has been well studied and carefully managed, the figures shown for the actual business may be but a poor guide as to what is possible. If approximation be desired and this method be used, operating expense may be taken at 60 to 70 per cent. of gross earnings.

Experience thus far shows that the substitution of electricity for horses is followed by an increase in gross receipts ranging from 25 per cent. to 300 per cent. over the original receipts. Even when the service with horses has been good considerable increase (rarely less than 30 per cent.) is marked. No better evidence could be had of the desirability of the electric railway. The expense per car-mile will generally prove to be less than by horses. We say *generally*, because in a city where horses and forage are exceedingly cheap, while coal and mechanical skill are unusually high, it might be impossible to secure the economy usually counted upon. As an illustration of the difference, exemplified on a large scale, we subjoin a statement made by the West End Railway Company of Boston, showing the greater economy per car-mile of electric traction. It is to be noted in this case that the figures for horses represent the result of years of effort toward the best attainable practice, while for electric traction the figures cover contract prices for repairs made very early in the history of the business and consequently at a higher price than would now be made. Moreover, the practice generally being new, it remains for time to develop those small economies which show a system at its best.

For both horses and electricity these figures of cost are high as compared with the rates attainable elsewhere. Coal, forage and labor are, in Boston, well above the average for the entire country. Other statements are also given in a

OPERATIONS OF THE WEST END STREET RAILWAY CO., BOSTON, MASS.

1891.	ELECTRIC.				
	APRIL.	MAY.	JUNE.	JULY.	AUGUST.
Gross receipts	$134,321.00	$144,638.00	$145,319.00	$144,553.00	$136,246.00
General expenses	8,193.00	7,796.00	7,465.00	6,956.00	6,027.00
Maintenance expenses	9,795.00	8,891.00	8,078.00	11,999.00	12,498.00
Transportation expenses	37,653.00	36,552.00	31,556.00	31,895.00	31,123.00
Motive power	30,194.00	30,924.00	26,360.00	26,399.00	25,630.00
Total operating expenses	85,835.00	84,163.00	73,459.00	77,249.00	75,278.00
Net earnings	48,487.00	60,475.00	71,860.00	67,304.00	60,968.00
Miles run	394,459	376,321	360,567	377,191	363,836
Ratio of mileage	26.68	25.58	25.15	25.19	24.92
Per cent. operating expenses	63.36	58.18	50.55	53.44	57.13
EXPENSES PER MILE RUN.	CTS.	CTS.	CTS.	CTS.	CTS.
Motive power per mile	07.65	08.22	07.31	07.00	07.04
Car repairs per mile	01.39	01.33	01.18	01.17	01.57
Damages per mile	00.75	00.89	00.16	00.12	00.23
Conductors and drivers per mile	07.33	07.36	07.25	06.92	06.85
Other expenses per mile	04.63	04.56	04.47	05.27	05.00
Total expenses per mile	21.75	22.36	20.37	20.48	20.69
Earnings per mile run	34.05	38.43	40.30	38.32	37.45
Net earned per mile	12.30	16.07	19.93	17.84	16.76

1891.	HORSE.				
	APRIL.	MAY.	JUNE.	JULY.	AUGUST.
Gross receipts	$344,396.00	$374,606.00	$404,225.00	$409,878.00	$392,474.00
General expenses	22,514.00	22,682.00	22,218.00	20,657.00	18,160.00
Maintenance expenses	22,692.00	17,748.00	18,560.00	27,228.00	18,669.00
Transportation expenses	114,001.00	110,154.00	106,836.00	108,732.00	105,953.00
Motive power	117,740.00	118,972.00	116,211.00	116,271.00	113,351.00
Total operating expenses	276,947.00	269,556.00	263,825.00	272,888.00	256,133.00
Net earnings	67,449.00	105,049.00	140,399.00	136,990.00	136,341.00
Miles run	1,083,887	1,094,683	1,073,218	1,120,377	1,096,376
Ratio of mileage	73.32	74.42	74.85	74.81	75.08
Per cent. operating expenses	80.62	71.95	65.27	66.58	64.52
EXPENSES PER MILE RUN.	CTS.	CTS.	CTS.	CTS.	CTS.
Motive power per mile	10.86	10.86	10.83	10.38	10.33
Car repairs per mile	00.93	00.60	00.61	00.61	00.03
Damages per mile	00.78	00.37	00.15	00.06	00.10
Conductors and drivers per mile	08.28	08.24	08.25	08.23	08.10
Other expenses per mile	04.70	04.55	04.74	05.07	04.81
Total expenses per mile	25.55	24.62	24.58	24.35	23.37
Earnings per mile run	31.77	34.22	37.66	36.58	35.80
Net earned per mile	06.22	09.60	13.08	12.23	12.43

Allowances should be made for the omission of cents in the above statement.

foot-note,* showing results obtained elsewhere on a somewhat smaller scale.

Those companies having severe snows to contend with must expend a considerable, but widely variable, amount of money in order to clear their tracks. In some of our most northern cities a period of inaction is forced by the magnitude of expense that would be incurred in clearing tracks, and also because such effort on the part of the railway company would render the streets unfit for the service of "runners"—wheeled vehicles being out of the question for any transportation.

The electric snow-sweepers now being put on the market will undoubtedly diminish the interruption to traffic. But they will demand a good supply of power. Even with their assistance the severe winter months must remain but slightly profitable. The average road lying north of 40° north latitude

---

* In February 1891 President D. F. Henry, of the Federal Street and Pleasant Valley Passenger Railway Company of Pittsburg, Pa., made a report of the operation of the company for the preceding six months, in which was embodied some interesting information in this line. We quote from it as follows.

Passengers carried, six months, 3,370,531; receipts, $168,526.55; mileage, 31 cars, averaging each 108 miles per day, total daily average, 3,348 miles. Cost per mile:

Conductors and motor-men	per mile,	6.80	cents.
Motor and electric repairs	"	1.68	"
Mechanical repairs	"	1.14	"
Motive power	"	1.54	"
Overhead system	"	0.45	"
Maintenance of way	"	1.08	"
General expense	"	0.88	"
Stables	"	0.46	"
Officers and salaries	"	0.84	"
Interests	"	2.71	"
Tolls	"	0.25	"
General labor	"	2.43	"
Total		20.26	cents.

as total cost per mile of operating expense, fixed charges, salaries and interest, as against receipts per car-mile of 27.55 cents; showing net earnings per mile of 7.29 cents.

Separating the above into strictly operating expense and fixed charges, we have operating expense, 12.74 cents per mile; and in comparing the same with the cost of operating the horse line, which was 10 cents per car-mile, we must remember that we then paid but one man on a car, where we now pay four, or 3 cents per mile, against 6.80 cents now. This increased cost per mile for conductors and motor-men is a necessary adjunct of rapid transit, and is not peculiar to the system. Allowing for this, we have a difference of 1.04 cents per car-mile in favor of electricity as against animal power.

I call attention to the fact that we have one of the most difficult roads in existence to operate, with streets having over one hundred curves, many of which have a short radius, and with heavy grades, leaving but little straight and level roadway. We have also six steam and six cable railway crossings.

President John N. Beckley of the Rochester Railway Company, at a recent meeting of the Street Railway Association of the State of New York, gave the following as the result of the experience of his company

In the month of May last the Rochester Railway Company operated 44 18-foot vestibule electric cars. The gross receipts from passengers riding on these cars during the month in the United States will not be wide of the mark in calculating that its profits must be made in nine months of the twelve—the other three just paying their own expenses.

As an assistance in the commercial management of an electric railway we give in Appendix D a set of accounting rules. These were very carefully prepared by Mr. H. I. Bettis, of Boston, while serving as auditor of the accounts of the Thomson-Houston Electric Company, in their contract work of maintaining electric equipment and lines of the West End Railway Company—the work being under the direction of Mr. W. E. Baker. Mr. Baker has kindly added to these rules a few words on the economical organization and administration of the working force of an electric railway.

A comprehensive statement of monthly operations may be made on a form similar to form "C." It will be seen on this

---

were $37,053, or 23.15 cents per car-mile for a mileage of 159,567 miles. The total expense of operation of these cars for that month was $18,332, thus leaving a net profit of $18,721. The total cost of operation per car-mile was 11.4 cents, and the profit per car-mile was therefore 12.11 cents It may be observed in passing that the operating expense was a trifle under 50 per cent. of the gross receipts The cost of operating was divided as follows:

Motive power	2.8	cents.
Car repairs	0.7	"
Conductors and motor-men	4.9	"
Other expenses	3.0	"

During the same period the company operated 62 horse cars, all of them without conductors. Most of the horse cars were one-horse or bobtail cars. The total cost of operating the horse cars, without conductors, during this period was about 10 cents per car-mile, but the total receipts per car-mile were but little above 12 cents.

In the month of June the Rochester Railway Company operated 54 electric cars and 60 horse cars. The electric cars earned each per day $23.60, or 22.77 cents per car-mile, and the total expense of operating them per day was $10.50, or 11.07 cents per car-mile. The cost of operation per car-mile was divided as follows:

Motive power	2.40	cents.
Car repairs	1.00	"
Conductors and motor-men	5.66	"
Other expenses	2.01	"
Making a total per car-mile of	11.07	cents

The cost of operating the horse cars during the same month per car-mile was 11.06 cents, and they earned 14.37 cents per car-mile. My experience in the operation of street railroads has convinced me that the most economical system of operation is the electric system. I have not, in the statements which I have now made, taken into consideration the greater fixed charge in the operation of an electric railroad as compared with a horse railroad, due to the much greater cost of the former; but in arriving at the conclusion which I have above expressed, due consideration has been given to this element of increased cost. We know that when a horse railroad is changed over and operated by electricity the receipts are very largely increased. It is safe in any case to say that the increase in gross receipts will be at least 15 per cent., and the average increase is probably as high as 30 per cent. Some of the increase is undoubtedly due to the greater mileage which the cars make, and still more is due to the cleaner, more rapid and more comfortable transportation of the people.

form that the final result of a month's operation is expressed in four different manners—namely, per cent. of expenses to earnings, net earnings, expense per car-mile, and expense per passenger. All these have a certain value for comparison. Neither, alone, tells the whole story.

The next matter is how to arrive at and keep those expenses sufficiently accurately and without too much clerical labor; and this brings us primarily to the stock or supply account. All material while in the stock room, and until used, evidently bears the same relation to the company as cash in the till or bank, except for availability, and should be so treated. Material should not be charged to the expenses of the road until used, and should be charged when used, regardless of the bill for this material being received or paid.

A careful record should be kept of all material received and all material used on blanks somewhat similar to forms "E" and "D," bound in book form. It is sometimes considered better to have the form "D" on loose sheets instead of bound, and there are apparent variations to be made in the blank to suit individual cases.

The "Material Received" book, or form "E," will be found to pay for itself in checking bills and preventing duplication. The material used is arrived at by charging out of the stock room on "Foreman's Requisition," in small quantities, as used, or obliging a foreman to make on a proper form a daily or weekly or monthly report of the material he uses. A simple form is that of form "F." One man should receive all the material for the road; he may, of course, be a storekeeper for this purpose if the road is sufficiently extensive; or, if not, this duty may be added to the other duties of the clerk.

An invoice book with copies of all bills should be kept, and a monthly summary of both books should be made—giving from the invoice book all the charges to stock, and from the material used all the credit to stock, and all the charges of material to expense properly classified. If in addition a stock ledger is kept, it will be found that all errors and mistakes in foreman's and clerk's pricing, etc., will correct themselves; and it will be easy at any time to check up the ledger with-the material actually on hand in the stock room.

In regard to the labor, there are many forms of time books used; the one here presented as form "P" is convenient, and the foremen soon learn to classify their time with little trouble and few mistakes. On the pay roll should be classified these expenses, and the total of all expenses is thus easily found.

The mileage sheets from "A" will be found very complete, and will give a record of the mileage of each car per month. The original records can be taken from the conductor's report most easily on a blank book, and transferred to these reports monthly. The conductor's reports should cover a report of the number of passengers carried on each trip, the number of trips made, and the route. There are many excellent forms for these.

To summarize, if the superintendent, as is frequently the case, has an office distinct from the office of the company, he would be required to have the following books: stock ledger, time book, pay rolls, material used and material received books. Copy books as follows will be found convenient: A letter book, a copy book for pay rolls, a requisition book (unless he buys his own supplies wholly, when this becomes an order book), a daily report of earnings book or deposit book, a motor report book for copying reports made on blank form "S," and a statement and bill copying book for bills made against other parties. Monthly reports should be made about as follows: mileage report, material used, balance of stock ledger, receipts report, invoice summary. The daily reports would be of daily earnings and of condition of motor cars.

On a road of any considerable size, say of 20 cars and upward, the superintendent will need the assistance of six good men—namely, a chief conductor, engineer at power house, line foreman, foreman of car repairs, foreman of track, and clerk; and, in fact, this organization will need only to be extended to operate a road of the largest size, when the foreman of car repairs becomes the master mechanic, the clerk the auditor, etc.

As to the force employed with an equipment of two 15 horse-power motors per car, we can allow about as follows:

One day mechanic to 12 cars.

One inspector to 50 cars.

One cleaner to 15 cars.

The motors ought to be cleaned thoroughly every third day. When less than 25 cars are run from one car house, the repair foreman may take the place of one mechanic, and in case less than 12 cars are run from one car house one mechanic may do the cleaning and repair work. The foreman will be in addition to the figures given, and these men will take care of the motors only; if additional work is required to any extent in repairs of the car bodies, etc., more men will be needed.

A proper manner to handle the bills for supplies and expense is as follows: provide a stamp with which every bill can be stamped on receipt, and which provides a place for the storekeeper to certify that the goods have been received, and have the bill checked and a place for a clerk to certify that any extensions and additions on the bill are correct; also a place for the same person ordering the goods to certify to the same; lastly, a place for the approval of the superintendent. It is important that all the names be written on the bill certifying to these facts, as then the whole history of the bill can at any time be ascertained.

CHAPTER XII.

HISTORICAL NOTES.

THE history of the electric railway presents some very curious features. Important though this particular application of electricity is to-day, it is almost impossible to trace out the various steps in its development with anything like certainty of giving credit to whom credit is due. For not only has the electric railway been a gradual evolution of ideas in the hands of many inventors working almost simultaneously, but it is almost unique in that it was worked out—so far as invention is concerned—during two distinct periods; first, at the time when the only source of electricity was the battery, and then over again after a commercial source of electric current had been found in the dynamo.

So long as the efforts of inventors were confined to the use of batteries, almost nothing could be done in the way of practical electric railway work; for its cost was evidently so prohibitive as to deter inventors from spending time and money on the solution of the various problems involved. Not until the modern dynamo had been invented and the great principle of its reversibility discovered was it possible to bring electric motive power to anything like a practical state.

This fact of there being two distinct periods of evolution in the history of the electric railway has had, perhaps, a happy effect on its modern development, in that the fundamental principles and methods had already become the common property of inventors, and hence the growth of the art has not as yet been checked by any basic patents of so formidable a description as to deter improvement on the part of others than some single individual or company.

The inception of the electric railroad and the first period of its history had its scene of action in the United States; the working out of the broad principles on which modern electric traction is based, and the first great steps toward the reduction of these principles to practice, took place abroad; and then again the scene changed back to America so completely that one may say, without risk of doing any injustice, that electric traction as a whole is of American development.

The electric motor, from which has sprung so many important practical applications, may probably be said to have had its origin in the Barlow wheel in 1826; and, oddly enough, this first effort was the forerunner of a rather advanced type of machine, being the motor counterpart of Faraday's disk machine and the unipolar dynamos of later years. Inventors were to pass through a long period of perplexity and roundabout expedients before they came back again to the principle of a reversed dynamo.

In 1830 the Abbé Salvatore Dal Negro, an Italian, professor of natural philosophy in the University of Padua, devised a reciprocating form of motor in which a permanent magnet oscillated between the poles of an electro-magnet commutated at each movement. A year or two later many inventors worked in the same line, but the first step toward the electric railroad was taken by Thomas Davenport, a blacksmith living in Brandon, Vt., who, quite independently of previous researches, devised a rudimentary electric motor working on the general principle of revolving an electromagnet by its attraction for fixed armatures or magnets, the current being commutated at the proper time to let the poles pass the first set of armatures and take up the work at the next. This he applied to an automobile electric car supplied by batteries carried upon it, and as early as 1835 he constructed a little circular electric road.

This was exhibited during the autumn of 1835 in Springfield, Mass., and also for two weeks in December of the same year in Boston. The road consisted of an automobile car carrying its own batteries. Of course nothing came of this rudimentary experiment, and the first extension of the same idea was at the hands of Robert Davidson, of Aberdeen, Scotland, who resides there to-day—perhaps the oldest living electrician. He began his experiments about 1838, at the period when wide attention was called to the possibilities of the electric motor by its use in the hands of Jacobi for propelling a small boat on the river Neva. The power in this case was furnished by primary batteries, and the speed reached by the boat—28 feet long and 7 feet beam—about 3 miles per hour.

Davidson was filled with the idea that the electric motor would find a place in the ordinary locomotive work of railways. He built with this in view a powerful motor, and fitted it on a truck of the ordinary railway gauge. The motor carried batteries for supplying the power; it was 16 feet long, 5 feet wide, and its total weight—including the batteries—was about 5 tons; 40 battery cells were employed, of a type claimed by Davidson as his own, although this was disputed by Mr. Sturgeon and J. Martin Roberts. The elements consisted of plates, 12 by 15 inches in size, of iron and of amalgamated zinc immersed in dilute sulphuric acid. The current obtained from these was delivered to an electric motor consisting of a pair of drums, each bearing two sets of iron bars parallel to the axles. Eight electro-magnets were placed on the bottom of the car in two opposite rows, and put the drums in rotation by attracting the armatures fastened upon them, while a commutator revolving with the armatures made the necessary changes of current in the magnets. Davidson's experiments are principally interesting on account of the large scale on which they were carried out. The locomotive just mentioned made several successful trips on some of the Scottish railways, and, finally, while the machine was being taken home to Aberdeen, it was found one morning in the engine-house at Perth shattered beyond repair by some mischief-making intruder. The railway engineers of the time felt very strongly that this innovation might be destined to supersede their own machines, and evidence pointed to the destruction of this first electric locomotive as having been at their hands.

After the Davidson experiments nothing on so large a scale was attempted for more than a decade, when Prof. C. G. Page, of the Smithsonian Institution, well known as a versatile genius in many fields of research, took up the subject of the application of the electric motor to railroading and performed some exceedingly interesting experiments.

Page, like Davidson, was thoroughly persuaded that the electric motor might be successfully used in regular railway work, and a machine of 10 horse-power was built by him in 1850. The general design followed was similar to a steam engine, two solenoids replacing the cylinder, and drawing backward and forward a soft iron piston rod which actuated a fly-wheel by means of an ordinary connecting rod and crank, while the motion of the piston shifted the connections from one solenoid to the other at the end of the stroke.

From this design he passed to that of two double solenoids acting on U-shaped magnet cores in a precisely similar way, and he eventually wound his solenoids in sections so that by a sliding commutator the current could be so applied as to produce both a long and a strong pull.

His first experiment with the electric locomotive made on this model was tried April 29, 1851, along the Washington and Baltimore Railroad to Bladensburg. This later machine was of 16 horse-power, and derived its supply of electrical energy from 100 large flat Grove cells stored on the locomotive, each having platinum plates 11 inches square.

The speed attained in this experiment was quite high, being at some points in the run at the rate of 19 miles an hour. Trouble was encountered, however, from breakage of the light earthenware porous cups, and after several subsequent experiments Prof. Page dropped the subject, having proved that considerable power and speed could be obtained at, however, a prohibitive cost.

Two other experiments on electric railroading, at about the same time, are worth mentioning, although on a very small scale. One was by Prof. Moses G. Farmer, who built a model locomotive, running on a track of 18-inch gauge and carrying a small car, which could accommodate two passengers. The power supplied was but small, furnished by 48 pint cells of Grove battery. In 1851 he built another little model road, which is important as showing a decided development in the art. It was constructed by Mr. Thomas Hall, of Boston, Mass., and was exhibited at the Charitable Mechanics' Fair. The track on which it ran was 40 feet long, and the gauge was but 5 inches. The motor was such an one as had been frequently used by Page and others, consisting merely of a permanent magnet, with an electro-magnet, provided with a suitable commutator, rotating between its poles. This armature shaft was provided with a worm, which engaged the teeth of an intermediate gear wheel that drove the axle.

The track was so arranged that when the car reached either end of it a pole changer would automatically be thrown over and reverse the direction of motion. The specially interesting feature of this road, however, is the fact that the current was supplied to the motor through the rails, and not from batteries carried on the car; further, the motor armature was geared to the driving axle with a considerable speed reduction, which embodied the principle of economical working that has since been generally followed.

This model of Mr. Hall's very plainly shows the combination of a moving motor car supplied from a stationary source of energy, and, with some similar anticipations, has gone far to prevent any fundamental patents from being granted during the second stage of the development of the electric railroad. In fact, the same idea had been disclosed four years previously by Mr. Lilley and Dr. Colton, of Pittsburg. Their device was simply a little model electric locomotive running around a circular track. The rails were insulated, and one was connected with each pole of the battery.

In 1840, however, a provisional patent had been granted in England to Henry Pinkus, in which not only was the general idea of supplying current to a moving train from a fixed source plainly shown, but it was even suggested that an electric motor might be used to drive the train itself. The patent had primary reference to ordinary railroad work, and its provisions are not very clear; but the idea was very distinctly put, and Pinkus even suggested the possible use of mechanical generators to replace the batteries shown in his plans.

It may also be mentioned that an English patent to Swear, in 1855, on telegraphy from moving trains, involved in a very obvious way the idea of taking current from an uninsulated conductor running along the line, thus forestalling the modern trolley system.

The same principles are shown and described in French and Austrian patents granted to Major Alexander Bessolo in 1855. The French patent shows what is now known as the third-rail system, with the return through the ordinary rails. It was proposed originally for train telegraphy, as in the case of Pinkus and others. Bessolo also stated that he proposed to use an overhead wire for one side of the circuit and the rail for the other, the current being taken from a stationary generator. In a supplemental specification to the Austrian patent the inventor explained that he proposed to use this system of distribution for driving vehicles on ordinary roads and on railways, and stated, as in the French patent, that the current may be conveyed to the locomotive machine by means of a circuit formed by a conductor insulated from the ground and suspended in a manner analogous to telegraph wires and by the rails of the track. Bessolo further suggests that such a system of locomotion would have great advantages on account of the ability of the current easily to reach independent locomotives or vehicles composing small trains, and that, besides, the generators were to be so located that they could be controlled from the stations.

All these early efforts at electric traction, however, resulted in nothing practical, for the simple reason that power had to be derived from batteries and was consequently too expensive for commercial use. It was not until the dynamo had appeared and the principle of its reversibility had been established that anything could be hoped for along this particular line of progress.

In 1864 Pacinotti made his now famous electro-magnetic machine, and, crude though it was, and unappreciated at the immediate time, it was the forerunner both of the modern dynamo and motor; for Pacinotti understood perfectly well that his machine was reversible, and that if current were supplied to it power could be obtained; while if power were supplied current could be obtained.

Three years later Wheatstone and Siemens, independently

and almost simultaneously, devised self-exciting dynamos—that of the former being shunt wound, that of the latter series wound; and at about the same time Siemens found that his machine could be used as a motor, although his discovery produced no immediate results.

He is said, however, at that period to have contemplated its adaptation to electric traction. Six years later the use of the dynamo as a motor was clearly demonstrated by Messrs. Fontaine and Breguet at the Vienna Exposition. Just previous to this time Siemens and Gramme had brought out commercial machines, and the latter had observed in his workshop that his dynamo could be used as a motor.

Fontaine and Breguet had intended to exhibit the principle by running one of the dynamos as a motor from a primary or secondary battery, but failing in this they coupled two machines together, and finding that the experiment was thoroughly successful, used the motor to drive a pump at the formal opening of the Machinery Hall, June 3, 1873.

The experiment, of course, created a great sensation, and during the next four years it became familiar in lectures, and was on a few occasions used for the transmission of a small amount of power. The scene of action now shifted to America, and the experiments of Mr. George F. Green, of Kalamazoo, Mich., began. Green was a poor mechanic, but possessed of more than ordinary skill and ingenuity, and for years had been interested in the study of electricity. About 1875 his ideas took shape, and he constructed a little model electric railway quite similar to that devised by Farmer, but on a somewhat larger scale.

Green used the track rails to transmit the current from the source of electricity—a battery in his experiments—to the moving car, which was driven by a small electric motor of the pole-changing type. He also proposed, as shown by his drawings of that date, to use an overhead wire as one of the sides of the circuit, but the experiment was not carried out.

During the next few years he was working at the problem as steadily as his occupation permitted, and in 1878 or 1879 built a larger model, with a car large enough to seat one or two people, propelled in substantially the same way.

He fully understood the advantages of employing a dynamo instead of batteries, but no dynamo was available for the experiments. Not until August, 1879, did he apply for patents; and then, through lack of funds, he was obliged to act as his own patent attorney, with the customary result of encountering great difficulties in the Patent Office. He went into interference with other inventors, and his claims, after being rejected on appeal to the Commissioner of Patents, were taken to the Circuit Court of the District of Columbia, which tribunal alone has the right to order a patent issued under such circumstances.

Here he was successful, but it was not until the latter part of 1891 that the decision was finally reached, and in December a patent was issued covering more broadly than would *a priori* be supposed possible the principles of the electric railway. His experiments formed the connecting link between the older and the newer period of inventions in the line of electric traction.

It was not until 1879, however, that the electrical transmission of energy was taken up on a serious scale. On Ascension Day, 1879, there was carried out at the little village of Sermaize, near Paris, by Chrétien and Félix, an experiment on plowing by electricity which marks so decided an epoch in the subject as to be worthy of more than a hasty description, particularly as it has been too lightly passed over in most historical notes on the transmission of power.

Plowing by steam power had already been tried in England and the United States, and it occurred to the engineers who instituted this notable experiment that the electric transmission of power offered great opportunities in facilitating the convenient application of mechanical power to such tasks. A rectangular plot of ground 220 meters wide had been laid out for the experiment, and the method employed was to establish at each end of a furrow a massive wagon carrying a motor, dragging a double-ended gang-plow by means of a cable and drum. A similar device had been previously used in steam plowing with tolerable success.

In this case there was mounted on each wagon a "type A" Gramme machine of the sort ordinarily used for electric lighting; these two were united by a conductor 250 meters long, forming part of a complete metallic circuit of copper wire 3 millimeters in diameter connecting them with two precisely similar Gramme machines for furnishing power. These two generators were distant, at various times during the experiment, from 400 to 620 meters from the motors.

The wagons themselves were automobile, as an axle of each could be driven by means of throwing the motor into gear with a large gear wheel connected to a wheel upon the axle by a sprocket chain. Probably about 3 horse-power was developed at the motors in this experiment, which was entirely successful. The speed reached by the plow was 40 to 50 meters per minute. The drums on which was wound the càble were driven by the motors through the intervention of friction gearing.

The power was supplied from the Sermaize beet sugar factory, and during the latter part of the previous season (1878) Chrétien and Félix had arranged at the river front, about 100 yards distant from the factory, an electric hoist operated from the machines in the factory and serving to transfer the beet root from the boats in which it was brought down the river to the factory wagons. Up to the time of the famous plowing experiment some 4,000 tons had been thus handled, at a cost of 40 per cent. less than unloading by hand.

The experiments just mentioned are in all probability the first examples of the transmission of power on a considerable scale either to fixed or movable motors.

During the summer of 1879 the first electric railway was put in operation, by the firm of Siemens & Halske, at the Industrial Exposition in Berlin. An oval track about 300 meters in circumference was laid down, and upon it ran a little electric locomotive dragging a single platform car. A third rail placed midway between the others served as the working conductor, and the current was taken from it by means of a sliding contact under the locomotive. The motor used was one of the regular Siemens dynamos, placed with its armature spindle parallel to the track; and the power was transmitted to the axle by a double-gear reduction, including, of course, a pair of bevel gears. The outer rails served as a return for the current through the wheels of the locomotive. Eighteen or 20 passengers constituted the full load of the little train, and the time required for a complete circuit was from one to two minutes, corresponding to a speed of perhaps 8 miles per hour.

This road is notable as being the practical starting-point of modern electric traction. It is beyond all question the first working electric railroad on a practical scale. Up to that time nothing had been constructed which bore even the semblance of a commercial electric railway, and such few plans as had existed before, both in Europe and America, were mainly on paper.

The success of the Berlin experiments stimulated invention in every part of the world, and the natural result was repetition of them under varied conditions and in different places. During the exposition at Vienna in the succeeding year Egger showed a model electric railway in which the third rail was dispensed with and the two working rails alone constituted the metallic circuit. Very early in the same year Siemens had been negotiating with the city authorities of Berlin for a franchise for an electric road on a considerable scale. About the same time some experiments were conducted in Paris with a view toward the utilization of a very small electric road as a substitute for pneumatic tubes, constituting a sort of forerunner of automatic electric traction.

Not until the middle of 1880 was any real work in electric railroading done in America. Mr. T. A. Edison was, so far as experimental work goes, first in the field, although in projecting such a system priority was awarded in the course of Patent Office litigation to Stephen D. Field, who filed a caveat on the subject May 21, 1879, and an application for a patent the succeeding March.

During the last months of 1878 and the first of 1879 Mr. Field had been doing some experimentation on electric motors, arranging among other things a small electric hoist; but so far as railroading is concerned his plans remained on paper until the latter part of the year 1880, a full year after the Siemens experimental road had been operating in Berlin.

Mr. Edison, too, only began work in 1880, and it was June 5 before he applied for a patent, nearly a year after the Siemens road was in successful operation, and some months after the same distinguished pioneer had been negotiating with the city authorities of Berlin for an electric road on a considerable scale.

Edison laid down a short experimental track near his laboratory at Menlo Park, N. J., and conducted some crude but interesting experiments; in his first locomotive the power was transmitted to the car wheels by means of belts, and the two rails were utilized as conductors.

The next year, 1881, saw decided changes and developments. Mr. Field had by that time put a small experimental road at Stockbridge, Mass., under way, and Edison was continuing his experiments. Meanwhile Siemens, the practical originator of modern electric traction, had not been idle, and plans of an elaborate character—never realized in their original form—were drawn up for electric rapid transit in Berlin. In May, 1880, these had taken shape, and were in the main for an elevated road of a meter gauge, the rails of which were to form the conductors, while the power was to be transmitted from the motor armature to the driving wheels by means of belts. At the same time Siemens was drawing up plans for electric rapid transit express service somewhat like that proposed in Paris some months before.

His scheme was for an automatic electric road of a little over a foot in gauge running completely incased in a sheet-iron tube of square section about 20 inches on a side. One especially interesting feature of this proposition, which was fully illustrated in the *Electrotechnische Zeitschrift* and *La Lumière Electrique* at that time, was that the armature of the motor was to be directly on the axle of the driving wheels, much after the plan that is being followed in the gearless motors of to-day. The speed proposed was about 40 miles an hour, and the service was to be entirely automatic.

It was not until the next year, however, that the first commercial electric road for regular service was put into operation. This was formally opened to the public May 12, 1881, at which time the Lichterfelde line—which is in operation to-day—was finally completed. In this, the third-rail method of supply is employed. The construction of the road was begun in the latter part of 1880. This Lichterfelde line was the first *commercial* electric road, as the little line of 1879 at Berlin was the first extensive experimental road.

It is worth noticing that July 20, *La Lumière Electrique* announced that the Lichterfelde line had up to that time been running without any serious accident, that a company for continuing it had already been formed, and in addition, that Siemens & Halske were just then changing the horse railroad that ran from Charlottenberg to Spandau into an electric road. In this installation—to avoid some of the inconveniences encountered in supplying the current through the rails—overhead wires were to be used, to which communication was made by a trolley towed by the car.

The experimenters were then hesitating between the use of the double trolley line and the single trolley using the rails as return, as practiced to-day. A month or two later, at the Paris Exposition, the system of overhead supply was used, in a somewhat different form, however, from that just mentioned. The Siemens road exhibited there was worked on the double trolley system, but the conductors, instead of being wires, were a pair of slotted tubes in which slid contact-makers connected to and dragged by the car.

Meanwhile the accumulator had not been forgotten. In 1880 a locomotive driven by accumulators was put into operation at the linen bleaching establishment of M. Duchesne-Fourmet, at Breuil en Auge, the total length of the track being over a mile. In June, 1881, an electric car operated by accumulators was tried upon the Vincennes tramway line, and from about that time the development of accumulator traction went gradually on, both in England and on the Continent.

The next noteworthy electric road to be put in operation was that at Portrush, in the north of Ireland. Soon after the franchises for a tramway line over the route had been obtained it was suggested to Dr. Werner Siemens that it was a good place for installing an electric road, and work was at once begun. The line was about 6 miles long. It was completed in the early summer of 1883, and the regular running of the trains began November 5 of that year. At the start the power was generated by steam, but a pair of turbines was soon installed at a convenient point on the river Rush, 1,600 yards distant from the nearest point on the line, and for the last seven years the road has been in steady and successful operation. The method of supply is by a third rail.

Up to 1883 the electric road in this country was practically undeveloped—all the advances had been made elsewhere; but soon the scene of activity was to shift. The conflicting interests of Field and Edison were consolidated after a couple of years of litigation, and the Electric Railway Company of the United States was organized early in May, 1883. The next month the Chicago Railway Exposition was to open, and by the greatest energy and push an electric locomotive was constructed and a road laid around the gallery of the main building, the total length being a little over 1,500 feet. It was of 3-foot gauge, and the third-rail method of supply was used, the two ordinary rails serving for the return. Auxiliary conductors were supplied to improve the conductivity of the rails.

The road began running early in June, and after a successful career of something less than a month was exhibited at the Louisville Exposition during the later part of the same year. By this time American inventors had settled down to hard work on the electric railway problem, and soon pushed on far in advance of their European competitors.

In the spring of 1883, too, Mr. C. J. Van Depoele began work on electric traction, and laid down a track about 400 feet long, on which a single car was operated for several weeks by means of a 3 horse-power motor. Later in the same year a little overhead line was constructed by Mr. Van Depoele for the Chicago State Fair, and remained in operation six or seven weeks.

Late in the autumn of 1883 still another effort at an electric road was made by Mr. Leo Daft. The experiments were tried on the Saratoga and Mt. McGregor Railroad in November, 1883, and, as in the case just mentioned, an electric locomotive was used. The motor was of about 12 horse-power, and its best actual performance consisted in hauling a 10-ton ordinary railway car loaded with 68 persons, at 8 miles per hour, over a track having a grade of 93 feet to the mile. The current was supplied by a third and central rail composed of iron rails of 35 pounds to the yard laid on blocks of hard wood saturated with resin. The current was taken off this rail by a pair of phosphor-bronze contact wheels pressed lightly down upon the rail by springs. The regulation of speed was effected by commutating the field coils.

The next year, 1884, saw still other inventors in the field, and before the end of the year considerable advance had been made. During the summer Mr. Daft equipped a short line on one of the piers at Coney Island, which ran during the latter part of the season and carried in the aggregate 38,000 passengers. At about the same time, however, two notable short roads were opened—one of them on the underground conduit principle, the other adopting the overhead trolley wire now generally used.

The former road, installed in Cleveland, Ohio. by Bentley and Knight, was actually thrown open to the public on July 27, 1884, and was the first electric system to be actually operated in competition with horses on street railway lines. The road thus equipped was about two miles long. The motor employed was placed between the wheels, being supported from the car body, and connection to the axles was made by means of wire cables. The slotted conduit was of wood placed between the rails, and of rather meager dimensions. The line was operated with quite uniform success for about a year, when the Bentley-Knight Company transferred its headquarters to Providence, R. I.

A little later in the autumn of 1884 the first practical overhead line on this side of the water was erected, the place being the suburbs of Kansas City, Mo.; and the inventor whose ideas were embodied in the short experimental road of a half mile in length was Mr. J. C. Henry.

This little road is notable as embodying some radical changes from previous methods. It was known as the Westport Electric Road, and was building during the summer of 1884. The overhead line consisted of two bare, hard-drawn copper wires supported from the top, as is usually the case with all trolley wires to-day. The size was No. 1 B. & S., and at that time such hard-drawn wire was only put on the market in 60-foot lengths, so that the experimenters were obliged to straighten it and braze the joints on the ground.

The motive power was supplied by a 5 horse-power Van Depoele motor, which was connected to the car of standard gauge by means of differential gearing not unlike that used on lathes. There were five changes of speed. The motor and gearing were supported together in a frame, one end of which was connected to the car axle, the other to the car platform. There was a bevel gear at the car axles and a clutch connecting it to the motor, so that the armature could run free if necessary.

Various experiments were tried in the way of different systems of current supply and different sorts of trolleys. The rails were temporarily bonded by driving horseshoe nails between the fish plates and the rails themselves, and in some cases one overhead wire was used merely as a feeder for the other and the rail as return. The two overhead wires were also sometimes used for separate tracks at turn-outs and the rails again used as a return. Various descriptions of trolleys supported by poles, and otherwise, were tried, but the form finally adopted was a trolley running on the line wires and with contact wheels bearing against their sides, the whole being connected to the car by a flexible cable.

The generator was a series-wound dynamo intended for arc lighting, of 10 horse-power capacity; and the electromotive force was several hundred volts—in some of the experiments; where a dynamo of double the capacity was employed, probably nearly 1,000. The steepest grade was 7 per cent., and the car was handled upon it without any special difficulty.

A few experiments were also tried with a motor of double the horse-power, fitted on an open street car, which was used for dragging a freight car on a short section of a neighboring steam railroad fitted up for the purpose. The highest speed attained was said to have been at the rate of 20 miles an hour.

This little Kansas City road was a nearer approach to the

forms familiar to us now than anything that had gone before it.

Unfortunately, however, nothing important came of the experiments, owing to the difficulty that inventors always experience in getting the public to adopt their ideas.

The fundamental difficulty with the electric road at that time was that it lacked public confidence; even electricians turned up their noses and looked askance at the new-fangled substitute for the horse. The demonstrations made on a small scale from time to time were not sufficiently convincing, so that it was an up-hill fight for those who were possessed of the courage of their convictions and determined to demonstrate the practicability of electric traction.

Nevertheless, the work went on, and in 1885 very perceptible headway was made. In the first place, Mr. Daft, in the early summer of that year, equipped electrically a portion of the lines of the Baltimore Union Passenger Railway Company. The branch selected for this purpose was that connecting Hampden and Baltimore. It includes considerable grades and curves, the heaviest grade being nearly 7 per cent., while the radii of some of the curves were as low as 50 feet. The system of supply selected was the third rail, as in Mr. Daft's earlier roads.

To this end a 25-pound T-rail was laid on special insulators and formed the outgoing lead, the return being through the two outer rails together with the ground. The electromotive force used was but 125 volts, which enabled a tolerable degree of insulation to be kept up. The electric locomotive contained a single 8 horse-power motor weighing 1,000 pounds. A single gear reduction was employed, and the 3-inch phosphor-bronze armature pinion engaged an internal gear 27 inches in diameter keyed on the axle of the driving wheels. The total weight of the locomotive was about 4,200 pounds. The regulation, as in Mr. Daft's previous motors, was secured by the commutated field method.

On August 8, 1885, the road went into operation, and the original motors were in service until quite recently. Four electric locomotives were installed at various times during the first year of operation. At street crossings the third rail was abandoned, and a bare overhead wire carried the current, from which it was obtained by a trolley pole placed on top of the car much as at present, carrying on its end the contact brush.

About the same time that this road went into operation Mr. Van Depoele equipped a short line at Toronto, Canada, with a single motor car drawing three trailers. Here an overhead line was employed all the way, the bare wire being supported over the car from brackets at the side of the track. An underrunning contact much like that just mentioned in connection with the Baltimore road was used.

Meanwhile Mr. F. J. Sprague entered the field, and, profiting by the experience of others, spent a large amount of time in working up the details of electric traction apparatus. His first effort was the application of electricity to the New York Elevated Railroad system, and in December, 1885, he publicly discussed the matter. In the early part of the succeeding year he was busily engaged on the problem and carried out an extensive series of experiments on a section of the elevated track.

Though this attempt at the elevated railroad came to nothing, it nevertheless taught many a valuable lesson regarding the handling of suitable power units and the precautions that would have to be taken in railway work.

During 1886 and 1887 Mr. Sprague was steadily at work on the electric traction problem, and the features which he brought into special prominence were the method of suspending the motors now generally used and the development of the overhead underrunning trolley into something like its present shape. The method of suspension employed was, as is well known, to pivot one end of the motor directly upon the car axle, sustaining the other by springs, so that the center of the armature shaft would preserve an invariable distance from the center of the axle, thereby securing the proper meshing of the gears. At first only a single gear reduction was used, but afterward the intermediate shaft made its appearance, to be displaced only in the practice of the last year.

This particular method of support was new in electric

motors, although it is said to have been anticipated abroad by the same method patented and used for gas motors.

It was not until almost the beginning of 1888 that the first Sprague road was actually put into operation—a small line at St. Joseph, Mo.

By January 1, 1888, there were in operation in the United States and Canada 13 electric roads, operating 95 motor cars over 48 miles of track. They were as follows: 6 Van Depoele roads, 3 Daft, 1 Fisher, 1 Short, 1 Henry, and the above-mentioned Sprague road at St. Joseph, the most extensive of these being the Daft road at Asbury Park, N. J., where eighteen cars were in operation. Electricians had not been idle, but the electric road was not yet fairly established.

During the spring of 1887 the Union Passenger Railway, of Richmond, Va., determined to adopt electricity as a motive power, and Mr. Sprague undertook the equipment of the line. This was successfully accomplished, and the road went into operation, after a brief period of experimental use, early in 1888. It was the first of the important electric roads equipped in this country or elsewhere, and gave an impetus to electric traction that is unique even in the progress of this nineteenth century.

Thirteen miles of line were equipped, and during the spring of 1888, on an average, 20 cars were employed. There were six 40,000-watt, 500-volt Edison dynamos; three 125 horse-power engines, and three 125 horse-power boilers. Although the apparatus was crude and the road underwent many vicissitudes, the present success of the electric road is very largely due to the perseverance and the energy that put the Richmond road in operation in the face of every sort of discouragement and gloomy prediction of failure. From that time on electric railways have grown in number and size so rapidly as to almost defy any attempt to record them. During the succeeding year a number of additional roads were installed, and the Thomson-Houston Company, which had acquired the Bentley-Knight and Van Depoele patents, entered the field.

Since that time other electricians have been busy. The

Sprague Electric Railway and Motor Company came into the possession of the Edison General Electric Company, new companies have arisen, and electric roads have been installed so rapidly and in such numbers as to occupy to the fullest the resources of the several companies that have been engaged in the work.

This brief record of the history of the electric railway makes no claim to completeness; indeed, a complete history would be hard to write until the position of the various inventors shall be officially determined by the courts. The name of electric railway patents is legion, but their value is as yet an indeterminate quantity—with a tendency, however, toward a zero value, in view of the early anticipations both of some of the most essential features of modern electric traction and the fact that many of the patented devices have existed only upon paper. Perhaps it has been well for the development of this branch of the electrical transmission of energy that such is the case.

We have made no attempt to mention all those who have taken a part in the development of the electric railway, or all the various attempts that inventors have made to solve the many difficulties that have been encountered. It has been our endeavor, however, to sketch out, with more attention to general accuracy of perspective than to minuteness of detail, the growth of the art that has been already productive of such important results and promises an almost immeasurable future.

APPENDIX A.

ELECTRIC RAILWAY *VERSUS* TELEPHONE—DECISIONS IN THE SUPREME COURT OF OHIO.

JANUARY TERM, 1891.

THE CINCINNATI INCLINED PLANE RAILWAY COMPANY,
Plaintiff in Error,

against

THE CITY AND SUBURBAN TELEGRAPH ASSOCIATION,
Defendant in Error.

ELECTRIC STREET RAILWAYS—SINGLE-TROLLEY OVERHEAD SYSTEM—RIGHTS OF TELEPHONE COMPANIES.

Syllabus.

1. The dominant purpose for which streets in a municipality are dedicated and opened is to facilitate public travel and transportation, and in that view, new and improved modes of conveyance by street railways are by law authorized to be constructed, and a franchise granted to a telephone company of constructing and operating its lines along and upon such streets is subordinate to the rights of the public in the streets for the purpose of travel and transportation.

2. The fact that a telephone company acquired and entered upon the exercise of a franchise to erect and maintain its telephone poles and wires upon the streets of a city, prior to the operation of an electric railway thereon, will not give the telephone company, in the use of the streets, a right paramount to the easement of the public to adopt and use the best and most approved mode of travel thereon, and if the operation of the street railway by electricity as the motive power tends to disturb the working of the telephone system, the remedy of the telephone company will be to readjust its methods to meet the condition created by the introduction of *electromotive* power upon the street railway.

3. Where a telephone company, under authority derived from the statute, places its poles and wires in the streets of a municipality, and in order to make a complete electric circuit for the transmission of telephonic messages, uses the earth, or what is known as the "ground circuit," for a return current of electricity, and where an electric street railway afterward constructed upon the same streets is operated with the "single-trolley overhead system"--so called—of which the ground circuit is a constituent part, if the use of the ground circuit in the operation of the street railway interferes with telephone communication, the telephone company, as against the street railway, will not have a vested interest and exclusive right in and to the use of the ground circuit as a part of the telephone system. (Decided Tuesday, June 2, 1891.)

But it is urged that the franchise of the telegraph association to construct lines of telephone is greatly impaired by reason of the single-trolley railway using a grounded circuit, whereby a large part of the electric current flows off from the rails to the surrounding earth, and to and upon all telephone wires which may be connected with the earth in proximity to the railway. The action is described as *conduction*, causing more or less of electric current to be poured into the earth and into all electric conductors connected with the earth, thereby reaching telephone wires in a grounded circuit, and creating loud and continuous noises upon the wires, which disturb telephonic communication. This disturbance, however, results not solely from the earth circuit of the railway company, but also from the fact that the defendant in error likewise relies upon the earth for its return circuit, by connecting with the earth the end of its wire furthest from its electric batteries.

It is claimed that in addition to this conduction or leakage disturbance the single-trolley electric railway introduces serious disturbances on telephone lines by *induction*, for the reason that such electric railways employ large wires to convey the current used for the propulsion of their cars, and this current is constantly and rapidly changing its strength; that these rapidly changing currents in the electric railway wires induce disturbing currents in parallel telephone wires near which the electric railways have been built, and thus prevent a successful transmission of telephonic messages.

These interferences with the telephone service may be obviated, it is stated, by the railway company giving up the single-trolley system with the ground circuit, and substituting the double-trolley system with its two trolley wires, two trolley wheels, and electric current passing from one wire through one trolley, through the motor, back through the other trolley to the other wire, and so back to the generator, without escaping to the earth. The grounded circuit, it is insisted, should be abandoned and surrendered to the sole use and service of the defendant in error. But it is admitted that other remedies of the telephone disturbances may be easily obtained by constructing the telephone with a complete metallic circuit, or by resort to what is known as the McCluer device, consisting of a single return wire, to which a number of telephone wires are attached.

Conceding that the mode adopted by the railway company of propelling its cars by electricity is an interruption to the telephone service of the defendant in error and calculated to impair its franchise in the manner contended, the inquiry is suggested whether the railway company must yield up a useful franchise that the same may be exclusively enjoyed by the telegraph association, or whether the association shall adapt its system to existing conditions; whether the company shall change from the single to the double trolley system, from the grounded to the metallic circuit, or whether the association shall either use a complete metallic circuit or resort to the McCluer device.

It is immaterial on which party the expense of the change may fall the more heavily. It is a question of legal right.

When the telegraph association erected its poles and lines in 1881 and 1882, with the design of conducting the business of a telephone company, it found the railway company operating its street railway, with authority under the statute to use other motive power than animals, to wit, electricity, cable, or compressed air, upon obtaining the consent of the Board of Public Works. The telephone business was not among the probabilities when the streets of Cincinnati now made use of by the telegraph association were dedicated or condemned for the public use. The primary and dominant purpose of their establishment was to facilitate travel and transportation; they belong from side to side and end to end to the public, that the public may enjoy the right of traveling and transporting their goods over them. The telephone poles and wires and other appliances are not among the original and primary objects for which streets are opened, for they may be placed elsewhere than on the highways and yet accomplish their purpose. In Taggart *v.* Street Railway Co., 16 R. I., it was said by Durfee, C. J., that telephone poles and wires are not used to facilitate the use of the streets for travel and transportation; "whereas the poles and wires of the railway company are directly ancillary to the use of the streets as such, in that they communicate the power by which the street cars are propelled." As a general rule, an occupation of the streets otherwise than for travel and transportation is presumptively inferior and subservient to the dominant easement of the public for highway purposes, for if not so, the primary object of their dedication or appropriation might be largely defeated. And the fact that permission is granted to occupy the streets or highways for a purpose other than travel, does not confer a prior and paramount right to occupy them to the exclusion of their use for travel in a mode different from what obtained when such permission was given.

To those improved agencies, devised for the convenience and advantage of the community in general, the franchise of the telephone company to occupy the streets for carrying on its business must be secondary and subordinate. Whether all who go upon the streets shall have the most convenient and expeditious passage and carriage of person and goods, has not been made dependent upon the manner in which the defendant in error has preferred to locate its poles, stretch its telephone wires, or form the electric circuit.

It is in recognition and maintenance of the superior easement of the public in the streets that city councils are required to "cause the same to be kept open and in repair and free from nuisance;" that the streets are graded and paved and proper regulations of police provided to govern the actions of persons using them; that the abutting owner, though having a peculiar interest and easement in the adjacent street appendant to his lot, has no right to place permanent obstructions in the street, nor do any act on his own land outside the limits of the street, that will make the way inconvenient or hazardous or less secure than it was left by the municipal authorities.

This paramount easement or estate which the public acquires in the streets, carrying with it a special interest in the adoption of the most approved systems of modern street travel, cannot be made subservient to the telegraph or telephone when admitted on the highway, without the clearest expression of the legislative will.

The demand made by the telegraph association is, not that the railway company shall so modify its existing electrical apparatus as not to interfere with the telephone service, but shall forever abandon the use of an essential part of its electromotive system, or be perpetually enjoined. In other words, the association claims the exclusive use of the grounded circuit, inasmuch as the mechanism of the telephone is so complex and the electric currents employed so delicate and sensitive that they cannot be used without disturbance from the heavier currents employed by neighboring electrical enterprises that operate with the grounded circuit. We find no foundation for such an exclusive franchise or right.

When the telegraph association began its operations under the telephone system, neither the statute authorizing it to erect and maintain poles, wires and other necessary fixtures, nor the ordinance under which it obtained the power to extend its lines in the streets, gave an exclusive right either to use the earth for a return circuit or a complete metallic circuit formed by double wires. The Legislature did not grant the right by general enactment, nor was the municipal corporation empowered by the Legislature to give the telegraph association the exclusive right to make use of its streets so as to create a monopoly.

It is contended, however, that the defendant in error, by virtue of its grants, acquired, before the railway company had a right to use electricity as a motive power, a vested interest in the telephone system as it now operates it, with a grounded circuit, and that not even the Legislature of the State could take away from it or injure this franchise, on the faith of which it has expended its capital and labor. Special privileges or immunities are under the control of the Legislature. If granted they may be altered, revoked, or repealed by the General Assembly. Art. 1, Sec. 2 of the Constitution. And while corporations with valuable franchises may be formed under general laws, all such laws may, from time to time, be altered or repealed. Constitution, Art. 13, Sec. 2. In view of these constitutional provisions, it is clearly within the power of the General Assembly to authorize one class of corporations to use, in the streets, electricity with the grounded circuit as a motive power, and another class to employ the same or a similar agency for the transmission of telegraphic or telephonic messages. And if the proper exercise of the rights granted to the one class under general law is irreconcilable and plainly interferes with a prior grant to a corporation of the other class, it may be construed as the intention of the Legislature to deny an exclusive franchise, if not to repeal the antecedent grant.

It is contended, however, in behalf of the defendant in error, that conceding the railway company and telegraph association to be upon an equal footing on the streets and highways in the enjoyment of their respective franchises, the company is bound to conform to the rule *sic utere tuo ut alienum non laedas.* In the view which we take of the relation to each other of the parties to the action, we deem it unnecessary to inquire whether there has been a want of conformity, and to what extent, if any, on the part of the railway company, to the requirements of the legal maxim. Nor do we think it necessary to determine how far an incorporated company making a lawful and careful use of its own property, or of a franchise granted to it by the proper municipal authorities, may be held liable for damages incidentally caused to another.

From the undisputed facts in the case, as disclosed in the record and printed arguments of counsel, it is evident, as we have already seen, that the railway company acquired from the State and from the city of Cincinnati authority to erect and maintain poles and wires in the streets or highways, and to use electricity as a motive power for its cars. Clothed with such authority, we have, upon weighing the allegations in the original petition and applying to them the well-settled principles governing the legal rights of the public in the highways, reached the conclusion that the facts set forth in the petition are not sufficient to constitute a cause of action. We are of the opinion that there has been no invasion of the rights of the telegraph association by the plaintiff in error, and that the telegraph association is not entitled to the relief prayed for in its petition. The judgment, therefore, of the Superior Court at general and special term must be reversed and the original petition dismissed. Judgment accordingly.

SUPERIOR COURT OF CINCINNATI.

THE CITY AND SUBURBAN TELEGRAPH ASSOCIATION

vs.

THE CINCINNATI INCLINED PLANE RAILWAY COMPANY.

Taft, J.—This is an action to enjoin the defendant from using a system of electric railway propulsion, known as the single-trolley system, in running its cars upon the streets of this city and the roads of the county, because, as plaintiff claims, it does a great injury to the conduct of the public business in which it is engaged—*i.e.*, the maintenance for profit of telephone communication between a large number of subscribers in the city and county.

Plaintiff alleges that it has been conducting its business under lawful grants of the Legislature of the State and the municipal authorities, and by virtue thereof has lawfully erected many lines of telephone wires on the streets upon which the street railway of defendant is constructed and operated; that the defendant has no lawful authority to use electricity as a propulsive power for its cars; that the single-trolley system in use by defendant conducts upon and induces upon the wires of the plaintiff erected on the same streets currents of electricity, which makes it impossible to use those wires for the purpose of telephone communication; that the result is that many of the subscribers of plaintiff have made complaint and threatened to discontinue the use of the telephone, and some have refused to continue payment for the same, all to the great injury of the plaintiff, that the damage is irreparable and continuous, and that plaintiff notified defendant before the erection of the single-trolley system that it would result in a serious injury to plaintiff, and of the objection to the system.

The defendant answers, denying plaintiff's lawful right to operate telephone lines; avers its own lawful right to conduct a system of electrical railway propulsion; avers that the single-trolley system is the only practicable one in use and secures advantages to the defendant not afforded by any other; that plaintiff's difficulties arise from the defects of its own telephone system, in that, for economy, it uses the earth as a return conductor, and so invites onto its lines many disturbing currents, both natural and artificial, although the mechanism of the telephone is so delicate that any proper system would make the return circuit metallic; that plaintiff's claim amounts to a claim to the exclusive use of the earth as a conductor, which is without warrant of law; that the use by the defendant of the earth as a return circuit is a material assistance in propelling cars up heavy grades; that it could not alter its system so as to avoid the use of the earth as a return circuit except at great outlay and loss of efficiency. The answer denies that the operation of its road produces interference with the telephone lines of plaintiff, and says that if any injury has been produced for which defendant is liable, the damages are capable of exact ascertainment, because they can all be remedied by an additional expenditure by the plaintiff, and that therefore there is no ground for equitable relief.

There can be no doubt that the plaintiff has a lawful right to occupy the streets upon which defendant's track is laid, with the poles and wires erected for the conduct of its business, and that the use of the earth as a return is not a violation of its franchise. It was organized as and still is called a telegraph company. It began its telephone business before Section 3,471 was passed, and perhaps some of its poles and lines along the streets occupied by defendant were erected for such use. But this circumstance is immaterial, for it has been held, both in England and in this country, that the telephone system, which is communication over long and short distances through the agency of currents of electricity on metal wires, is really a magnetic-telegraph system, and that an exclusive right to operate the telegraph includes the exclusive right to operate telephone lines. Attorney General *v.* Edison Telephone Co., 6 Q. B. D., 244; Wisconsin Tel. Co. *v.* Oshkosh, 62 Wis., 32.

Having determined, then, that plaintiff's use of the streets on which defendant operates its road is lawful, the next question which arises is whether defendant has the right to use electricity for the propulsion of its cars. If it has not and it thereby interferes with plaintiff's franchise, it is very clear that such interference gives plaintiff a right of action.

On the whole, then, I am of opinion that the Legislature conferred the right upon defendant to use any other motive power than animal whenever the Board of Public Works should consent. Now, the board did consent, on October 24, 1885, that defendant should use either a cable, compressed air, or electricity. It has chosen electricity, and has produced the necessary authority to erect its poles and string its wires.

Such being the condition of the franchises which the plaintiff and defendant are lawfully entitled to enjoy, considered each without reference to the other, it becomes necessary to inquire, first, whether any loss has been inflicted upon plaintiff by defendant, and if so, how has it occurred; second, whether such loss, if any, is justified by defendant's franchise, so as to be *damnum absque injuria*, which involves the preliminary question whether the Legislature, after having given the plaintiff the right to construct its telephone system, on the faith of which right it has expended large amounts, can confer a franchise on another, the exercise of which will seriously impair the plaintiff's franchise as heretofore enjoyed.

First, a current of electricity cannot be produced without a circuit—that is, unless the negative and positive poles of the generating battery or instrument are connected by a continuous substance capable of conducting the current. Such a substance may be a metal wire, or if both poles of the battery be connected with the earth by metal wires the current will find a circuit through the wires and the earth. The earth, by reason of its immense mass, makes an excellent conductor. By what paths the current discharged into it over the wire from one pole finds its way throughout to the wire from the other pole is not capable of determination.

The telephone is a mechanism by which the sound of human speech is reproduced over long distances. Without attempting to describe the exact mode in which this result is brought about, I may say that the sound waves of the human voice produce vibrations in a thin ferrotype plate, which, by means of a magnet and an induction coil, are converted into corresponding vibrations in an electric current on the connecting wire, and these vibrations are, in turn, by means of the inducing coil and magnet at the other end, converted into exactly corresponding vibrations on a plate there, reproducing the sound waves of the voice of the speaker in such a manner as to enable the receiver to understand. The current on the connecting wire is a slight one, and the circuit is completed, not by a return wire, but by a ground wire brought into contact with the earth. This contact is usually made by attaching the wire from the negative pole of the single cell battery in each telephone to a gas pipe or water pipe running down into the earth.

In the Sprague system of electric railways, which is the kind used by the defendant, the electricity used to operate the motors under the cars is conveyed to them by a single overhead wire suspended over the middle of the track, along the under side of which runs a trolley wheel on a single mast attached to the car, making electric connection between the overhead wire and the motor of the car, and allowing the current to pass down through the motor and onto the track, whence some of it returns directly to the dynamo generator at the power house. A large part of the electricity leaves the track, however, and by other and various paths also finds its way through the earth back to the dynamo.

In addition to the overhead trolley wire, which is supported by guide wires from iron posts erected on the curb at regular intervals, there is what is called a feed wire strung along on these posts for the purpose of keeping up the required quantity of electricity on the trolley wires. On the streets where are telephone wires and electric railway wires their general course must be parallel. The evidence in the case establishes beyond a doubt that since the defendant began the operation of its line by electricity in June, 1889, down to the time of trial, the usefulness of all the telephones along the line of the railway has been more or less impaired; that in many cases the buzzing noise which seems to be the chief form of the disturbance has been so loud and continuous that communication over the lines has been impossible. Nor has the disturbance been confined to telephones on the line of the railway. Telephones several miles from the city whose connecting lines ran parallel for any distance with the railway were similarly affected and the buzzing noise in many of these seems to have been quite as deafening as in telephones along the line. Altogether more than two hundred lines have been affected in this way, to a greater or less degree. The cause of the trouble is undoubtedly the operation of the electric railways, and the way by which it is brought about is twofold. First, the escape of the electric fluid from the rails, which is called earth distribution or leakage, near where the wire from the telephone is connected with the earth, brings upon this earth-connection wire of the telephone varying currents of electricity of much greater quantity than that necessary for the telephone current, and produces upon the magnet and inducing coil an effect which results in vibrations of a very different character from those produced by the human voice, and makes a noise like the buzzing of a saw. Second, a similar noise is made by induction. It is a physical fact of much importance in electric mechanism that where two wires of two circuits are parallel to each other, and there is a current of varying intensity on one of them, this will produce in the other, in the opposite direction, a current of electricity of similar variation. The insulation of the wires has no effect to reduce the current produced in this way. The amount of induction depends upon variation in the current, the distance of the wires from each other, and the length of the parallelism of the wires. The current upon the trolley wire and the feed wire of the railway is quite variable in quantity and intensity, owing to the drain upon the store of electricity by the moving and stopping of the car. Nor is the electricity, as generated, exactly uniform in its flow from the dynamo. The result is that wherever the telephone wire is parallel with the trolley wire and the feed wire, there is induced upon the telephone wire a current whose variations correspond with the variations of the electrical current on the electric railway wires, and this, acting upon the inducing coil and magnet, produces vibrations of the pin plate, which makes the buzzing sound. It is not possible, in listening to the sounds produced by the electric railway, to say whether it is the result of induction or earth leakage. Some of plaintiff's subscribers, notably Proctor & Gamble, have loud buzzing sounds upon their telephones, the ground wires of which are at such a distance from the railway track as to make it quite unlikely that the disturbance could have been caused by earth leakage, while, on the other hand, their wires are for some distance parallel with the trolley and feed wire of the railway company. Other telephones are disturbed on the line of the road, though their wires are not parallel with the trolley wire. Expert evidence attributed the disturbance about one-half to induction and one-half to conduction or earth leakage. This, of course, is only a rough estimate, and the fact may vary much in particular instances. The injury to plaintiff's service is produced, then, speaking in a general way, one-half by the defendant pouring electricity into the earth, which finds its way into the property of plaintiff's subscribers, and thence into the wire of the telephone, and one-half by a creation of a current on plaintiff's wires in the street by the parallelism of defendant's wire and the varying character of the current. The result has been a very substantial interference with the plaintiff's business, and if the present condition continues it will end in a serious loss.

Is it a loss for which defendant is liable? The contention on behalf of the defendant is that because it has full power to operate by electricity under the law, the loss resulting to the plaintiff is *damnum absque injuria*, and if the plaintiff wishes to avoid the loss, it must adopt safeguards in the shape of a metallic circuit to avoid the difficulty. To this plaintiff replies that by virtue of its grants it acquired, before the defendant had a right to use electricity as a motive power, a vested interest in the telephone system as it now operates it, with a grounded circuit, and that not even the Legislature of the State could take away from it or injure this franchise, on the faith of which it has expended so much capital and labor.

I am inclined to think that under the constitutional provision that all laws for the forming of corporations may be altered or repealed (Sec. 2, Art. 13), it would be in the power of the Legislature to grant a right to other corporations for a public use to so use the street as to require the plaintiff company, if it wished to continue in the telephone business, to change its system, and that without any right of action against such corporations. The case of Ry. Co. *v.* Ry. Co., 30 Ohio St., 604, shows that where one railway company condemned a right of way across the track of another, that other cannot recover for an injury to its franchise as a railroad or for the increased expense entailed upon it in obeying the laws of the State with reference to railroad crossings. However this may be, I am very clear that no intention on the part of the Legislature to abridge the granted rights of one corporation by a new grant to another will be recognized by the courts unless such intention plainly appears in the law. In England, the power of Parliament is unlimited, and it may even confiscate private property, and *a fortiori* may abridge and destroy chartered rights and franchises. Nevertheless, we find in that country that where one corporation is granted a right which may be so exercised as to injure or interfere with a right previously granted to another, the presumption of law is that Parliament intended only such uses as were consistent with the rights of the first corporation. In Gas Light and Coke Co. *v.* Vestry, 15 Q. B. D., 1, plaintiff was a gas company which had laid its gas pipes, by virtue of a public grant, under a street which the defendant, a public corporation, was charged with keeping in repair, and upon which it used such heavy rollers as to injure the pipes of the plaintiff. The rolls used were economical and well fitted for the purpose, but it was held that unless the defendants were expressly authorized by statute to use rollers of the size and weight of those which did the injury, the defendant could not justify under a duty to keep in repair which might be discharged by rollers of less weight and without breaking the pipes.

This case is peculiarly applicable to the case at bar, because here was a case of a public grant to the gas company enjoyed in a certain way, followed by a grant to the defendant to exercise another right, which, if exercised in one way, would injure plaintiff's enjoyment of its right, and which, if exercised in another, would not. The same principle is here applied that courts recognize wherever private property is injured by the exercise of authority granted by the Parliament. See Hammersmith, etc., Ry. Co. *v.* Brand, L. R. 4 H. L., 171; Queen *v.* Bradford Navigation Co., 6 B. & S., 631; Geddis *v.* Proprietors of Bann Reservoir, 3 App. Cases, 430; Att'y Gen'l *v.* Colney Hatch Lunatic Asylum, L. R. 4 Ch. App., 416; Att'y Gen'l *v.* Gas Light and Coke Co., 7 Ch. D., 217; Managers Metrop. Asylum District *v.* Hill, 6 App. Cases, 193.

Even, then, if franchises to occupy the public streets conferred by the Legislature may be subject to modifications of their use and enjoyment by other public grants of the Legislature, it is certain that unless the legislative intent to make such modification clearly appears, either by express words or by necessary implication arising from the impossibility of enjoying the second grant without such modification, it will not be inferred.

But it is said that this principle can have no place here because the right to occupy the street for the purpose of travel is a superior right to that of using it for the telephone poles. Defendant's counsel, to establish this, rely on The Spring Grove Ave. *v.* Cumminsville, 14 Ohio St.; Smith *v.* Tel. Co., 2 C. C. R., 259; Mt. Adams & E. P. R. Co. *v.* Winslow, 3 C. C. R., 425; R. R. Co. *v.* Williams, 35 Ohio St., 171; Pike's Ex'rs *v.* Western U. Tel. Co., decided by this Court.

These cases have no bearing upon this case at all, as it seems to me. They involved a discussion of whether the erection of certain structures in the street could be considered a new burden and use, not included in the easement which the abutting property holder had originally granted to the public, and whether therefore he was entitled to compensation. It is unquestionably within the power of the Legislature, so far as the public is concerned, to enlarge the benefit to be derived from the streets, so as to include other public purposes than those of mere travel (Ry. Co. *v.* Lawrence, 38 Ohio St., 45). If this takes more from the abutting property holder than he originally gave, then he may have compensation, but the public cannot complain, for their representative, the Legislature, has spoken and granted the use. When the telephone company is granted the right to use the streets, its right is as well founded as that of a street railway company, and in the absence of express legislative direction to the contrary, there is to be no yielding to any other. The provision that the telephone and telegraph lines shall not incommodate the public in the use of the street, in Section 3,461, does not help the defendant. The inconvenience must be determined when the enjoyment of the franchise is entered upon.

After rights have been acquired by the outlay of capital and user, there must be express legislative sanction, at least, to warrant a court in finding a use of the street to be an interference with public travel which was not so when it began. It should be noted in this connection, too, that the plaintiff performs a very important *quasi* public duty, and is in fact a common carrier of messages. It is given the power of condemnation on that ground alone. Pierce *v.* Drew, 136 Mass., 75; Hackett *v.* State, 105 Ind., 250.

Coming now to apply the principle just under discussion to the case at bar, what do we find? For ten years the plaintiff has exercised the franchise of occupying the streets along defendant's lines with its poles and wires, conducting a telephone business with a single wire circuit and an earth return. This mode of return was universally employed when it began, and is to-day in general use. It has constructed a valuable plant, many parts of which will have to be changed at great expense if it is to adopt the only system which will obviate the difficulty it now encounters from the operation of defendant's railway. I refer to the metallic circuit. It is admitted by every one that if the telephone company were to make every one of its lines a complete metallic circuit, with the return wire parallel with the outgoing wire, the disturbance both from induction and earth leakage would be completely removed. It is obvious that if the circuit never came into contact with the earth, the electricity dumped into the ground by defendant could not reach the telephone wire, and so no disturbance could arise there. And it is also a well-ascertained fact that if the two wires of the circuit, the outgoing and the return, are parallel and the same length, no effect will be perceptible from induction by a third parallel wire of another circuit, however variable may be the current on that wire. This is because the induction which actually takes place upon each of the wires of the circuit results in currents of equal intensity and variability in opposite directions, which being on the same circuit exactly neutralize each other.

To construct such a circuit for every subscriber would entail an expense upon plaintiffs, perhaps, even greater than the original outlay. Nor does it seem practicable to have metallic circuits in disturbed districts and allow the other lines to remain grounded, because of the difficulty in arranging the switchboards at the exchange by which subscribers are connected with each other, and for the further reason that in order to connect a metallic circuit with a grounded circuit, the return wire of the metallic circuit must be grounded at the exchanges. The result is that the circuit then made between the two subscribers is a grounded circuit with a long leg and a short leg of wire parallel in the disturbed district. The difference in the length of the two wires prevents the neutralizing effect of the induced current in one wire upon the opposite induced current in the other, because, by reason of the different lengths, that part of the induced currents said to be produced by static induction is unequal. Moreover, to put up metallic circuits through the disturbed districts and to make the necessary alterations in the switchboard would entail an expense, not nearly so heavy, of course, as a complete conversion into a metallic circuit, but which would be quite substantial. There remains to consider what is called the McCluer device, which consists in a large wire carried into the disturbed district with which are connected all the return wires of the telephones. This main is grounded at a point away from danger of earth leakage. It is likely that such a device would get rid of earth leakage, but it would have no practical preventive effect upon the disturbances by induction, which, as I have said, make up about half of the troubles. The expense of such a device would not be large, still it would be something. It follows, from what has been said, that if the defendant is permitted to use its present single-trolley system, it will require the plaintiff either to lose the business of many subscribers, or it will have to go to an enormous expense to obviate all difficulty or to a less expense to get partial relief.

The plaintiff invested large capital on the faith of its present mode of enjoyment of its franchise. It has continued in such enjoyment for some ten years. To change involves loss and expense. Clearly, then, it cannot be deprived of what so nearly approximates a vested right without clear legislative intent.

Is it impossible, then, for the defendant to run an electric railway along its streets without making these disturbances? It is practically conceded, although there was some slight evidence to the contrary, that if instead of a single trolley wire and an earth return two trolley wires were used, one for the positive and the other the negative current, the difficulties would be just as completely obviated as if the telephone company used a metallic circuit. There is such a system in use in a number of cities of the country. There are two in operation in this city and two or three more are projected here. In such a system the electricity is carried from one wire down through one trolley wheel and mast to the motor of the moving car, and returns from the motor to the other wire by means of a second trolley wheel and mast. It is said it is wholly impracticable.

The single-trolley system is in use on nine-tenths of the electric railway systems in the United States. This arises doubtless from two facts: First, that it is perhaps one-fourth cheaper in its outside construction; and, second, because in single-track railways, of which there are many more than double track, where it is necessary to have many switches and turn-outs, the complications of wires overhead increase much more rapidly with a double trolley than a single trolley. But in the case at bar we have a double track from one end of the road to the other, so that no switches are required. Mr. Short, who is a skilled mechanician engaged in a company which bears his name and which erects both the single and the double trolley systems, says that he recommends the single trolley for single tracks and the double trolley for double tracks. He seems the only witness on either side that is free from suspicion of bias.

It is admitted by many of defendant's witnesses that electrically the double-trolley system is exactly as good as the single-trolley, and this must be so, because a return by wire makes quite as good a current as by the earth. The claim that electricity in the wheels passing to the track gives additional traction is conjectural, and is not apparent in a practical comparison of the grades ascended by the cars on the two systems. The objection to the double-trolley system is the mechanical difficulty in making proper switches and in supporting the superstructure. In the double-track road, like the defendant's, the first difficulty does not arise, and the second difficulty has in fact been overcome in every double-trolley system which has been erected. On the whole, then, it seems to me that there is no serious obstacle to defendant using the double trolley.

The defendant cannot rely on an estoppel. It was notified by the plaintiff as early as March 12, 1889, some three months or more before its electrical plant was put into position, that the use of the single-trolley system would interfere with and injure the use of the plaintiff's telephone lines.

But it is said that the injuries here occasioned are not cognizable in the courts, even if the telephone company is to be regarded as a private property owner in the enjoyment of the telephone franchise, and the case of Frazier *v.* Siebern, 16 Ohio St., 614, is relied on in this connection. That was a case between adjoining proprietors, where the defendant dug a well on his premises and knowingly diverted percolating waters which made a spring on the plaintiff's premises and dried the spring. This was held to be *damnum absque injuria*. There is no parallel between that case and the one before us. There the spring owner had no ownership in the percolating waters until they appeared on his property, and the injury he suffered arose simply from his neighbor making his own that to which, while it was on his premises, he was lawfully entitled. In the case at bar, the disturbances, in their twofold origin, present slightly different circumstances, but call for the application of the same principle. The earth leakage arises from the defendant pouring electricity into the earth, which, following a course as natural for it as the seeking of a lower level is for water, comes upon the premises of plaintiff's subscribers, and by getting upon the wires of the telephone does harm. The plaintiff, by contract with its subscribers, has a right to have its wire where it is, and for the purposes of this discussion has exactly the same right to object to the presence of electricity on its premises caused by defendant and resulting in damage as would the subscriber himself. It is impossible to distinguish such an injury in principle from the case of one discharging filthy water into the ground so that it shall percolate into another's premises and there do damage, or of one causing smoke and cinders to be constantly thrown into one's windows and injuring one's enjoyment of one's property. These are all examples of nuisances which have been held to give a right of redress.

Ballard *v.* Tomlinson, 29 Ch. D., 116; Ry. Co. *v.* Gardiner, 45 Ohio St., 309.

As was said in Reinhardt *v.* Mestasti, 42 Ch. D., 685, "the principle governing the jurisdiction of the court in cases of nuisance does not depend on the question whether the defendant is using his own, reasonably or otherwise. The question is, Does he injure his neighbors?"

The disturbance by induction is of similar character, except that the force is applied from one wire to the other through the air instead of the earth. If, as we have found, defendant is not entitled to interfere with plaintiff's enjoyment of the telephone franchise, here is a direct act of interference.

Then it is said no claim can arise for disturbances by the defendant, because there are similar disturbances from other causes, for which the defendant is not responsible, and Wood *v.* Sutcliff, 2 Simons N. S., 163, is cited upon this point. That was a case where the injury sought to be enjoined was the pollution of a stream from which plaintiff had, in time past, drawn water for his business. One of the reasons given by the vice-chancellor for refusing the injunction was that others than the defendant were polluting the stream. The real basis of the decision was the laches of the defendant in asserting his right, for he had provided a new supply of water and had not used the river for some time. This case presented no difficulty in considering the case at bar, first, because there are no substantial disturbances shown other than those caused by defendant, and, second, because the case itself is not sound authority. (See Crump *v.* Lambert, 3 Eq. Cas., 414; Crosky *v.* Lightowler, 2 Ch., 478; Walter *v.* Selfe, 4 DeGex & S., 315.) In the last case the Court said that because one man was maintaining a nuisance against the plaintiff was no reason why defendant should set up an additional nuisance.

We find, then, that defendant is inflicting a legal injury upon the plantiff in the nature of a nuisance from which has already arisen loss, and which must inevitably cause loss in the future, constantly recurring. It is said that the damage is not irreparable, because the plaintiff can expend money and avoid it, and in the same way can arrive at its exact loss, and that, therefore, its remedy is not by injunction, but at law. Neither of these claims can be sustained. The most frequent exercise by a court of equity of the power of injunction is to prevent the continual recurrence of injuries from nuisance. The ground is that the plaintiff should not be put to a multiplicity of suits and endless litigation. To say that, in order to entitle a man to obtain an injunction against such injury, he should not be able, by the expenditure of even vast amounts, to avoid the injury, is to say that no injunctions can ever issue for such a cause. The point is that he is not obliged to expend money to protect himself when a neighbor injures him, but he may appeal to the courts; and when upon the law side he finds that the remedy there afforded is inadequate because of the necessity for bringing a separate suit for each injury, he may fairly say that by appeals to legal tribunals he cannot repair his damage, and therefore his loss will be irreparable. As to the ascertainment of damages, it is by no means true that in each suit the entire cost of introducing a metallic circuit or a McCluer device would be the measure of damages for this sort of interference, and the very reason for going into a court of equity is to get into a forum where all the injuries can be considered together.

The order of the court will therefore be that the defendant be enjoined perpetually from the use of the system of electric railway propulsion as now operated by them, or any other which will occasion similar disturbances to those now caused by defendant's single-trolley system. The order of injunction will not take effect until six months from this day, with leave to the defendant to apply to the court hereafter for further time, if necessary, in a *bona fide* effort to make the necessary changes. The costs of the action must be paid by the defendant. Decree accordingly.

APPENDIX B.

INSTRUCTIONS TO LINEMEN.

WIRE.

For the trolley wire, hard drawn copper is used to a very great extent, its size depending on the service required. No. 0 B. & S. gauge is a very convenient size and is generally used, although the wire may be smaller with sub-feeders as explained. Silicon-bronze wire has also been largely used on account of its great tensile strength. Copper is quite strong enough, however.

Span wires should be No. 1 B. & S. gauge, galvanized, wiped, Swedish iron wire if the trolley wire to be used is No. 0. They may be of smaller size if a smaller trolley wire is used.

Pull-offs and anchor guys should be the same size as the span wires.

For guard-wire spans, use No. 8 B. & S. gauge, galvanized, wiped, Swedish iron wire.

For longitudinal guard wires, use No. 10 B. & S. gauge, galvanized, wiped, Swedish iron wire.

In case of extra heavy or long stretches, two or more strands of the above iron wire may be twisted together. Stranded wires of galvanized iron have been used for ordinary lengths, and are preferred by some.

POLE CLAMPS.

The poles should be fitted with suitable pole clamps, so that the wire may be readily adjusted to the required weight. It is very important that the height of the trolley wire from the track should be uniform, and by using pole clamps this uniformity can easily be secured. Great care should be taken in this respect. Fig. 127 shows one style of pole clamp which may be used, and Figs. 128–132 several insulator holders for substantially the same purpose. There should be two clamps on each pole, one for trolley spans, the other for guard spans. In certain cases where a number of crowfoot wires diverge from one pole it may be found necessary to make use of more than two clamps on the pole. The clamp to which the guard span is to be attached may be smaller and lighter than that intended for the trolley span.

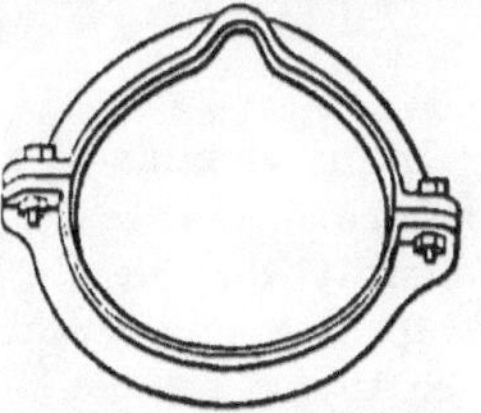

FIG. 127.—POLE CLAMP.

After the wires have been erected and lined up they should be more closely adjusted by loosening or tightening the bolts in the clamps which hold the spans.

In some cases where the spans are extremely long or heavy it may

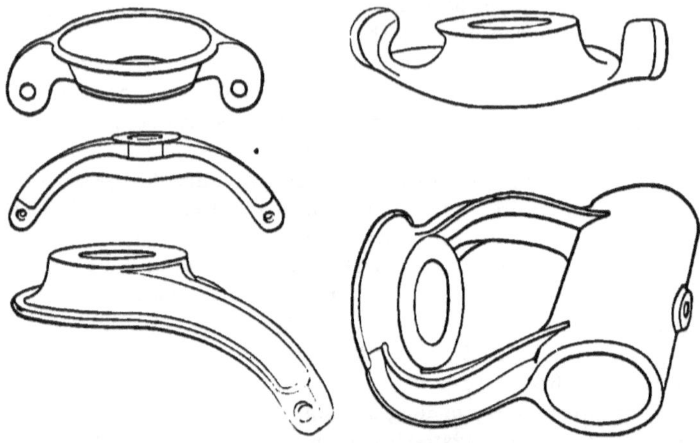

FIGS. 128-132.—INSULATED HOLDERS.

be advisable to use a third span fastened to clamps and allowed to sag toward the trolley span, which should be secured to it for additional support.

SPANS.

In the erection of a line the trolley spans should be the first wires stretched, and should be pulled up into their permanent position. Ordinarily, they should be as tight as two men can pull them with the help of a single 3-inch block and fall. The tension should be from 300 to 600 pounds. The guard spans should be as nearly perfectly insulated as possible from the trolley spans and the poles. This is accomplished by stretching them from pole clamps to which nothing is attached but the guard spans. Anchor guys should be attached to the trolley-span pole clamps and not to the guard-span clamps.

Next, guard spans should be stretched, and should ordinarily be pulled tight with the vise and strap, and should have about 100 pounds tension on them. In extra long and heavy stretches more tension should be placed upon both trolley and guard spans than that above mentioned, but care should be taken that the spans in every case shall be as loose as steadiness and appearance will allow, since the more tension there is on them the greater liability is there of breakage. The attachment of the wires to the eyebolt of the clamps should be made by a long and close half-connection.

EAR BODIES, OR SPAN-WIRE CLAMPS.

After the spans are up the next work is to locate and put on the ear bodies. On straight line work the ear bodies, if similar to the ones shown in Figs. 133–146, are "sprung on" the wire. Figs. 147–149 show several of the many kinds of ear bodies that may be used on curve

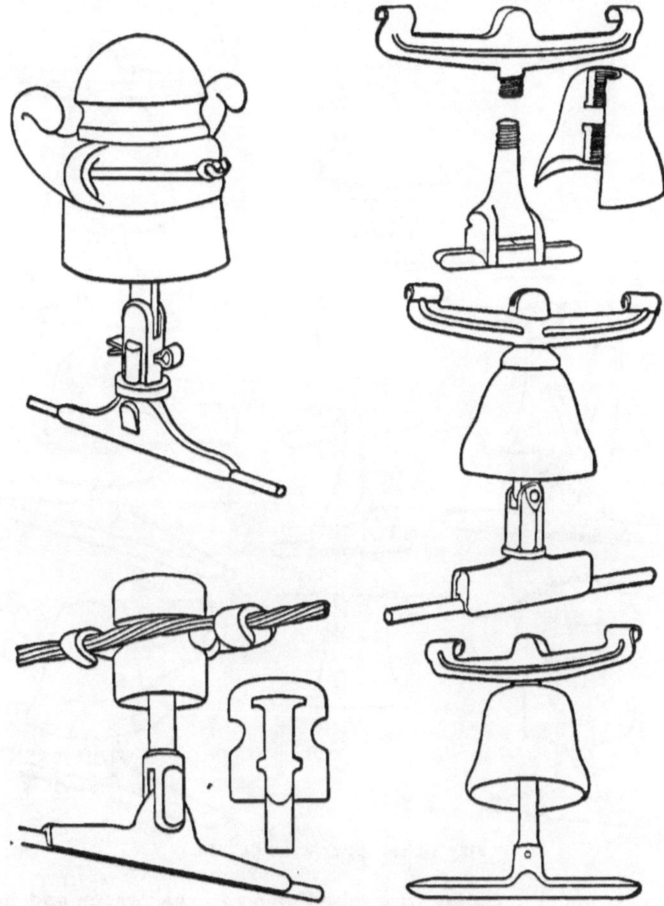

FIGS. 133-137.—SPAN-WIRE CLAMPS.

work. These should be "cut into" the span, that is, the proper point on the span wire having been determined by means of a plumb line, the wire is cut at this point and the ear inserted, as shown in Fig. 150. Sometimes insulating holders, such as shown in Figs. 151–153, are used in this work.

Figs. 154–157 show some of the types of "pull-offs," which are also intended for curve work, and Figs. 147–149 several bracket clamps used in special cases.

"RUNNING" TROLLEY WIRE.

Next, the trolley wire should be anchored securely at one end of the line and reeled out for a distance of 1,000 feet, or as far as is found convenient; temporary slings of iron wire should be placed over the

FIGS. 138-146.—SPAN-WIRE CLAMPS.

span wires and the trolley wire raised up on a tower wagon and hung in these slings.

Hauling clamps should be placed on the trolley wire and a temporary anchor made, the wire being pulled up tight by four or five men with double "block and tackle." The permanent anchor should then be pulled up tight and fastened to a pole so as to securely hold the tension placed upon the trolley wire.

On reaching a curve the wire should be pulled up tight to this point, anchored, and then run around the curve, allowing as much slack as may be needed to make the curve. On the curves the slings should be attached to the span wires or pull-offs in such a way that they cannot slip along them. The trolley wires should be placed a little inside

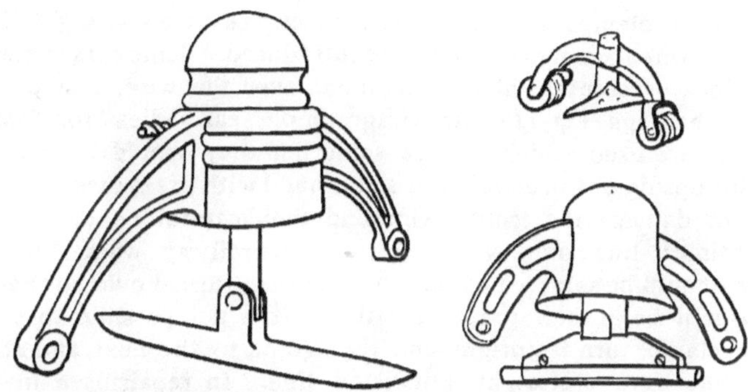

FIGS. 147-149.—CLAMPS FOR CURVE WORK.

FIG. 150.—"CUTTING IN" A SPAN-WIRE CLAMP.

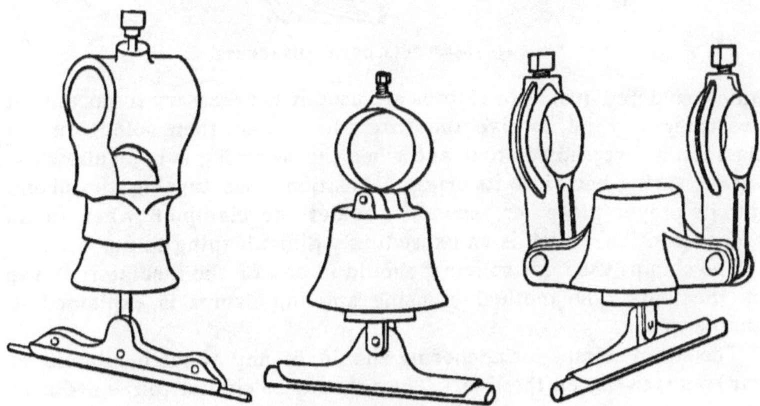

FIGS. 151-153.—INSULATOR HOLDERS.

of their usual position, as judgment may dictate, and pulled comparatively tight. The wire should then be anchored both ways at the end of the curve, so as to make a fixed point.

The construction may be continued in the same manner as before from this point on.

After running and hanging up the trolley wire the ears or clamps should be put on.

"EARS" OR INSULATOR CLAMPS.

The same clamps should be used on curves as on straight lines, excepting when anchors or splices are introduced. Some ears or clamps have sides or points, which, being bent over the wire, clamp them to it. (See Figs. 133-146 with paragraph on "ear bodies" for "ears.") If clamps are used which require soldering they should be fitted to the wire upside down, closely pressed around with great care, so as not to cut or damage the trolley wire, and then carefully soldered with the ordinary lineman's soldering iron for railway work. All the clamps should be soldered on and the line then turned over so that the clamps will be in their proper position. It is not possible to solder on one clamp, turn it upright, and then go on to the next, and at the same time have a straight, untwisted line. In repairing a line on

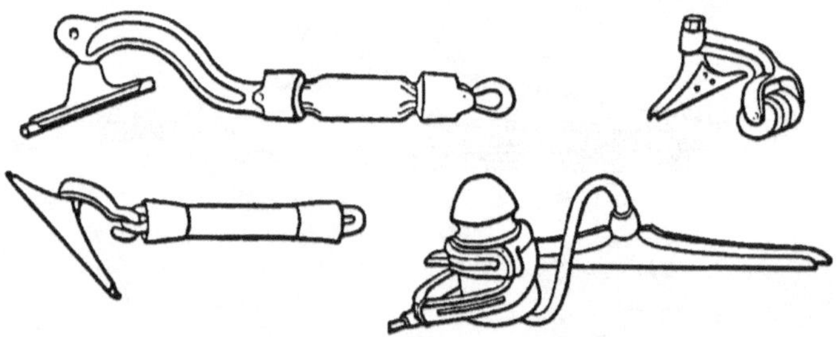

FIGS. 154-157.—"PULL-OVER" BRACKETS.

which soldered insulator clamps are used it is necessary to do one of two things: First, to give the wire a half twist, then solder on the clamp in a reversed position, and when the soldering is done allow the wire to spring back into its original position, thus turning the clamp into its proper place; or, second, to solder the clamp on when in its proper position, which is an exceedingly difficult thing to do.

The clamps used for splicing should fit any of the insulators in use on the line. The method of using splicing clamps is explained in another place.

The strain clamps for anchoring should fit any of the insulators or ear bodies in use on the line. They should be similar to the ordinary clamps so far as the groove for holding wire is concerned, and should be put on in the same way.

GUARD-SPAN HANGERS.

In the same way that insulators are used the guard-wire hangers should be employed, several styles of hanger being shown in Figs.

158-160. For straight line work the straight line guard hanger should be "sprung on" the wire with the insulator down.

The longitudinal guard should be suspended from below, not over the insulator, by means of the common insulator tie of No. 12 wire.

FIGS. 158-160.

On curves single and double guard-wire curve suspensions should be employed, the longitudinal guard wire being secured to the insulator by a stout tie.

ANCHORS.

Double anchors should be put in every few hundred (500 to 1,000) feet on straight line work, and should be put in at each end of every

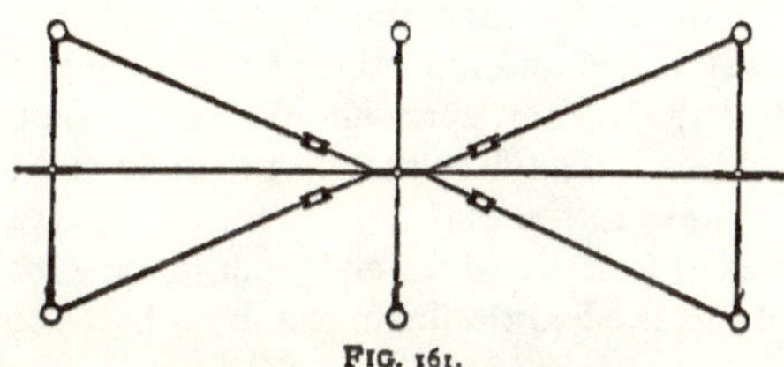

FIG. 161.

curve. By double anchors are meant anchors which hold the line both ways. Figs. 161-162 show how these anchors should be installed.

A single anchor must never be put in except at the end of a line. Anchoring is half the line construction, and if not well done will cause endless trouble.

The following instructions for anchoring should be carefully noted:

For the immediate connection to the trolley wire, use a short strain clamp.

The same wire that is used for spans should be used for anchors,

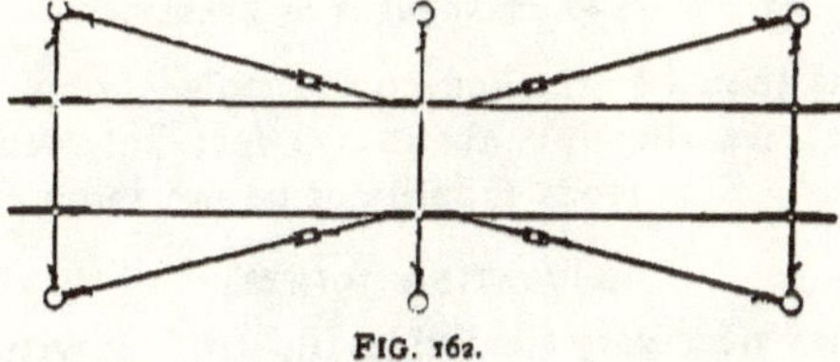

FIG. 162.

and should be pulled about one-half as tight. Some insulator should be put into each anchor wire about 2 feet from the anchor clamp.

For double-track construction of anchors, see Fig. 162 above.

SPLICING.

When the trolley wire requires splicing a "splicing ear" should be used in connection with an insulator. If a splicing ear like the one shown in Fig. 163 is used, the end of the wire after being passed up

FIG. 163.—SPLICING "EAR."

through the proper holes and soldered should be turned back over the top, as shown in the figure, and not left sticking up straight.

FEEDERS.

In "cutting in" feeders, measurements should be made of the width of street, distance between trolley wires, etc., and the feeder span made up in the shop ready to install, since trouble will be found in making up a span on the street. All feeder spans should be put up as represented in Fig. 164.

The wire used for connecting the feeder with the clamp or ear should be of the same size and material as the trolley wire. The wire on the dead end of the feeder span should be the same size and kind as used for span wires. The insulator on the dead end should be close to the ear and not next to the pole.

The insulator at the other (live) end should be close to the pole, and the side feed fastened to the insulator by a half connection. The

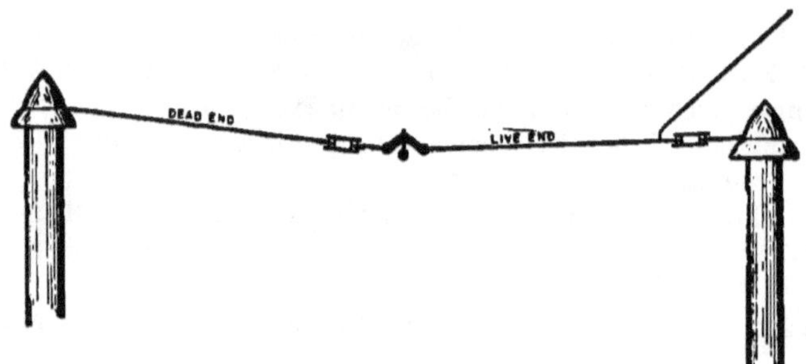

FIG. 164.—ARRANGEMENT OF FEEDER.

insulator should then be attached to the pole clamp by the regular wire used for spans, allowing about two feet between the pole cap and the insulator. Side feeds should not be put in on curves.

INSULATING JOINTS.

It is sometimes necessary to divide the trolley wire into sections insulated from each other. This is done by the insertion of "insulating joints."

Insulating joints should be inserted at spans, in the same manner as a splicing ear.

If such an insulating joint as the one shown in Fig. 165 is used, the span wire should be attached to its central segment.

Connection can be made with a feeder by the ordinary side feed

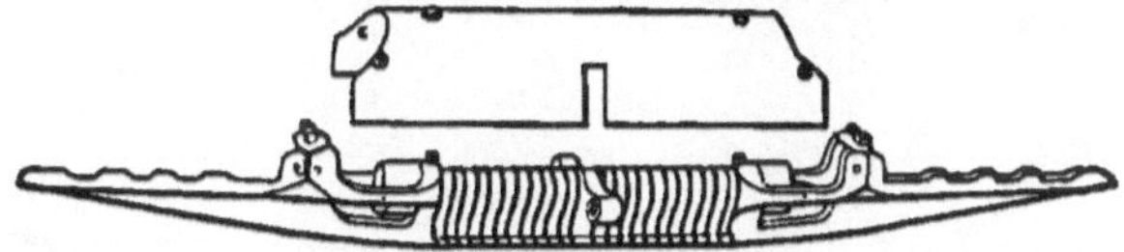

FIG. 165.—INSULATING JOINT.

simply by clamping the side feed wire into the tongue of the insulating joint, by means of a screw provided for this purpose.

It is sometimes advisable in the construction of overhead lines to run a wire from each side of an insulating joint to a switch on the pole, so that the insulating joint may be short-circuited in case it is desirable to throw the two sections of line together. One method of doing this is shown in Fig. 166.

The kind of insulating joint shown in Fig. 165 may be used as a lightning arrester, if supplemented by a special device having a series of small breaks of about one-sixteenth of an inch each, which should be "cut into" the span wire attached to the ear and dead grounded at the other end on the pole, assuming that iron poles are used. If wooden poles are used this ground should be made by leading a wire down the pole and fastening it to the rails.

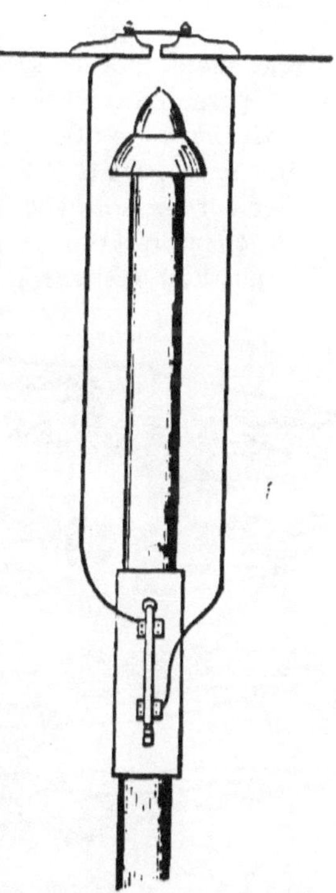

FIG. 166.—SWITCH AROUND INSULATING JOINT.

LOCATION OF TROLLEY WIRE.

The trolley wire should be located as follows:

On straight line work the wire should be over the center of the track.

At curves the wire should be placed on the inside of the curve, its distance from the center of the track depending on the degree of curvature.

FROGS.

Where lines branch from each other frogs are used for directing the trolley wheels to the proper line. When the line branches off to the right use a right-hand frog, and when to the left, a left-hand frog, and when the track splits symmetrically a symmetrical frog should be used.

In case the angle between the diverging tongues is not the same as that which the trolley wires make, the tongues should be slightly bent to make the two angles the same. Some frogs are made with lugs on the sides for the attachment of an insulating truss, so that guy wires may be fastened thereon to steady the frog and prevent its tilting.

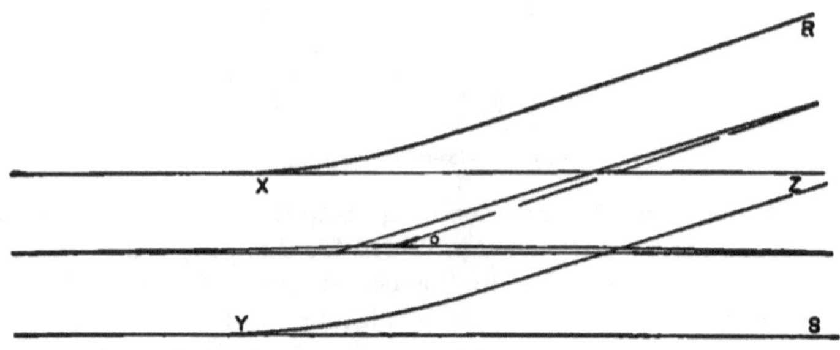

FIG. 167.—PLACING FROGS.

Frogs should not be secured to the trolley wire until the proper location is assured. With the line at its usual height, 18′ 6″ to 19′ 6″, the proper place may be found as follows:

In Fig. 167 let R X Z Y S represent the position of the track.

Let o represent the center of the triangle, X Y Z; that is, the point of intersection of right lines bisecting the three angles X, Y, Z. The position of the frog should be directly over the spot (o). Observe

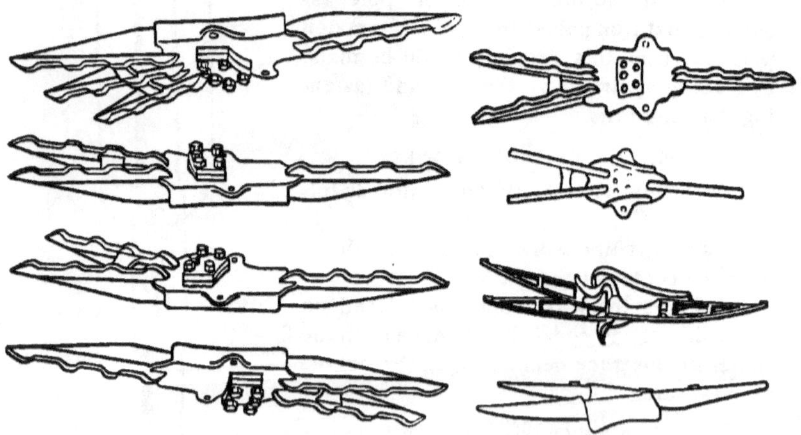

FIGS. 168-175.—FROGS.

that this throws the main line a few inches out of center, and likewise the branch line a little out of center. This is all right, however, and is the secret of making the frog work successfully. Figs. 168–175 show a few types of frogs.

DIAGONAL AND RIGHT-ANGLE CROSSINGS.

Where trolley wires cross there should be used: First, a diagonal crossing, if the lines make an acute angle of less than 60 degrees; second, a right-angle crossing, if the acute angle is between 60 and

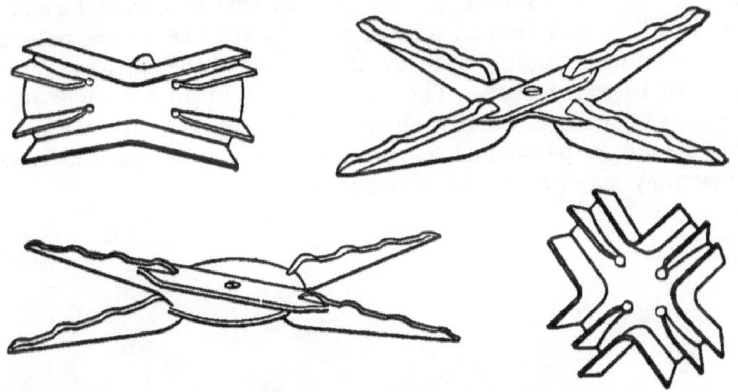

FIGS. 176-179.—TROLLEY-WIRE CROSSING.

90 degrees. The method of putting up diagonal and right-angle crossings is the same as that for frogs. See Figs. 176-179.

GUARD WIRES.

Guard wires should be insulated from everything else. They should be cut into sections at the same points as the trolley wire, and at every 1,000 feet besides. The guard wires should be directly over the trolley wires, that is, one guard wire to each trolley wire.

Before the line is left as completed all insulators, pole clamps, and other parts of the line material needing it should be painted neatly and with care, that the parts may be kept from rusting.

POLE SPECIFICATIONS.

If wooden poles are used they are generally of cedar, chestnut, or Georgia pine, the latter being usually squared and pointed at the top. Round poles should be trimmed, smoothed and pointed, and the butts of all wooden poles should be painted with some preserving compound. The length should be not less than 28 feet, with mean diameters of at least 7 and 9 inches at the top and bottom respectively.

Iron poles should be at least 28 feet long, and if made in sections, as the Walworth pole, the respective diameters of the sections should be generally 6, 5, and 4 inches. There are many pole manufacturers, but the Walworth Co. is perhaps most widely known. Iron poles should have insulated caps and cast-iron trimmings, though these latter are not absolutely necessary. All poles should be set six feet in the ground, with tops on a level, about 20 feet above the level of the rails. For iron poles, the rake should be about five-sixteenths of an inch to one foot of length, and for wooden poles at least twice as much, this depending of course upon circumstances. The object aimed at is that when the strain of the span and the other wires is upon the poles they shall stand straight. All poles should be well grouted with broken stone and cement, and it is well to place a large stone at the heel of the pole, also one in front of the pole just below the surface of the earth, so that the pole shall be well braced. The distance apart of the poles will depend upon circumstances; usually about 125 feet apart will answer. All poles should be painted once before being set, and again afterward, so that they may present a good appearance.

APPENDIX C.

ENGINEER'S LOG-BOOK.

January 17, 1892.

COAL USED, *5,650* LBS. WATER USED, *36,560* LBS.

CONDITION OF TRACK.

Muddy.

CARS IN USE.

Standard, *8.*
Double truck, *1.*
Trailers, *2.*

NOTE TIME AND NUMBER OF EXTRA CARS IN LOG.

MACHINES IN USE.

Engines, *A.*
Boilers, *1 and 2.*
Dynamos, *1, 2, 3.*

NOTE ANY MACHINES THROWN INTO USE IN LOG.

Started, *5.50.*
Shut down, *11.20.*

TIME.	INDICATOR CARDS.	AMPERES ON LINE.	VOLTS ON LINE.	WEATHER.	SPECIAL REPORTS.
8.00	*H 48.2* *C 47.9*	*112*	*550*	*Cold clouds.*	
9.17		*182*	*505*		*Heavy current for about six minutes. Dynamo 1 sparked badly. Cut it out and threw in No. 4.*
12.50				*Snow.*	
1.32		*125*	*540*		
2.40		*117*	*546*		
2.45		*113*	*548*		
4.50				*Snow ceased.*	
6.10		*95*	*550*		
10.00		*80*	*550*	*Clear.*	

APPENDIX D.

CLASSIFICATION OF EXPENDITURES OF ELECTRIC STREET RAILWAYS.

COMPILED BY H. I. BETTIS.

(By permission)

An authority upon railway accounts has said that "the most remarkable diversity exists among railway companies in reference to the classification of their operating expenses."

Undoubtedly the statement was correct as applied to steam railroad accounts, but could the same writer have had his experience in street railway management he probably would have declared that there is no similarity whatever in their treatment of disbursements.

The electric street railways of to-day are in their infancy, and we have but a limited knowledge of the practical utility of the various parts and the manner in which they are operated, but with careful attention to the accounts the benefits derived from the various combinations and methods are readily seen.

Also comparative statements can be made from month to month, by which it can be seen at a glance wherein we have not paid the proper attention to certain parts of our system, and demonstrate that our neglect results in dollars and cents on the wrong side of the income account.

At the time of the Buffalo Convention our managers and superintendents made statements concerning their operating expenses which actually staggered one another. One would state that on his line operation cost so much per mile; another would state that it cost his company a much different sum per mile.

The first had included, perhaps, the repairs of cars with wages of motor-men and conductors, while the second might have omitted repairs of cars and included the cost of operating the power house, with or without the cost of repairs on the dynamos.

There was, and is now, no way in which a comparison can be made between the costs of operating our various street railways, and it is with this object in view that this manual has been compiled, hoping that although it might not suit all, it might at least be a guide to those who have not yet decided upon any classification, and for those who are at all desirous of having a standard or universal system.

Such a classification should be as simple as possible, and yet with sufficient detail to afford the management and owners an easy means for determining in what manner the business of the road could be conducted with better success, where the failures had been, and what had been the causes.

The classifications as given here are not subdivided to the extent that some might think desirable, but sufficiently, I think, for any practical purposes, as any further division would be purely statistical.

It has been the author's intention to follow the idea of the Interstate Commerce Commission when designing the standard classification for steam roads, to separate the expenses which add to the present or future value of the property from those which do not, as the reader will see by comparing the general expenses and transportation expenses with the maintenance of way and buildings and maintenance of equipment, the latter accounts adding materially to the value of the property.

There is a short classification of construction expenses added, but too much care cannot be exercised in charges to these accounts, bearing in mind, however, that nothing should be charged to construction and equipment except that which adds to the first or original cost of the property.

I wish to express my thanks to Mr. W. E. Baker, Superintendent of the West End Street Railway contract for the Thomson-Houston Company, Boston, whose generosity has enabled me to profit by his experience and advice upon the subject treated.

OPERATING EXPENSES.

GENERAL EXPENSES.

1. Salaries of General Officers.—In this account are included the salaries of the general officers; the heads of departments connected with the supervision and management of the general business of the road. Salaries of division superintendents and assistants may also be charged to this account. By general officers are meant officers in charge of departments and whose jurisdiction extends over the entire road.

2. Salaries of Clerks in General Offices.—This account embraces the salaries of all clerks in the general offices, clerks for heads of departments, and all clerks not hereinafter mentioned.

3. Miscellaneous Expense General Offices.— The expense of heating, lighting the general offices; wages of porters, messengers, etc.; telephone service, and all miscellaneous supplies and expenses of the general offices are charged to this account.

4. Stationery and Printing.—Includes the cost of all stationery, books, paper, stamps, pens, pencils, etc.; also cost of all printing of blanks, circulars, statements, tickets, etc., and the cost of advertising.

5. INSURANCE.—Includes cost of insurance on property of the company, and against injuries to employees, and all expense of collection.

6. LEGAL EXPENSE.—In this account are included the salaries, fees and expenses of attorneys, witnesses' fees and other court expenses.

7. INJURIES AND DAMAGES.—Expenses on account of persons injured and property damaged, with payments of claims, are all chargeable to this account. Wages of persons while disabled, medical attendance, and funeral expenses; also wages of claim agent and others connected with the claim department. Lawyers' fees and other court expenses *are not chargeable to this account;* nor are damages to property belonging to the company.

8. CONTINGENT EXPENSES.—This account includes the miscellaneous expenses not otherwise provided for; traveling and other expenses of general officers and assistants, etc., etc.

TRANSPORTATION EXPENSES.

1. CAR SERVICE.—This account includes the wages of conductors, motor-men, starters, aids, inspectors, and switchmen; cost of punches, ticket registers, sign sticks, switch sticks, and miscellaneous supplies for car service. The wages of the superintendent of time-tables and chief of conductors, with such clerks as may be under them, should also be charged to this account.

2. CAR-HOUSE EXPENSE.—This account includes the wages of shed foremen, shifters, cleaners, oilers, wipers, laborers, inspectors and watchmen, except such as are employed on repairs of cars. The cost of fuel and lighting the car houses and sheds, lanterns and oil for watchmen, and tools used by workmen on cars (cleaning and oiling, other work except repair) are chargeable to this account.

3. LUBRICANTS AND WASTE FOR CARS.—Oil, grease, tallow and other lubricants, with waste used upon car journals and motors, are included in this account.

4. ELECTRIC SUPPLIES.—This account includes such supplies as are constantly needed for the operation of the electric cars, but cannot be charged to repairs, such as lamps, fuses, carbon brushes, trolley cord, etc.

5. WRECKING.—Wages of those employed in getting derailed cars on the track and removing obstructions and wrecks; tools used and all other expenses incurred on the same account. Expense of getting cars back to the car house when broken down on the line is also chargeable to this account.

6. OPERATION OF POWER HOUSES.—This account includes wages of engineers, firemen, coal shovelers, dynamo-men, oilers, cleaners and others employed in the power houses, except when employed upon repairs. Also the cost of water, water rates, or cost of pumping where the company furnishes its own water-works; carbon brushes,

fuses, lamps and other supplies necessary for the daily operation of the power houses, and not otherwise provided for; cost of heating and lighting power houses. Repairs and renewals of engines, boilers, dynamos, switchboards and station fixtures *are not chargeable to this account.* Fuel and lubricants are also chargeable to separate accounts.

7. FUEL.—This account includes the cost of all fuel used in the power houses, with transportation charges on the same.

8. LUBRICANTS AND WASTE FOR POWER HOUSES.—Oils, greases, tallow and waste for use in the power houses, for engines, shafting, dynamos, pumps, etc.

MAINTENANCE OF WAY AND BUILDINGS.

1. REPAIRS OF ROADWAY AND TRACK.—This account includes all expenditures on account of the road-bed and track, except the cost of rails and ties used, and the cost of repairs and renewals of paving and the supplementary wire. It includes tracks laid in buildings, yards, on turn-tables, wharves, and over bridges; wages of roadmasters, track foremen, laborers, watchmen, and others, while engaged in track repairs and renewals. It includes cleaning, oiling and sanding track, repairs and renewals of drains under the track, repairs and renewals of planking over bridges, repairs and renewals of frogs and switches, joint fastenings, etc., etc.; removing snow and ice, repairs of snow plows and sweepers. It also includes repairs of rails and all work on rails, cutting and drilling, *except drilling for tie wires;* also labor expended in taking up track. The cost of tools implements and all supplies used in connection with the track is included in this account. The expense of removing snow and ice, with the cost of repairs on snow plows and sweepers, may be made a separate account if so desired, but comes under the head of Maintenance of Way.

2. RENEWALS OF RAILS.—This account includes the cost of new rails laid in the track, with the transportation charges on the same, less the value of old rails taken up. The expense of loading, unloading, drilling, cutting, laying and repairing rails *is not included in this account.*

3. RENEWALS OF TIES.—This account includes the cost of new ties laid in the track and the freight on the same. The expense of loading, unloading and laying ties *is not included in this account.*

4. REPAIRS AND RENEWALS OF PAVING.—This account embraces all expenditures on account of the paving. It includes the cost of paving blocks and sand and the cost of transportation of the same; the wages of pavers, laborers and others engaged in repairs and renewals of paving; also the cost of tools and other supplies for the same work. The expense of taking up and relaying paving when necessitated by the repairs on the road-bed, the track and the supplementary wire *is not chargeable to this account,* but to the account for which such expense was incurred.

25

5. Repairs and Renewals of the Supplementary Wire.—This account includes all expenditures on account of the supplementary wire and its connections. It includes the cost of the wire, tie connections, channel pins, solder, and other supplies, also tools and implements used in connection with the work of repairing the supplementary wires; wages of solderers, laborers and others engaged upon this work. The expense of drilling rails for channel pins and tie wire rivets is also chargeable to this account. Expense on the supplementary wires, necessitated by the taking up of rails, ties, switches, frogs, etc., *is not chargeable to this account*, but to repairs of roadway and track. Expense on the supplementary wires necessitated by the taking up or laying of paving should be charged to the account of paving.

6. Repairs and Renewals of Buildings, Docks and Wharves.—This account includes the cost of repairs and renewals of all buildings, docks and wharves and of the stationary fixtures and furniture of the same not otherwise provided for; car houses and sheds, store houses, car shops, repair shops, blacksmith and machine shops, power houses, coal sheds and bins, stations, etc., etc. Repairs of pits in car houses and shops, cranes in power houses and coal sheds, etc., are embraced in this account. Repairs of tracks in buildings and on wharves *are not chargeable to this account.*

7. Repairs and Renewals of Poles and Overhead Lines.—This account includes the cost of repairs and renewals of poles and brackets, with trolley, span, guard and feed wires, with all appliances for suspension and insulation of the same.

MAINTENANCE OF EQUIPMENT.

1. Repairs of Cars.—This account includes the cost of all repairs on car bodies, painting, varnishing, upholstering, relettering cars and car signs; repairs and renewals of the trucks, brakes, brake shoes, axle boxes, springs, track brushes, snow scrapers, pilots, sand boxes, etc.; repairs and renewals of wheels and axles. The cost of new cars taking the place of old to make the number good is also chargeable to this account. On roads using both motor and tow cars it is advisable to keep each kind separately.

2. Repairs of Electrical Equipment.

A. Armatures and Fields.—This account includes the cost of repairs and renewals of armatures and fields, labor of removing from the cars those damaged, also cost of replacing them and making all connections. New armatures and fields put in to take the place of old, to keep the number good, are chargeable to this account.

B. Gears and Pinions.—This includes the cost of repairs and renewals of gears and pinions, labor removing and replacing in the cars; also the cost of new gears and pinions put in to take the place of those damaged or destroyed.

C. Trolleys.—This account embraces all repairs and renewals of trolleys and their parts, labor of taking off and replacing, and the cost of new trolleys replacing those damaged or destroyed.

D. Sundry Repairs.—This account embraces the repairs of motors and their connections, *excepting armatures and fields, gears and pinions, and trolleys.* It includes the repairs of wiring, cables, repairs and renewals of lightning arresters, reversing switches, cut-out switches, main motor switches, rheostats, and pans, brush holders, motor pans, etc., etc.

3. REPAIRS OF STEAM PLANT.—To this account should be charged all repairs and renewals of the steam plant in the power house, including the boilers, engines, pumps and shafting, repairs and renewals of belts, piping, steam fitting, etc., etc.

4. REPAIRS OF ELECTRICAL PLANT.

A. Dynamos.—Repairs and renewals of dynamos and their parts; armatures, fields, pulleys, commutators, oilers, bearings and boxes, brush holders, etc., are all chargeable to this account; also labor removing and replacing damaged parts.

B. Switchboard Equipment.—Repairs and renewals of the switchboard equipment are charged to this account, such as repairs and renewals of station switches, rheostats, circuit breakers, ammeters, wiring and connections.

5. TOOLS AND MACHINERY.—Repairs and renewals of tools and machinery, shafting, boilers, engines, etc., in the shops of the company; also cost of lubricants for the same. Small tools (not shop fixtures) are chargeable to the account most benefited by them.

6. MISCELLANEOUS EXPENSE.—To this account should be charged all miscellaneous expense of maintenance of equipment not otherwise provided for.

CONSTRUCTION AND EQUIPMENT EXPENSES.

Too much care cannot be exercised in charges to this account. Nothing should be charged to construction and equipment except that which adds to the first or original cost of the property.

1. SUPERINTENDENCE AND GENERAL EXPENSE.—Salaries of superintendent of construction, assistants, wages of clerks and others employed in the offices of this department. Expense of the office, furniture, fuel, lighting, supplies for office, miscellaneous and personal expense of superintendent and assistants while on business. Includes stationery and printing for this department.

2. ENGINEERING.—Wages and expenses of engineers and draughtsmen on preliminary work and construction.

3. RIGHT OF WAY.—Salaries and expenses of right-of-way agent, together with payments for rights of way, easements, franchises, payments for land for buildings, shops, etc.

4. BUILDING CONSTRUCTION.—Cost of buildings—car houses, stations, offices, store houses, power house, repair shops, wharves, coal sheds, etc., etc.; also furniture and fixtures for the same.

5. TRACK AND ROADWAY CONSTRUCTION.—Includes the expense of grading, surfacing, ballasting, ditching and paving; the cost of rails, rail chairs, ties and stringers, tie rods, joint fastenings, track spikes, frogs and switches, supplementary wire, tie wires, channel pins, solder and miscellaneous track material; also the cost of distributing and laying the same, with the supplementary wire and its connections.

6. OVERHEAD CONSTRUCTION.—Cost of poles and setting, putting up trolley, feeder and guard wires, including cost of wire and all devices for overhead construction.

7. CAR EQUIPMENT.—Cost of cars built or purchased, including the cost of trucks, motors, upholstering, painting, lettering, varnishing, etc.

8. SNOW PLOWS AND SWEEPERS.—Cost of snow plows and sweepers built and purchased, including the electrical equipment of the same.

9. POWER STATION EQUIPMENT.—Cost of steam plant, engines, boilers, pumps, piping, shafting and belting, dynamos and switchboard equipment, together with installation of the same.

10. TOOLS AND MACHINERY.—Cost of tools and machinery for repair shops, car houses, etc., and expense of setting and placing in running order.

11. IMPROVEMENTS AND BETTERMENTS.—All expenditures which improve the *original plant*, and of which a portion should be charged to operating and a portion to construction expenses.

ACCOUNTING FORMS.

Form A.

Form A is a sheet 9½ inches wide by 12 inches long, and printed on one side only. All the solid lines in the form, as shown below, both horizontal and perpendicular, are ruled in red ink. The dotted horizontal lines are all faint blue ruled. The upper set of dotted lines are crossed over each side of the double perpendicular lines by eleven perpendicular heavy blue lines, making thereby twelve columns, each set of twelve columns being separated by double perpendicular red ink lines. The figures under the columns entitled Date run from 1 to 31, inclusive. The space above the double horizontal lines at the top of the page is 1⅛ inches; the space devoted to *Summary of Mileage* is 2¼ inches high. The dotted lines, nine in all, are faint blue ruled. The short lines at the bottom of the page are printed heavy black, being in the original form one inch long.

FORM A.

Daily account of trips run and monthly report of revenue mileage of S............... D...............

Motor Electric Car No.............

and cars towed by it for the month of............................ 189..

Date.	Routes	Revenue Motor Trips.	Revenue Towed Trips.
1			
2			
3			
29			
30			
31			
Total..			

SUMMARY OF MILEAGE.

Motor trips on Route No....@ =	Towed trips on Route No....@ =
"	"
"	"
"	"
"	"
"	"
"	"
"	"
Total motor mileage,	Total towed mileage,

Form C.

Form C is printed on a four-page sheet 14 inches high by 8½ inches wide. The width of each column of the first page appears under its respective title, the Endorsement taking one-quarter of the page, and both sections entitled Pay Rolls appear on the first page. The following two subdivisions appear on the second page, and the next two on the third page. The section entitled Earnings, and all that follow, are on the fourth page. The dotted horizontal lines through the entire form as printed are ruled faint blue, spaced ¼ inch apart. All the solid lines are ruled in red, except the heavy perpendicular lines, which are ruled dark blue. Each of the four sections on the second and third pages is given half a page, each separate item being placed on a single line. The columns are uniform, being, on the second and third pages, ¾ inch and 1 inch wide alternately, while on the fourth page they are 1 inch and ½ inch wide alternately. On all three pages there are four sets of columns.

Form C.

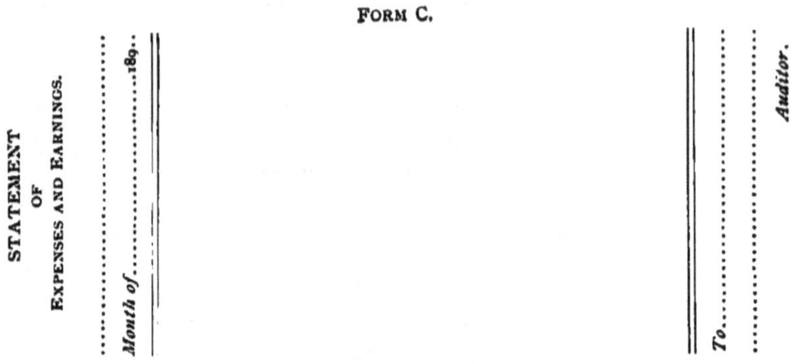
STATEMENT
OF
Expenses and Earnings.
....
Month of..............................189..
To..............................
....
Auditor.

RECAPITULATION.

Earnings, Expenses and Net Earnings.

	Motor Mileage.	Earnings.	Expenses.	Per cent. of Exp. to Earn.	Net Earnings.	Exp. per Car Mile.	Exp. per Pass.
	1	1 ¼	1 ¼	1	1 ¼	¾	½
Month of							
Same month of......							
Increase..............							
Decrease..............							

Earnings, Expenses and Net Earnings.

Current Year Compared with Corresponding Period of Last Year.

	Motor Mileage.	Earnings.	Expenses.	Per cent. of Exp. to Earn.	Net Earnings.	Exp. per Car Mile.	Exp. per Pass.
	1	1 ¼	1 ¼	1	1 ¼	¾	½
Jan. 1st to..........							
Same period of......							
Increase							
Decrease							

PAY ROLLS.

	18..	18..	Increase.	Decrease.
	1 ¼	1 ¼	1 ¼	1 ¼
General Office..............				
Motor Men and Conductors..............				
Power House..............				
Maintenance of Roadway and Track..............				
" " Equipment..............				
....				
Total				

PAY ROLLS.

Current Year Compared with Corresponding Period of Last Year.

	18..	18..	Increase.	Decrease.
General Office				
Motor Men and Conductors				
Power House				
Maintenance of Roadway and Track				
" " Equipment				
Total				

GENERAL EXPENSES.				
Salaries of General Officers				
" " Clerks in General Offices				
Miscellaneous Expenses, General Offices				
Stationery and Printing				
Insurance				
Legal Expenses				
Injuries and Damages				
Contingent Expenses				
Total				
TRANSPORTATION EXPENSES.				
Car Service				
Car House Expenses				
Lubricants and Waste for Cars				
Electric Supplies				
Wrecking				
Operation of Power House				
Fuel				
Lubricants and Waste for Power House				
Total				

DETAILS OF EXPENSE ACCOUNT.

MAINTENANCE OF WAY AND BUILDINGS.				
Repairs of Roadway and Track				
Renewal of Rails				
" " Ties				
Repairs and Renewal of Paving				
" " " " Supplementary Wire				
" " " " Buildings				
" " " " Overhead Line				
Total,				

DETAILS OF EXPENSE ACCOUNT—*Continued.*

MAINTENANCE OF EQUIPMENT.				
Repairs of Cars				
" " Electrical Equipment				
" " Armatures and Fields				
" " Gears and Pinions				
" " Trolleys				
Sundry Repairs				
Repairs of Steam Plant				
" " Electrical Plant				
Dynamos				
Switch Board Equipment				
Tools and Machinery				
Miscellaneous Expenses				
Total,				

Comparative Statement of Earnings, Expenses, Net Earnings and Mileage.

Statistics..189..,

and for Current Year Compared with Corresponding Period of Last Year.

EARNINGS.

	18..	18..	Increase.	Decrease.
Passenger	1 ½	1 ½	1 ½	1 ½
Tickets				
Mail				
Miscellaneous				
Total,				

EXPENSES.

	18..	18..	Increase.	Decrease.
General				
Transportation				
Maintenance of Way and Buildings				
" " Equipment				
Total,				

MILEAGE STATISTICS.

	18..	18..	Increase.	Decrease.
Double Motor Box				
Single " "				
Double " Open				
Single " "				
Tow Car				
Long Box				
" Open				
Total,				

EARNINGS.

Current Year Compared with Corresponding Period of Last Year.

	18..	18..	Increase.	Decrease.
Passenger				
Ticket				
Mail				
Miscellaneous				
Total,				

EXPENSES.

Current Year Compared with Corresponding Period of Last Year.

	18..	18..	Increase.	Decrease.
General				
Transportation				
Maintenance of Way and Buildings				
" " Equipment				
Total,				

MILEAGE STATISTICS.

Current Year Compared with Corresponding Period of Last Year.

	18..	18..	Increase.	Decrease.
Double Motor Box				
Single " "				
Double " Open				
Single " "				
Tow Car				
Long Box				
" Open				
Total,				

Form D.

Form D is a sheet 12 inches wide by 17¼ inches long, and is printed on both sides, Form E being on the reverse side, so that when the book lies open, the *Material Used* page is opposite the *Material Received* page. Single red ink perpendicular ruled lines divide the columns, and the width of each column is given under its respective title. There are 59 horizontal faint blue lines on the page. The height above the double horizontal red lines at the top is 1½ inches.

FORM D.
MATERIAL USED.

Date.	Kind and amount of material.	Price.	Amount.		Charge account.	Where used.	To whom delivered.
1¼ in.	3½	½	¼	¼	¼	2½	2½

Form E.

For explanation, see remarks descriptive of Form D.

FORM E.

MATERIAL RECEIVED.

Date.	Kind and amount of material.	Price.	Amount.		From whom.	How received.	Bill Checked.
1¼ in.	3½	½	¾	¼	2½	1¾	1½

Form F.

Form F is a slip 7¼ inches wide by 4 inches high, and printed on one side only. The word Foreman is at the foot of the slip on which appears 8 lines running the full cross-width of the paper; with a double line above and below the single lines. There is no colored ruling on the slip.

FORM F.

...., 189..

Storekeeper:

Please furnish the following material for use on

....

.... Foreman.

Form P.

Form P is a book 4 inches wide by 6¾ inches long, bound with a stout paper cover. It contains six sheets, which when folded in the centre make a book of 24 pages.

The blank space above the double ruled line at the top of the page is about 1 inch high.

The widths of the several columns appear under their respective titles. The dotted lines in the form below are ruled faint blue in the original, and are spaced ¼ inch apart, of which there are 19 in all. All the solid lines are ruled in red.

FORM P.

TIME WORKED by........................ Occupation,............................

for week ending.............................., 189.. Rate per day or month, $........

Specification of Work.	Sun.	Mon.		Tues.	Wed.	Thur.	Fri.	Sat.	Sun.	Mon.	Total hours.	Amount.		Charge Account.
3½ in.	¼	¼	¼	¼	¼	¼	¼	¼	¼	¼	3/16	3/16	½	1
....	..	..	..	..	..	..	..	..	..	..	......	....	...	....
....	..	..	..	..	..	..	..	..	..	..	......	....	...	....
Total,														

On the title-page of cover appears the following, properly spaced, of course:

TIME BOOK.

....Crew,

week ending.., 189..

certify the time and rates as shown herein are correct.

....

Foreman.

On the last page of cover appears the following:

The time of each man must be entered in this book at the close of each day's work, and the book forwarded to the General Office at the end of each week. Do not fill in the last two columns.

Form G.

Form G is printed on a sheet 8 inches wide by 7 inches high. An endorsement appears on the back of the sheet. All of the solid lines in the following form are ruled in red ink, and the width of the columns appears under each heading. There are 22 ruled lines, in faint blue, on the sheet, and the height of the space at the top is 1 inch.

FORM G.

....Department.

Foreman's report of material used in month of.., 189..

Date.	Location and description of work.	Quantity and kind used.	Foreman will leave these columns blank.			
			Cost.		Total cost.	
1 in.	2¼	2¼	¼	½	¾	½

On the back of Form G appears the following endorsement:

....Department.

Foreman's Report of material used
in month of..., 189..
I certify the within report to be correct.

....

Foreman.

Foreman will keep a careful and accurate account of all materials used or issued, and make report of same on this blank daily.

Form S.

Form S is a sheet 9¼ inches wide by 12 inches long, and is printed on one side only. There are 22 horizontal lines on the page, ruled faint blue. The lines that appear in the following printed form are red. The width of each column appears under its title.

FORM S.

THOMSON-HOUSTON ELECTRIC COMPANY, RAILWAY DEPARTMENT.

On account West End Street-Railway Company.

Daily Report of Condition of Motor Cars on...Division.

...., 189..

....Div. Clerk.

Car Number.	In Shops.		Out Shops.		Remarks.
	Date.	Time.	Date.	Time.	
1¼ in.	1	1	1	1	4

APPENDIX E.

CONCERNING LIGHTNING PROTECTION.

BY PROFESSOR ELIHU THOMSON.

The discharges of lightning as affecting railway and other installations may be divided into four classes.

The first and probably the most dangerous of all discharges of lightning is a stroke falling on the line from the clouds and seeking ground through the conductor itself and through the cars connected to it.

Second, inductional discharges produced by discharges in the clouds moving in a parallel or approximately parallel direction of the line of the track. These discharges are of course less severe than the former, but may introduce complicated actions owing to the induction on the track being the same in direction as that on the line.

Third, leakage discharges, which may be of the nature of lateral branchings from a portion of earth which has received a heavy stroke, and which lateral branchings will of course divide themselves over the return conductors, and may reach the station.

Fourth, electrostatic induction from cloud layers, which when discharged of course give rise to changes of electric potential in all metallic masses under them. This latter is perhaps the most common action occurring during thunder storms and the one which gives rise to the minor discharges tending to equalize the potentials over the system and which have to be guarded against by the arrester devices. The character of the discharges which reach the line may be that of a sudden or instantaneous action or may occupy a slight period of time, and may therefore be characterized as slower inductive discharges.

As to the character of the discharge, of course the more sudden and violent the action the more dangerous it is likely to be to the insulation of the plant. The slower discharges having time to reach an equalization are not so apt to produce those great differences of potential between one point and another which lead to perforation of insulation, etc. It is questionable, indeed, whether any devices whatever will be sufficient to take care of the plant and avoid all damage where the lightning falls directly on the trolley line.

The only safeguard in such a case is to seek out the exposed sections of such a line—places, for example, where the line passes over high ground and stands alone unsurrounded by metallic wires or high or metallic buildings—and protect such places by the introduction of lightning rods on the suspending poles run into the ground and extending high enough above the general trolley line to divert any discharge therefrom.

It is pretty safe to say that, so far as the station itself goes, damaging discharges will mostly enter on the feeder lines; though discharges which do damage may of course come in on the ground return, especially if the stroke should fall to the ground not far away from the station and if the ground itself be not of a very moist and conducting character. It is conceivable that in the case assumed the discharge from cloud to earth may find the earth so poor a conductor as to seek the rails and scatter itself along them. As the trolley line and the whole structure of the electrical installation is practically only a metallic extension of the rail conductor, there may, of course, be danger from the current scattering itself over the whole system and breaking down the insulation in its path. The remedies which seem most applicable to the case of lightning discharges reaching the stations by the feeders, whether these discharges are of an inductive character or direct strokes, are the employment of lightning arresters on the feeder lines in such a manner that the earth connection from them is as short and direct and of as little self-induction as possible.

To this end each station should have heavy, broad ground connections, such as good-sized piping led to the best metallic grounding conductor within reach, and this ought to be as short as possible, the feeder lines being brought down to it rather than that it be brought up to the feeders. The type of arrester which seems to be the best adapted to use here is one in which the spark gap is as small as it can be made safely and in which there is a means for disrupting or blowing out the arc, which will, of course, form to ground and short circuit the system when a lightning discharge takes place.

There should also be a heavy self-induction coil on the dynamo side of this arrester, placed in such a manner that it tends to divert any discharge to ground and oppose a reaction of great force to any heavy discharge moving in the direction of the dynamo. A self-induction of this kind is also useful in preventing the short-circuiting action through the arrester from attaining undue proportions in an instant.

It acts, in other words, to secure a little time for the blowing-out action to be accomplished before the current in the ground branch has become equivalent to a short circuit on the system. It would seem in all cases preferable wherever the grounded dynamo terminal exists to bring that terminal back to the ground plate of the arrester and make the ground at that spot.

In this way the spark gap in the lightning arrester becomes a shunt to the dynamo terminals, the gap possesses the less opposition to currents going to ground, and this tends to divert the discharges from the dynamo loop. A counter-inductive protector should be introduced into the circuit to protect the dynamo by the counter-inductive effect set up in it by the discharge coming to earth. This arrangement of devices in ordinary cases it would seem ought to be sufficient, and it is only conceivable that in the case of very heavy discharges they might not serve their purpose fully.

A station constructed with an iron framework which in itself was made the ground for the arrester would, of course, be an ideal arrangement. An open iron cage or framework of iron around the dynamo would be an addition of value, the openings in the cage being made of convenient size, such as a foot square, and therefore made so as not to interfere with the manipulation of the machine. In case such a cage were employed it should have a spark gap and blow-out device between each terminal leading from the commutator to the cage. The self-induction of the blowing-out apparatus should be on the side of the dynamo circuit. This arrangement might be an inconvenient one to adopt in stations, and while effective, the objections to it might be sufficient to rule it out.

An effective substitute for such an arrangement may be to use a spark gap between the field magnet and the conductors leading to the commutator brushes of the dynamo, so that a stroke tending to jump through the insulation of the armature to the core would be prevented. The principle of this arrangement would be to establish a weak point, as it were an artificial weak point, in insulation, between the iron frame of the dynamo and its wiring—such that the discharge on reaching the dynamo from either terminal would be led to take its path across the discharge plates (or weak point provided) instead of through any of the insulation of the machine, which insulation should, of course, be superior at every point to the insulating power of the spark gap.

This, it would seem, ought to be a very effective arrangement in saving injury to dynamos, and if its action could be coupled with self-induction in the lines beyond the spark gap on the dynamo side, of course the effectiveness would be increased.

There still remains the condition of the machine acting as a condenser, taking in a certain static discharge, suddenly reaching the limit of its capacity, backing up the discharge and causing perforation—an action which occurs in condensers and frequently leads to their destruction by perforation. A counteracting condenser might possibly be introduced, which would cause the spark gap to be called into service as the weak point, instead of leaving it to the condensing action of the machine. If this reasoning be correct, and it appears that it must be, then the spark gap should be supplemented by a certain amount of condenser surface adjoining it, and leading, as it were, the discharge in a sort of blind alley for building up the potential across the spark gap.

As to continuance of the use of the pipe arrangements for discharges tending to reach earth at stations, it would seem that if the precautions above mentioned were caried out consistently there would be no real reason for the pipe being retained, though without question it does no harm if present, and may do some good.

Concerning the protection of dynamos from discharges reaching them from a ground line, it would seem that the precautions above stated would be sufficient to protect them, the greater danger being from the discharges reaching the machines from the overhead lines, for which all the precautions are taken; while the discharges coming from earth would necessarily be of rather lower intensity. and with the device of the spark gap attached to the dynamo it would seem that these discharges would not necessarily puncture the insulation.

It may be well to reiterate here that the best station arrangement would be one constructed of an iron framework connected to the ground, and it is further to be understood that if the station be itself exposed to strokes of lightning none of the wiring leading from the station overhead should exceed in height the highest point of the station which has a conductor, but that, on the other hand, a solid conducting rod should be placed in connection with the framework of the station and pass upward to a sufficient height to cover electrically, or at least protect from discharges, a considerable extent of area of the surrounding territory. This would prevent any lightning stroke falling on the lines very near the station, the object being to interpose as much of the self-induction of the lines as possible between the point of stroke and the station.

Whether the lightning-arrester devices be put upon the feeders or upon the wires leading to the dynamo separately does not seem to matter very much. This remark applies to the general lightning arresters which are applied to the lines entering the station. If applied to the feeders alone it would seem to be sufficient when supplemented by the dynamo spark gap shunting arrangements above described. As to the line protection, it is recommended that a survey of the trolley line be made in relation to the chances of its being struck by lightning, or in relation to its exposure to cloud effects directly. Note if it should pass over a hill more or less open and which is not built upon, as the line would be exposed in an exceptionally favorable way to strokes of lightning, being, as it is, an excellent conductor of electricity and having connection to ground though the cars, the ground also being an excellent ground on account of the extent of surface covered by the connected track.

A survey of the line, then, should discover the points of danger from lightning discharge, and precautions should be taken along the line and particularly at those points liable to receive strokes of lightning. This protection might be accomplished by connecting the foot of the poles to the track and extending the poles upward by a rod to a sufficient height properly to protect the trolley line. It is not to be expected, however, that in such case a lightning discharge falling upon any pole or poles would fail to disturb the electrical equilibrium of the line. It would, indeed, set up a complicated set of inductive disturbances, if it did not leak directly to the line. The line, on account of its extent, is also subject to extraordinary inductive actions from cloud discharges and to electrostatic induction from discharging cloud masses.

To take care of these as far as possible there should be placed along the line a number of grounded lightning arresters, well protected from rain so as to avoid leakage, whereby a free discharge to earth may be secured without allowing the line current to follow. These arresters should be placed, perhaps, every few hundred feet apart and in exposed locations more frequently. They should be found, also, at every angle of the line where, for example, the line takes a turn at right angles or in some other direction, the inductive action of the clouds being expected to accumulate or be more pronounced at such places, and therefore more apt to seek earth through arresters placed at these points.

The track connections of the line may of course be regarded, in a sense, as extensions of the general conducting system, and the track itself may become subject to inductive disturbances giving rise to trouble. To avoid this there does not appear to be any better means than to have a thorough grounding at points along the track, such grounding being obtained by driving tubes into the ground, iron bars or pipes, until a water stratum is reached, or by making attachments to existing pipes leading deep into the ground.

In regard to the protection of motors on cars, the problems are of a similar nature to those existing in stations, and the car itself may be regarded for the time being as a small moving station. The lightning arresters already existing may be retained with advantage and arranged so that they become efficient spark gap shunts of the motor. I am inclined to think that the field-magnet cores in motors will be best protected by being insulated thoroughly from the ground as well as the armature cores, as by that means there will be less tendency for discharges of lightning to leave the wire and jump through the iron of the machine. Where they are not so protected by being insulated, that is, where they are really grounded, it would seem that the lightning arresters should be able in most cases to take care of the motors.

There comes in, however, in this connection a condenser action between the field-magnet coils and their cores and between the armature winding and its core which might divert at low potentials a discharge insufficient to jump the spark gap, but which by the condenser action existing would suddenly give rise to local potentials at weak points, breaking through the insulation and spoiling the machine. The only remedy available seems to be such a one as has been suggested for the dynamos, to establish, as it were, a sort of blind alley for the discharge, and force it to leap an artificial weak spot which virtually takes the place of the insulation between the winding and the cores of the dynamo.

Upon this point experiments are needed to determine just what the danger is and how to meet it. I am inclined to think that here again the precautions used on the dynamo would be effective, that counter-inductive devices of simple character might be introduced along with the ordinary self-induction of the arresters, and that possibly a properly disposed condenser and spark gap combined might be introduced. In fact the spark gap of the lightning arrester itself might be utilized as the protecting spark gap, especially should the sparking be assisted by the condenser action just indicated.

APPENDIX F.

MOTORS WITH BEVEL GEAR AND SERIES-MULTIPLE CONTROL OF MOTORS.

SINCE the first edition of this book was published there have been two noticeable improvements in the street railway apparatus, somewhat antagonistic in their result, it is true, but decidedly ingenious, and each having an important field of action.

One of these is the improved arrangement of bevel gear transmission that appears in the Sperry street railway motor shown in Figs.

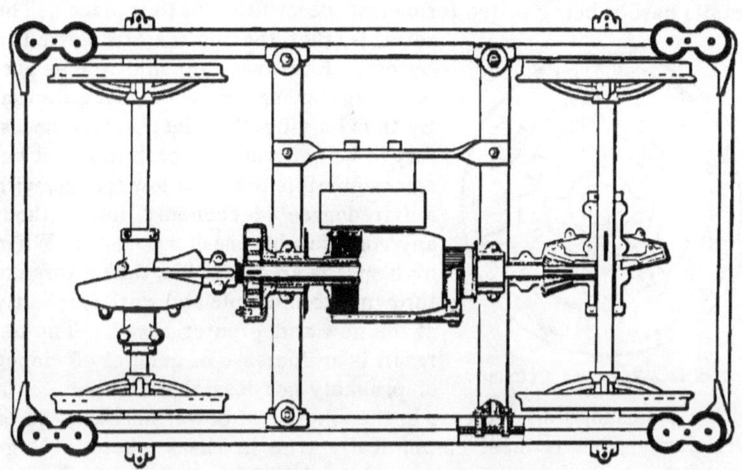

FIG. 180.—SPERRY MOTOR WITH BEVEL GEAR.

180 and 181. Its essential principle consists in supporting bevel gears by rigid bearings in oil-tight cases, so that the teeth always mesh accurately, and in driving one or both of these gears from the motor shaft through a peculiar form of flexible clutch which seems to give excellent results in practice. It is shown in section in Fig. 181, and needs little explanation, consisting as it does of a pair of concentric spiders with projections interlocking through rubber cushions. The outer spider has teeth inwardly projecting which engage a toothed wheel on the motor shaft. These two sets of teeth admit of considerable play, which is permissible because there is no marked rubbing action as in ordinary gears—only a slight sliding and steady pressure.

The result of this device is to give the bevel-geared street-car motor a much better place in the art than it has ever taken before, rendering it possible to drive both axles effectively from a single motor flexibly supported. In certain classes of work the advantage of this procedure is very decided.

The other improvement noted is the evolution of a successful series multiple controller. The scheme has already been mentioned and somewhat discussed; the device for accomplishing it, however, has necessitated considerable experimentation, and only very recently has it come into a thoroughly practical form. The difficulties that had to be overcome have been of a mechanical rather an electrical nature, the details hardly being suited for a close description in this place. The result is, that the motors are started in series and thrown into multiple as part of a regular speed-varying regulation. By thus halving the electro-force necessary to be generated in each motor it becomes possible to run at low speeds, with a fair degree of economy and without any considerable use of a rheostat. When high speeds are desirable, the motors are thrown into multiple and work efficiently at the new and greater speed. The net result is an increase of general efficiency of probably not less than 15 per cent., with a corresponding decrease in the amount of power necessary to be furnished. This is most emphatically true in cases where a large range of speeds is necessary, as in roads which start in the centre of a city and then run out into the suburbs, where the limitations imposed by street traffic no longer exist.

FIG. 181.—FLEXIBLE CLUTCH.

This series multiple control appears to avoid most of the difficulties which have heretofore been met in running motors economically at low speed. It may be mentioned in passing that this method would perhaps find a very favorable application in enabling the gearless motors to be used with a much greater degree of economy than has been heretofore possible. It goes without saying that to a certain extent the series multiple control decreases the advantages to be gained from the use of a single motor, by raising the efficiency of the ordinary combination to a considerably higher point than heretofore. It seems probable, however, that the single motor geared to both axles may prove very serviceable where speeds are comparatively uniform, as in certain classes of tramway work, while the series multiple control secures for street railway purposes a good efficiency at very widely varying speeds.

APPENDIX G.

METHOD OF MEASURING INSULATION RESISTANCE OF OVERHEAD LINES.

In operating an electric railway plant efficiently, it is necessary that the loss of current through those devices supporting the overhead construction should be a minimum, but no method is generally known by which the insulation resistance of any individual insulator

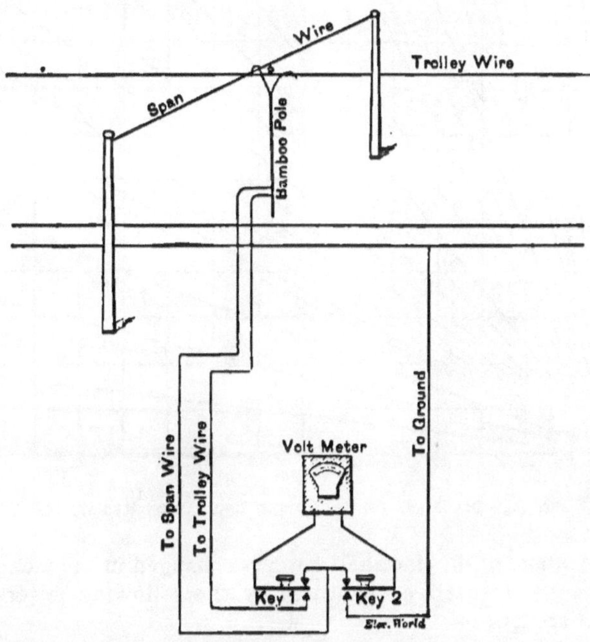

FIG. 182.—CONNECTIONS FOR MEASURING INSULATION RESISTANCE.
Depress key 1 to measure x. Depress key 2 to measure y.

can be measured without removing it from the line. Such a method is here described.

Its use depends upon the line being insulated at each point of support by two resistances in series, of which there is commonly one between the trolley wire and the span wire, and the insulation of each pair of poles between the span wire and the ground.

The method consists in measuring the voltage from trolley wire to span wire, and also voltage from span wire to ground. These measurements should preferably be made in immediate succession, in order

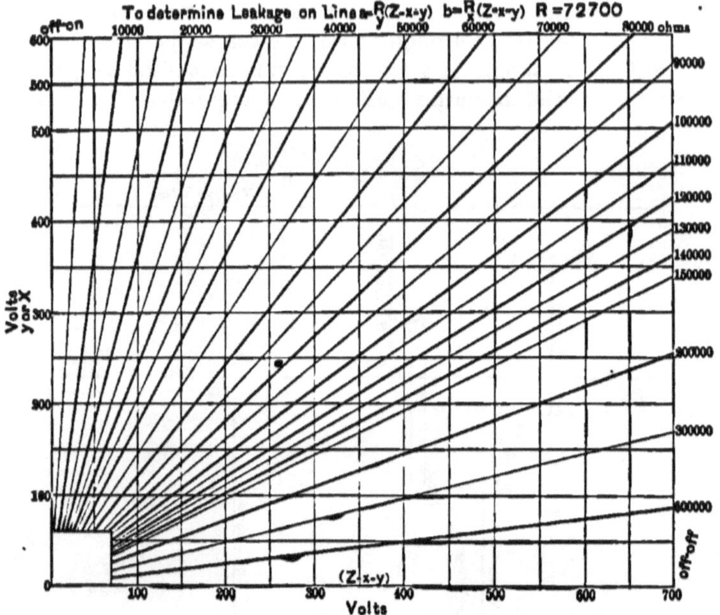

FIG. 183.—DIAGRAM FOR COMPUTING INSULATION RESISTANCE.

that the voltage of the line shall not have changed in the mean while. Indicating the respective quantities by the following letters, these equations are true:

Let R equal resistance of the volt meter.
A be the insulation resistance of the line insulator.
B " " " " pole or poles.
x " measured voltage from trolley to span wire.
y " " " " span wire to ground.
z " " " " trolley wire to ground.

$$A = \frac{R}{y}(Z-x-y). \qquad B = \frac{R}{x}(z-x-y).$$

For rapidity in securing these results, it is convenient to have a bamboo pole about fifteen feet long (weighing only about 18 oz.), with two metallic hooks on the upper end thoroughly insulated from each other. One hook should be caught over the trolley wire and the other hook over the span wire, and, by means of suitable commutating keys in metallic connection with each of these hooks, the volt meter, and the ground, the three necessary measurements of voltage at which point of support can be made within say five seconds. The adjoining cut shows the diagrammatic connections.

These measurements can be systematically entered in a note-book, and the required value of insulation resistance secured without calculating, by the use of a chart with radial lines drawn thereon, as shown in the cut.

By proper adding or dropping of ciphers from either or both ordinates, and the radial lines, results can be secured from this chart with as great accuracy as if it were drawn on a scale ten or one hundred times greater. This method is due to Mr. Theodore Stebbins.

INDEX.

www.ingramcontent.com/pod-product-compliance
Lightning Source LLC
Chambersburg PA
CBHW030557300426
44111CB00009B/1011